Energy, Environment, and Sustainability

This books series publishes cutting edge monographs and professional books focused on all aspects of energy and environmental sustainability, especially as it relates to energy concerns. The Series is published in partnership with the International Society for Energy, Environment, and Sustainability. The books in these series are editor or authored by top researchers and professional across the globe. The series aims at publishing state-of-the-art research and development in areas including, but not limited to:

- Renewable Energy
- Alternative Fuels
- Engines and Locomotives
- Combustion and Propulsion
- Fossil Fuels
- Carbon Capture
- Control and Automation for Energy
- Environmental Pollution
- Waste Management
- Transportation Sustainability

More information about this series at http://www.springer.com/series/15901

Dhananjay Kumar Srivastava
Avinash Kumar Agarwal
Amitava Datta · Rakesh Kumar Maurya
Editors

Advances in Internal Combustion Engine Research

Springer

Editors
Dhananjay Kumar Srivastava
Department of Mechanical Engineering
Indian Institute of Technology Kharagpur
Kharagpur, West Bengal
India

Amitava Datta
Department of Power Engineering
Jadavpur University
Kolkata, West Bengal
India

Avinash Kumar Agarwal
Department of Mechanical Engineering
Indian Institute of Technology Kanpur
Kanpur, Uttar Pradesh
India

Rakesh Kumar Maurya
Department of Mechanical Engineering, School of Mechanical, Materials and Energy Engineering
Indian Institute of Technology Ropar
Rupnagar, Punjab
India

ISSN 2522-8366 ISSN 2522-8374 (electronic)
Energy, Environment, and Sustainability
ISBN 978-981-10-7574-2 ISBN 978-981-10-7575-9 (eBook)
https://doi.org/10.1007/978-981-10-7575-9

Library of Congress Control Number: 2017959921

Printed on acid-free paper

This Springer imprint is published by Springer Nature
The registered company is Springer Nature Singapore Pte Ltd.
The registered company address is: 152 Beach Road, #21-01/04 Gateway East, Singapore 189721, Singapore

Preface

Energy demand has been rising remarkably due to increasing population and urbanization. Global economy and society are significantly dependent on the energy availability because it touches every facet of human life and its activities. Transportation and power generation are major examples of energy. Without the transportation by millions of personalized and mass transport vehicles and availability of 24×7 power, human civilization would not have reached contemporary living standards.

The first international conference on 'Sustainable Energy and Environmental Challenges' (SEEC-2017) was organized under the auspices of 'International Society for Energy and Environmental Sustainability' (ISEES) by the 'Center of Innovative and Applied Bioprocessing' (CIAB), Mohali, from February 26–28, 2017. ISEES was founded at IIT Kanpur in January 2014 with the aim of spreading knowledge in the fields of energy, environment, sustainability, and combustion. The society's goal is to contribute to the development of clean, affordable, and secure energy resources and a sustainable environment for the society and to spread knowledge in the above-mentioned areas and spread awareness about the environmental challenges, which the world is facing today. ISEES is involved in various activities such as conducting workshops, seminars, conferences in the domains of its interest. The society also recognizes the outstanding works done by the young scientists and engineers for their contributions in these fields by conferring them awards under various categories.

This conference provided a platform for discussions between eminent scientists and engineers from various countries including India, USA, South Korea, Norway, Malaysia, and Australia. In this conference, eminent speakers from all over the world presented their views related to different aspects of energy, combustion, emissions, and alternative energy resources for sustainable development and cleaner environment. The conference started with four mini-symposiums on very topical themes, which included (i) New Fuels and Advanced Engine Combustion, (ii) Sustainable Energy, (iii) Experimental and Numerical Combustion, and (iv) Environmental Remediation and Rail Road Transport. The conference had 14 technical sessions on topics related to energy and environmental sustainability and a panel discussion on 'Challenges, Opportunities and Directions of Technical Education & Research in the Area of Energy, Environment and Sustainability' to wrap up the three-day technical extravaganza. The conference included two plenary talks, 12 keynote talks, 42 invited talks from prominent scientists, 49 contributed talks, and 120 posters. A total of 234 participants and speakers attended this three-day conference, which hosted Dr. V K Saraswat, Member, NITI Aayog, India, as a chief guest for the award ceremony of ISEES. This conference laid out the road map for the technology development, opportunities, and challenges in this technology domain. The technical sessions in the conference included Advances in IC Engines and Fuels; Conversion of Biomass to Biofuels; Combustion Processes; Renewable Energy: Prospects and Technologies; Waste to Wealth—Chemicals and Fuels; Energy Conversion Systems; Numerical Simulation of Combustion Processes; Alternate Fuels for IC Engines; Sprays and Heterogeneous Combustion of Coal/Biomass; Biomass Conversion to Fuels and Chemicals—Thermochemical Processes; Utilization of Biofuels; and Environmental Protection and Health. All these topics are very relevant for the country and the world in the present context. The society is grateful to Prof. Ashok Pandey for organizing and hosting this conference, which led to the germination of this series of monographs, which included 16 books related to different aspects of energy, environment, and sustainability. This is the first time that such a voluminous and high-quality outcome has been achieved by any society in India from one conference.

The editors would like to express their sincere gratitude to the authors for submitting their work in a timely manner and revising it appropriately at short notice. We would like to express our special thanks to all those who reviewed various chapters of this monograph and provided their valuable suggestions to improve the manuscripts. We acknowledge the support received from various funding agencies and organizations for the successful conduct of the first ISEES conference SEEC-2017, where these monographs germinated. These include Department of Science and Technology, Government of India (special thanks to Dr. Sanjay Bajpai); TSI, India (special thanks to Dr. Deepak Sharma); Tesscorn, India (special thanks to Sh. Satyanarayana); AVL, India; Horiba, India; Springer (special thanks to Swati Mehershi); CIAB (special thanks to Dr. Sangwan).

Kharagpur, India — Dhananjay Kumar Srivastava
Kanpur, India — Avinash Kumar Agarwal
Kolkata, India — Amitava Datta
Rupnagar, India — Rakesh Kumar Maurya

Contents

About the Editors

Dr. Dhananjay Kumar Srivastava is currently working as Assistant Professor in the Department of Mechanical Engineering, IIT Kharagpur, since April 2015. He has done Doctor of Philosophy from Engine Research Laboratory, IIT Kanpur, in 2013. He was a Research Fellow at University of Birmingham, UK, in 2014. He was also a Visiting Researcher at University of Vienna, Austria, in 2004. His areas of expertise include laser ignition of engine, combustion visualization, emission control, engine calibration, gasoline direct injection. He is the recipient of Gandhian Young Technological Innovation Award in March 2013, Pool Scientist Fellowship by CSIR, India, from 2010 to 2013.

Prof. Avinash Kumar Agarwal joined IIT Kanpur in 2001 and is currently a Poonam and Prabhu Goyal Endowed Chair Professor. He was at ERC, University of Wisconsin, Madison, USA, as a Postdoctoral Fellow (1999–2001). His areas of interest are IC engines, combustion, alternative fuels, hydrogen, conventional fuels, lubricating oil tribology, optical diagnostics, laser ignition, HCCI, emission and particulate control, and large bore engines. He has published more than 160 peer-reviewed international journals and conference papers. He is Associate Editor of ASME Journal of Energy Resources Technology and International Journal of Vehicle Systems Modelling and Testing. He has edited 'Handbook of Combustion' (5 volumes; 3168 pages), published by Wiley VCH, Germany. He is a Fellow of SAE (2012), Fellow of ASME (2013), and a Fellow of INAE (2015). He is the recipient of several prestigious awards such as NASI-Reliance Industries Platinum Jubilee Award-2012; INAE Silver Jubilee Young Engineer Award-2012; Dr. C. V. Raman Young Teachers Award-2011; SAE International's Ralph R. Teetor Educational Award-2008; INSA Young Scientist Award-2007; UICT Young Scientist Award-2007; INAE Young Engineer Award-2005. He is the recipient of prestigious Shanti Swarup Bhatnagar Award-2016 in Engineering Sciences. He is the first combustion/IC engine researcher to get this honor.

Dr. Amitava Datta is a Professor in the Department of Power Engineering, Jadavpur University. He completed his graduate education in Mechanical Engineering from Jadavpur University and his Ph.D. from IIT Kharagpur. He is a recipient of Alexander von Humboldt Fellowship in Germany in 2000 and worked at Lehrstuhl fuer Technische Thermodynamik, University of Erlangen-Nuernberg. His research interests include the areas of combustion, atomization, energy, thermodynamic modeling and application of CFD in reacting flows, microfluidics, and biological flows.

He is instrumental in setting up the Combustion Laboratory in his department. He is an active researcher and has completed guidance of 12 Ph.D. theses and several master's theses. He has published 83 peer-reviewed research papers in various international journals and also presented and published several papers in national and international conferences. He has undertaken several sponsored research projects as principal investigator and co-investigator. These include projects funded by DRDO, ISRO, BRNS, UGC, and AICTE. He was a nodal person in the world bank-sponsoredTEQIP-I programme of Jadavpur University. He is also actively involved in various administrative responsibilities in his university. Presently, he is serving his second term as Head of Power Engineering Department. He is also the Director of the Internal Quality Assurance Cell of his University.

He has also served in various national committees in his professional capacity. He was in the executive board of the Combustion Institute, Indian Section. Presently, he is in the Joint Working Group on combustion setup by GTRE and DRDO. He has served as an expert for Russian Science Federation and also in various national bodies. He is also the reviewer of many leading international journals in the areas of combustion, energy, and fluid mechanics and acted as examiners of Ph.D. and M.S. theses of various IITs and universities.

Dr. Rakesh Kumar Maurya is Faculty Member in the Department of Mechanical Engineering, IIT Ropar, since August 2013. Before joining IIT Ropar, he was working as Senior Research Associate (Pool Scientist at CSIR) at IIT Kanpur. He received his bachelor's, master's, and Ph.D. degrees in Mechanical Engineering from IIT Kanpur, India. He received Early Career Research Award from Science and Engineering Research Board (SERB), Government of India, New Delhi. He is also a recipient of Young Scientist Award (2016) from International Society for Energy, Environment and Sustainability. He has served as a journal referee and committee member/session co-chair of international conferences on multiple occasions. He teaches and conducts research in the area of internal combustion engines. His areas of interest are low-temperature engine combustion, alternative fuels, engine combustion diagnostics, engine instrumentation, combustion and emission control, particulate matter characterization, engine management systems, and philosophy of science.

Part I
General

Horizons in Internal Combustion Research

Dhananjay Kumar Srivastava, Avinash Kumar Agarwal, Rakesh Kumar Maurya and Amitava Datta

Abstract With the major concern to increase the efficiency of internal combustion (IC) engines, various technologies and innovations have been implemented to improve efficiency and reduction of emissions. This monograph gives a detailed description of advanced IC engine concepts. The monograph is divided into advanced technology for IC engines, exhaust after-treatment and its heat recovery, simulations in the field of IC engine such as HCCI and GCI. Toward the end of this monograph, an overview has been presented related to future mobility solutions of Indian automotive industry. The latest research topics are included in this monograph which will be very useful to students, research scholars as well as industries working in IC engine.

Keywords LTC · HCCI · GCI · Indian automotive industry · Exhaust after-treatment

Internal combustion (IC) engines today represent a class of heat engines marked by their high power-to-weight ratio, making them the suitable choice for portable power solutions. Worldwide scientists and researchers are concerned about climate change and global warming. Automotive vehicles are a major source of emission of

D. K. Srivastava
Department of Mechanical Engineering, Indian Institute of Technology Kharagpur, Kharagpur 721302, India

A. K. Agarwal (✉)
Engine Research Laboratory, Department of Mechanical Engineering, Indian Institute of Technology Kanpur, Kanpur 208016, India
e-mail: akag@iitk.ac.in

R. K. Maurya
Advanced Engine and Fuel Research Laboratory, Department of Mechanical Engineering, Indian Institute of Technology Ropar, Ropar 208016, India

A. Datta
Department of Power Engineering, Jadavpur University, Salt Lake Campus, LB-8, Sector III, Salt Lake, Kolkata 700 098, India

D. K. Srivastava et al. (eds.), *Advances in Internal Combustion Engine Research*, Energy, Environment, and Sustainability,
https://doi.org/10.1007/978-981-10-7575-9_1

greenhouse gases (GHGs) and particulate matter (PM). This monograph has been designed after careful assessment of challenges faced by engine community.

This monograph is divided into five parts. The second part is on advanced technology for IC engines and has six chapters. One of the chapters is on low-temperature combustion which is an advanced technology for IC Engines. This chapter reviews fundamental aspects of development of LTC engines and their evolution. Another chapter is on HCCI engine concept. The homogeneous charge compression ignition (HCCI) strategy is an advanced engine combustion concept having higher thermal efficiency while maintaining the NOx and soot emission to an ultra-low level. Intense ringing operation in HCCI engine is one of the major challenges at high load conditions, which limits the HCCI engine operation range and can also damage engine parts. This chapter will be of great interest to the industry. Another chapter is on various strategies and methods used in commercial vehicles. This chapter discusses on fully flexible camless valve actuation systems explaining the working of some popular actuation systems, highlighting their advantages and limitations. Few experimental results from the literature have also been cited to substantiate the utility of variable valve actuation systems. Oxygenated fuels are a solution to emission and also play a critical role in IC engine. In this context, one of the chapters is on the effect of butanol addition in the diesel fuel on the combustion and emissions characteristics of a diesel engine. Combustion stability is also discussed with respect to diesel engine operation with butanol blends. Carbon monoxide (CO), unburned hydrocarbon (HC), and nitrogen oxides (NOx) emissions characteristics of diesel engine using butanol blends are discussed in this chapter. Special emphasis is placed on the discussion of particulate emission (soot particle numbers) in diesel engine with butanol blends. Another chapter is on cycle to cycle variations in conventional diesel engine using wavelets. Higher cycle to cycle variations in combustion engines lead to efficiency losses, engine roughness, and lower power output and higher exhaust emissions. Cycle to cycle variations in combustion engines are typically characterized by several techniques such as statistical method, symbol sequence statistics, chaotic methods, and wavelet analysis. This strategy has better temporal and spectral resolution, and thus wavelet analysis can be used to analyze the periodicities as well as magnitude of variations in the engine combustion cycles. This chapter presents the characterization of cycle to cycle variations in conventional diesel engine fueled using statistical as well as wavelets technique. Another interesting and innovating chapter is on development of online mixing of gaseous fuel such as hydrogen and CNG. In this chapter, gaseous fuel mixing system was developed by which one can change the proportions of hydrogen and CNG of the HCNG blends dynamically without necessarily stopping the engine. Validation of the system developed was done by theoretical methods and experimental investigations.

The **third** part of this monograph is on exhaust after-treatment and its heat recovery, and it covers three important topics. One of the chapters provides overview on recent advancements in after-treatment technology for IC engines. Recent advances include reduction with and without filter, reduction with catalyst and without catalyst, and some other after-treatment techniques such as plasma-assisted techniques, NOx, soot combined reduction. One chapter is on calcium oxide nanoparticles and its potential toward purification of vehicle gas exhaust. Calcium oxide nanoparticles and its potential toward purification of vehicle gas exhaust were investigated in this work. Calcium oxide nanoparticles were synthesized, and its efficiency in absorbing constituents of vehicle gas exhaust has been estimated. There is a review chapter on exhaust heat recovery using thermoelectric generators. Engineers and researchers are basically working for improving the conversion efficiency of TEG modules by developing and doping semiconductors and optimization of the AETEG system to utilize and recover maximum heat available from the exhaust line by designing efficient heat exchanger systems, thus trying to improve its feasibility. This article covers the wide spectrum of feasibility of application of TEG modules in diesel engines with possible ways to utilize the generated power.

The **fourth** part is on numerical and computational work, and it covers four important topics. One of the authors presented results from a series of computational fluid dynamics (CFD) simulation studies performed by us to understand the impact of design features and operating conditions on GCI, focusing on idle to low loads, where igniting gasoline purely through compression is challenging. These simulations are based on experiments performed at Argonne National Laboratory (Argonne) on a four-cylinder diesel engine modified in GCI mode. We studied the impact of factors like injector nozzle inclusion angle, injection timing, injection pressure, boost level, and swirl ratio. The preignition reaction space from the results was analyzed to understand the interplay between these factors and the overall reactivity. We also delve into the impact of uncertainties in CFD model inputs such as model parameters and initial and boundary conditions on simulation results by performing a global sensitivity analysis (GSA), based on thousands of CFD calculations run on a supercomputer at Argonne.

Another chapter deals with design of IC engine using CFD. The purpose of this chapter is to make the reader to familiarize with the complexities involved in the working of a four-stroke engine. The five events which are completed in four strokes are suction, compression, combustion, expansion, and exhaust. It is hoped that readers may be benefited in understanding the application of CFD for fluid flow analysis and engine design by reading this chapter. Therefore, the main aim of this chapter is to make the reader appreciate how exactly CFD can be applied for design of an engine. As it is application-oriented, we are not going deep into the equations, modeling, etc. A number of case studies are presented and discussed.

The **fifth and last part** of this monograph is on next step to be taken for Indian automotive industries. A chapter focuses on strategies and policies to be followed by government. Recently, government planned to debar gasoline and diesel vehicle by 2030. In this scenario, Indian automotive industry has to be future-ready. The future disruptions in Indian automotive would include implementation of hybrids, electric, and fuel-cell vehicles. Government has started working in infrastructure development of hybrid and electric vehicles such as charging units, battery development, charging infrastructure development. However, currently hybrid and electric vehicles are significantly costlier and are required to be economically feasible. It can be assumed that conventional gasoline engines will be used in hybrid vehicles. This monograph presents both fundamental science and applied emerging technologies for IC engine research. Specific topics covered in the manuscript include:

- Low-temperature combustion;
- Characterization of ringing operation in ethanol-fueled HCCI;
- Variable valve actuation systems;
- Use of butanol blends in a diesel engine;
- Cycle to cycle variations in conventional diesel engine using wavelets;
- Hydrogen-enriched compressed natural gas;
- Recent advancements in after-treatment technology;
- Calcium oxide nanoparticles as an aid for purification of vehicle gas exhaust;
- Exhaust heat recovery using thermoelectric generators;
- Chemical kinetic simulation of syngas-fueled HCCI engine;
- Gasoline compression ignition;
- Application of CFD for analysis and design of IC engines;
- Future mobility solutions of Indian automotive industry.

Part II
Advanced Technology for Internal Combustion Engines

Low-Temperature Combustion: An Advanced Technology for Internal Combustion Engines

Akhilendra Pratap Singh and Avinash Kumar Agarwal

Abstract Universal concerns about degradation of ambient environmental conditions, stringent emission legislations, depletion of petroleum reserves, security of fuel supply, and global warming have motivated R&D of engines operating on alternative combustion concepts, which have the capability of using renewable fuels. Low-temperature combustion (LTC) is an advanced combustion concept for internal combustion (IC) engines, which has attracted global attention in recent years. LTC is radically different from conventional spark ignition (SI) combustion and compression ignition (CI) diffusion combustion concepts. LTC technology offers prominent benefits in terms of simultaneous reduction of both oxides of nitrogen (NOx) and particulate matter (PM) in addition to reducing specific fuel consumption. However, controlling ignition timing and heat release rate (HRR) are primary challenges to be tackled before LTC technology can be implemented in automotive engines commercially. This chapter reviews fundamental aspects of development of LTC engines and their evolution, historical background, and origin of LTC concept and its future prospects. Detailed insights into preparation of homogeneous charge by external and internal measures for diesel like fuels are discussed. Combustion characteristics of LTC engines including combustion chemistry, HRR, and knock characteristics are also touched upon in this chapter. Emission characteristics are also reviewed along with insights into PM and NOx emissions from LTC engines.

Keywords Low temperature combustion · Oxides of nitrogen · Particulate matter
Combustion control

A. P. Singh · A. K. Agarwal (✉)
Indian Institute of Technology Kanpur, Kanpur, India
e-mail: akag@iitk.ac.in

D. K. Srivastava et al. (eds.), *Advances in Internal Combustion Engine Research*,
Energy, Environment, and Sustainability,
https://doi.org/10.1007/978-981-10-7575-9_2

1 Introduction

Transport is an essential component by which people not just connect with each other but also progress. To fulfill increasing demand for safe, reliable, environmental friendly, economical and efficient transport system, development of novel automotive technologies has become crucial. Automotive technology refers to the technologies incorporated in vehicles for their design and prototyping. These technologies are essential for evolution and adaptation in existing vehicles. With rapidly increasing demand as well as expectations of consumers for higher safety standards, low-carbon future is embodied aggressively in evolving fuel economy standards and stringent emissions norms. According to World Energy Outlook (2011) factsheet of International Energy Agency (IEA), global demand for primary energy is expected to increase by one-third between 2010 and 2035 and energy-related CO_2 emissions are expected to increase by 20% (reaching up to 37 $GtCO_2$ by 2035) [1, 2]. Rapidly dwindling petroleum reserves are another major concern for the automotive sector. Therefore, research efforts have focused on exploration of alternative energy resources, including renewable fuels such as biofuels, solar energy, and hydrogen.

Direct injection compression ignition (DICI) and spark ignition (SI) engines are the main technologies, which have reputed and established applications in the automotive sector. Over the past several decades, diesel engines have become more efficient, durable, quieter, and vibration-free. Diesel vehicles have undergone dramatic changes in the last decade with the advent of common rail direct injection (CRDI) technology. CRDI technology offers unprecedented flexibility, which was previously not available for DICI engines, and it delivers 25% higher power output compared to baseline DICI engines. Apart from engine performance, control of harmful pollutants like oxides of nitrogen (NOx), particulate matter (PM)/soot, carbon monoxide (CO), and hydrocarbons (HC) is essential in order to preserve the environment and protect the human health. Diesel engines are one of the major sources of PM, which mainly consist of soot laced with polycyclic aromatic hydrocarbons (PAHs), trace metals, and sulfates. PM is known to have adverse impact on human health and the environment through inhalation pathway, toxic contamination of different environmental media, visibility reduction (due to smoke), and global climate change (due to black carbon emissions) [3].

Different emission reduction techniques are being developed to comply with prevailing emission standards. These techniques can be broadly classified into active and passive emission reduction techniques. In the active emission reduction, polluting species are prevented from formation during combustion. The governing principle to avoid pollutant formation in the combustion chamber is to optimize the combustion. Combustion and pollutant formation in diesel engines are dependent on in-cylinder conditions, which are primarily governed by fuel injection parameters, in-cylinder air temperature, pressure, and charge motion. Numerous techniques such as turbo-charging, exhaust gas recirculation (EGR), high-speed direct injection (HSDI), modifications in engine configuration, and design of flow control valves have been implemented to control the formation of pollutants during combustion. In passive emission reduction, combustion products are neutralized before their exit from the tailpipe into the atmosphere. These methods include exhaust gas after treatment, which have been developed over decades. Three way catalytic convertors (TWC), diesel particulate filters (DPF), diesel oxidation catalysts (DOC), selective catalytic reduction (SCR) technique, and lean NOx traps (LNT) are some of the popular exhaust gas after-treatment techniques. Emission control hardware is located in the exhaust system of vehicles, where they oxidize and reduce pollutants before the exhaust is released into the atmosphere. Current and emerging after-treatment techniques promise significant emission reduction; however, the cost and complexities involved in their implementation threaten their application to the diesel engines.

Additionally, use of alternative fuels and hybrid vehicles can also be considered for emission control. In spite of the fact that fossil fuels are the backbone of the energy supply, depletion in their reserves and stringent emission norms have motivated researchers to develop advanced vehicles such as battery-operated vehicles, fuel-cell vehicles. Another concept for emission reduction is the use of hybrid vehicles, which are powered by electric motors along with IC engines. The electric motors provide the benefit of high power and efficiency but no emissions. According to driving conditions, in hybrid vehicles, power source can be switched between an electric motor and an IC engine, resulting in higher fuel economy and lower emissions. However, hybrid vehicles suffer from the limitation of higher cost, scarcity of raw materials (rare earth metals), large size and weight, which has resulted in very small share of hybrid vehicles among new vehicles sales. New vehicle technologies collectively are expected to account for 6% of new passenger vehicle sales by 2020 and 19% by 2035, bulk of which would be hybrids. Figure 1 shows the relative growth of different automobile technologies among new vehicle sales.

In such a scenario, automotive industry desperately requires technologies, which are cleaner and efficient, improve ambient air quality in an efficient manner, reduce greenhouse gas emissions, and contribute to energy security [5–9]. Considering stringent emission regulations and scarcity of primary energy resources, development of new highly efficient, and environment friendly combustion concepts and systems capable of utilizing alternative fuels in addition to conventional fuels have become increasingly important. Several experimental studies have been carried out

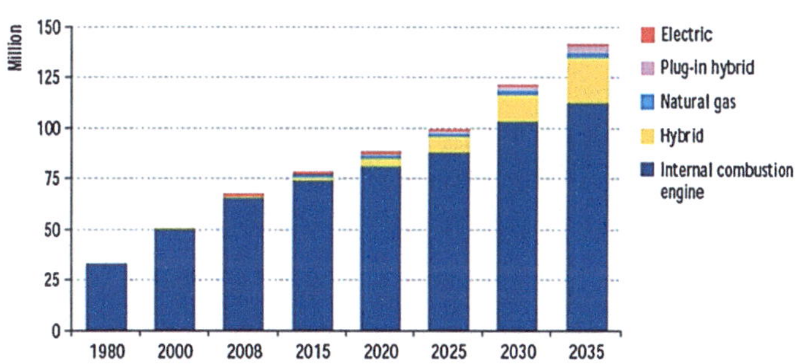

Fig. 1 Present and future status of light-duty vehicle sales [4]

to develop novel combustion concepts such as low-temperature combustion (LTC), which demonstrate the prospects of meeting stringent environmental challenges faced by automotive engines. In recent years, researchers have focused on LTC technology development primarily due to its extremely low NOx and PM emissions and high efficiency potential.

1.1 Advanced Combustion Techniques

LTC engines have great potential to achieve high thermal efficiency and ultra-low emissions of NOx and PM. This has attracted attention of researchers and automotive industry alike. Significant efforts are being made to understand the physical and chemical processes involved in LTC, which affect engine performance and emissions. LTC engines operate on the same fundamental principle as a four-stroke engine and use basic elements of CI and SI engines. The LTC principle is shown in Fig. 2.

In a LTC engine, during the intake stroke, a nearly homogeneous fuel–air mixture is introduced. After intake valve closing (IVC), the piston starts to compress the fuel–air mixture, which increases the in-cylinder temperature and pressure. As the piston approaches TDC, charge attains auto-ignition conditions. Chemical kinetics of the charge can be accelerated by increasing the charge temperature in the beginning of compression stroke by preheating the intake air or by retaining a fraction of hot exhaust gas from the previous engine cycle in the cylinder. In both strategies, chemical reactions occurring in the homogeneous fuel–air mixture accelerate due to relatively higher charge temperature and pressure of residuals [11]. Start of combustion (SoC) in LTC mode can be controlled by a combination of variables such as compression ratio, inlet charge temperature, and pressure. As soon as the auto-ignition temperature is attained during the compression stroke, fuel starts

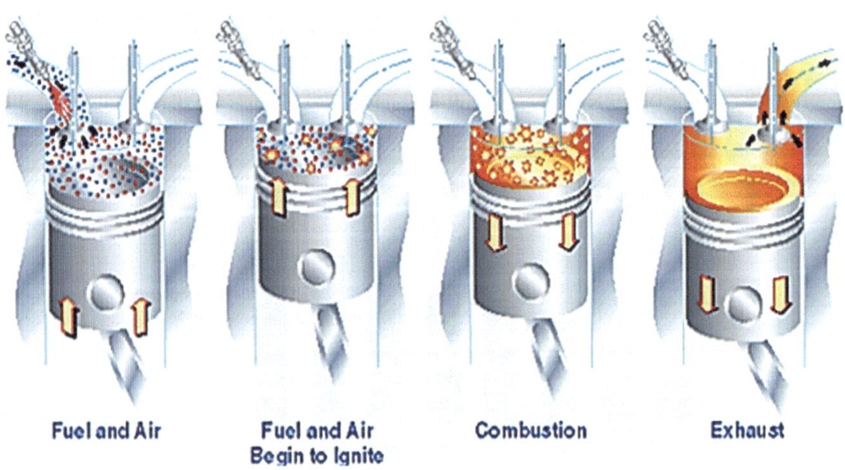

Fig. 2 Homogeneous charge compression ignition combustion principle [10]

oxidizing rapidly and its chemical energy is released instantaneously. Auto-ignition in LTC engine occurs simultaneously at several locations throughout the engine cylinder, and these locations are called hot spots. This quick heat release causes pressure rise in a significantly shorter time span compared to conventional combustion, while the peak cylinder local and global temperature still remains significantly lower. The fuel–air mixture temperature and pressure, therefore, increase further during combustion. During the expansion stroke, work is done by the expanding gases on the piston to produce a net positive torque, which is available at the crankshaft. The cycle is completed after the piston ascends to TDC during the exhaust stroke, forcing products of combustion out of the cylinder. In summary, LTC consists of the following steps:

(a) Preparation of a highly dilute fuel–air mixture using EGR to control combustion and the heat release rate.
(b) At the end of the compression stroke, fuel–air mixture temperature approaches auto-ignition temperature, leading to simultaneous spontaneous ignition of entire charge in the cylinder at several locations.
(c) Precise control of heat release rate (HRR) to achieve trade-off between combustion efficiency and emissions [12].

1.2 Advantages of LTC

This section describes potential advantages and scientific challenges in exploiting full advantage of this novel LTC concept, referred as homogeneous charge compression ignition (HCCI) combustion. LTC offers several advantages over the conventional combustion modes [13, 14]:

- LTC approximates a constant volumetric combustion in a very short combustion duration, and it can be achieved for high compression ratio; therefore, it results in higher thermal efficiency. Relatively lower peak combustion temperature in LTC leads to better energy utilization due to lower radiation losses. Throttling losses are also absent in a LTC engine in comparison to a SI engine.
- LTC has potential of significantly lower emissions compared to DICI and DISI engines with simultaneous reduction in NOx and PM emissions. There is no flame front, i.e., there is complete absence of localized areas of excessively high temperature and rich mixtures due to superior homogeneity of the fuel–air mixture. Therefore, there is no soot production. Further, there are low-temperature conditions and more uniform distribution of bulk gas temperature in the cylinder; therefore, NOx emissions are restricted to ultra-low levels. Hence, LTC is not affected by Soot–NOx trade-off.
- Main advantage of LTC is its fuel flexibility. LTC can be achieved by a wide range of fuels including gasoline, mineral diesel, biodiesel, alcohols, etc. Fuel flexibility of LTC engines enables the use of various alternative fuels that could reduce rapid depletion rate of petroleum reserves [15–17].

- LTC engines are suitable for the replacement of conventional SI and CI engines. These engines can also be coupled with advanced hybrid engines, i.e., to combine the advantages of highly efficient IC engines with electrical series hybrid powertrains.

1.3 Challenges in LTC

The main challenges that hinder the realization of LTC are as follows:

- Difficulties in vaporization of diesel hinder the development of diesel LTC because the fuel injection timings of LTC are significantly advanced compared to direct injection diesel combustion. When diesel is injected, the cylinder pressure and temperatures are close to atmospheric conditions. Viscous diesel fuel does not vaporize under these conditions. Therefore, preheating of intake air is required for vaporization of mineral diesel.
- LTC does not offer precise control over the start of combustion across wide range of engine speeds and loads. This issue becomes more important at the time of transient engine operation. Although several attempts have been made to resolve this issue by EGR control techniques such as variable valve timings; however, there is scope for further improvement in this area.
- LTC regime suffers from load limitations and can be implemented only at low-to-medium loads. For real-world application of LTC, the engine operation modes have to be switched back and forth between LTC mode at lower loads and conventional CI mode at higher loads.
- LTC is characterized by leaner air–fuel ratios. However, rich fuel–air mixtures at higher loads cause deterioration in engine noise, very rapid heat release rates, reduction in engine power output, etc.
- Lower in-cylinder temperature in case of LTC impedes post-oxidation of HCs and conversion of CO to CO_2. Thus, LTC basically suffers from the problem of high HC and CO emissions.

2 Premixed Charge Compression Ignition (PCCI): An Advanced LTC Technique

Different LTC systems can be differentiated by control over combustion phasing. Adaptation of HCCI technique results in reduced NOx and PM in diesel engines; however, HCCI combustion results in overall inferior thermal efficiency, and it also suffers from the lack of control over the combustion phasing. In case of a HCCI engine, combustion phasing is dominated by chemical reaction kinetics and decoupled from injection timing, which further results in lack of control over combustion events. Researchers have observed either too advanced or too retarded combustion phasing in case of HCCI combustion, which results in lower thermal efficiency. Poor combustion phasing results in relatively higher HC and CO emissions due to lower in-cylinder temperatures. These issues motivated researchers to develop a new LTC strategy known as premixed charge compression ignition (PCCI) combustion.

PCCI combustion is an advancement over HCCI combustion in terms of combustion stability. In PCCI combustion, control over combustion events with injection strategy is retained and a greater percentage of the total charge is premixed prior to ignition relative to conventional diesel combustion. In PCCI engine, fuel is injected at an intermediate timing between that of HCCI and conventional CI engine. This results in a partially premixed fuel–air mixture, which auto-ignites similar to HCCI combustion mode. PCCI combustion seems better compared to HCCI combustion due to better control over the combustion events, which leads to relatively superior engine performance. Various injection strategies are available for PCCI combustion, but commonly advanced injection timings are used to increase fuel–air mixing time. Despite the charge not being completely premixed (homogeneous), the same principles are applied to obtain low emissions as with HCCI. The factors used to promote combustion conditions toward the PCCI mode of operation are primarily the injection strategy and EGR rate. PCCI combustion is a single-stage combustion technique in which most fuel is burnt in premixed combustion phase. There is very small or practically no fuel remaining in the combustion chamber for diffusion combustion, which results in a relatively lower bulk temperature inside the combustion chamber. PCCI engines at high loads are facilitated with high boost pressure, which helps in oxidation of fuel resulting in relatively lower CO and HC emissions. Therefore, PCCI engine not only offers benefits of LTC with lower NOx and PM emissions but also results in significantly lower HC and CO emissions.

2.1 Charge Preparation for PCCI Combustion

Charge preparation, i.e., mixing of fuel and air is a very important feature of IC engines. The combustion process and its control are very much dependent on quality of mixtures and the technique employed for charge preparation. One of the major challenges of PCCI combustion is to prepare the premixed charge (highly diluted fuel–air mixture to give reasonable burn rates) before auto-ignition temperature is attained and combustion starts in the combustion chamber. In PCCI combustion engines, effective mixture preparation technique is required for achieving high efficiency, low HC, and PM emissions and preventing lubricating oil dilution. There are several techniques employed for charge preparation in PCCI combustion engines, depending on test fuel properties and control strategies being used.

The PCCI charge preparation techniques can be divided into three main categories, including external charge preparation (port fuel injection), internal charge preparation (in-cylinder direct injection), and concepts using both types of charge preparation techniques (dual-fuel mode). These strategies differ from each other in

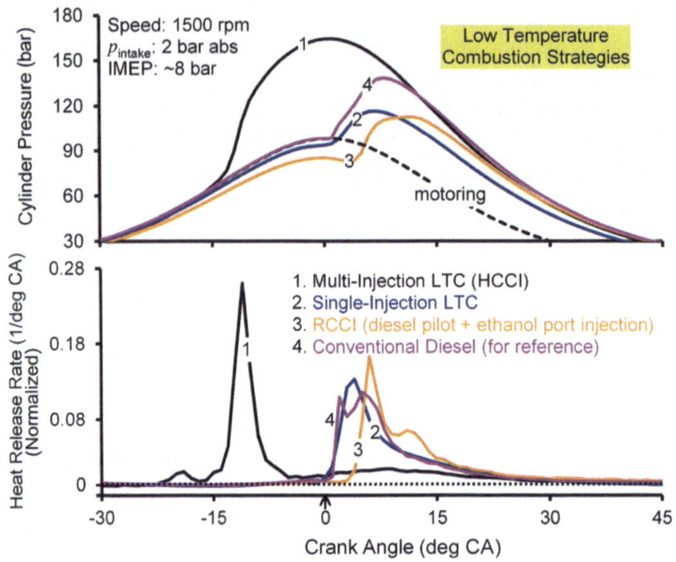

Fig. 3 Common LTC strategies for diesel engines [18]

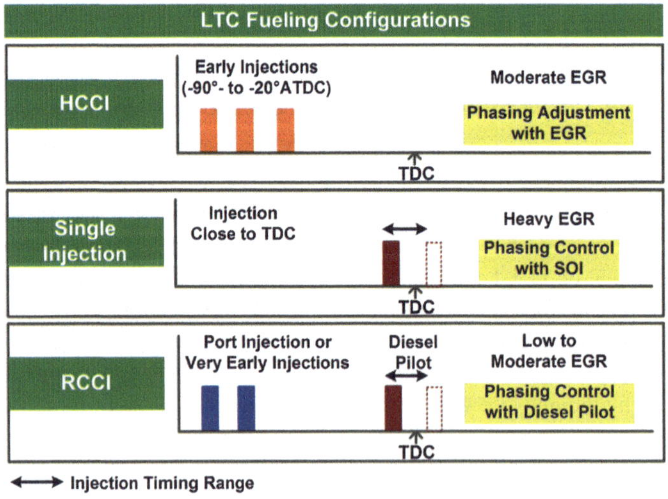

Fig. 4 LTC fueling configurations [18]

terms of time available for mixing and in degree of mixture homogeneity, which is achieved at the time of SoC [18].

Common strategies for achieving LTC include HCCI [19–22], single-injection LTC [23–26] and reactivity controlled compression ignition (RCCI) [27, 28], as shown in Fig. 3. An overview of fueling configurations is shown in Fig. 4. To achieve the diesel-fueled PCCI combustion, a multiple-fuel injection strategy is preferred due to its capability of generating nearly homogeneous fuel/air mixture [18].

2.2 *Fuel–Air Mixture Homogeneity*

Main requirement of LTC is the availability of homogeneous fuel–air mixture before SoC. In an IC engine, fuel–air mixing is governed by many parameters such as fuel properties, fuel injection strategy, in-cylinder conditions, fuel spray characteristics, in-cylinder flow patterns. These parameters are closely related to each other, i.e., in-cylinder conditions can be optimized by controlling fuel injection parameters and in-cylinder flows. Air motion redistributes dense liquid core of fuel throughout the combustion chamber. In-cylinder flows have strong impact on fuel evaporation; hence, they affect both the physical and chemical delay part of the ignition delay. Mainly in-cylinder flows near the intake and exhaust manifolds have been the focus for investigations of fuel–air mixing and its cyclic variations. Most researchers have operated engines in motoring mode. A common observation from these studies has been that there were large-scale dissipation of in-cylinder turbulent flow structures of the intake jet into small eddies with progression of an engine thermodynamic cycle. Among different flow structures that get formed inside the combustion chamber, "swirl," "tumble," and "squish" are of particular interest because of their significant impact on fuel–air mixing. Swirl is an organized rotational motion of air around the cylinder axis. Swirl is generated during the intake stroke due to specific intake manifold geometry and during compression stroke because of geometry of the piston and cylinder.

Fuel spray characteristics also significantly affect the fuel–air mixing inside the combustion chamber; therefore, it has been also been studied by many researchers using experiments as well as numerical techniques [29, 30]. They emphasized on the fuel spray atomization, which is required to enhance the in-cylinder fuel–air mixture homogeneity. They suggested that spray breakup processes become more complex due to higher ambient air density, which results in a dominant aerodynamic and viscous effects. Therefore, optimized fuel injection parameters are required to achieve a trade-off between in-cylinder conditions and fuel spray characteristics.

3 Combustion Characteristics of LTC

For any combustion mode, combustion characteristics inside the combustion chamber affect overall performance parameters of the engine such as power output, emissions. Considering the fluid mechanics, PCCI combustion can be divided into three distinct phases, namely pre-combustion, combustion, and post-combustion (Fig. 5).

(a) Pre-combustion: Flow dominates; small change in species.
(b) Combustion: Chemistry dominates; heat release occurs so rapidly and globally that turbulent mixing does not have time to be a significant influence.
(c) Post-combustion: Chemistry and turbulent mixing are likely to have some coupling, but no chemical heat release occurs (Fig. 6).

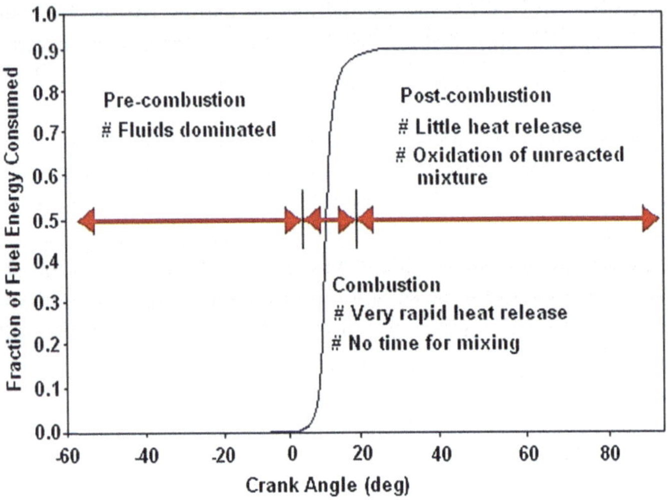

Fig. 5 Phases in LTC [31]

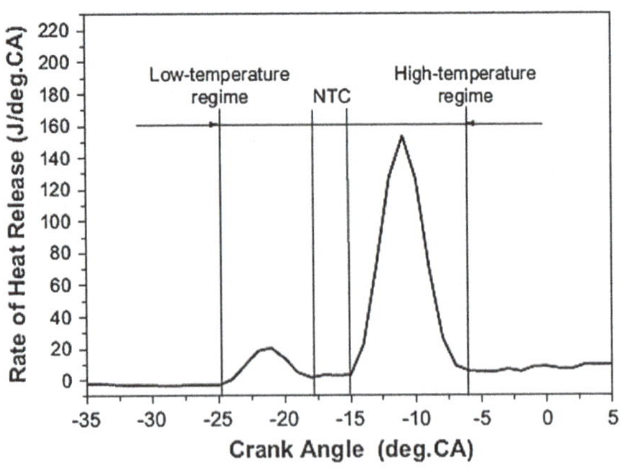

Fig. 6 Typical heat release curve from LTC of *n*-heptane [32, 33]

Due to small heat release, during early compression stroke, in-cylinder pressure, and temperature increases slightly. The low heat release (LHR) is associated with low-temperature chemical kinetics reactions. After LHT, the high heat release (HHR) occurs, which is attributed to negative temperature coefficient (NTC) regime [33]. NTC is the intermediate temperature region between LHR and HHR and is regarded as being the general characteristic in fully or partially premixed charge. Increasing pressure and cool flames result in lower NTC duration because these accelerate the activation of hot flame region [34].

The temperature range of the NTC area is between 700 and 950 K [35, 36]. In the NTC regime, overall reaction rate decreases with increasing in-cylinder temperature, resulting in a lower mixture reactivity. Heat release from low-temperature reactions is mainly related to the fuel and the engine operating conditions. In particular, in case of mineral diesel with a relatively high auto-ignition tendency or CN [37], LHR was observed. In case of gasoline-like fuels with low CN (high octane number), LHR was lower compared to mineral diesel like fuels under the same condition. Consequently, heat release from low-temperature reactions is too little to be observed expectedly in the heat release profile at most conditions for gasoline-like fuels [38]. In case of HCCI-DI, diffusive combustion is also noted due to directly injected mineral diesel [34]. Oxidation of hydrocarbon fuels mainly takes place via two general routes shown in Fig. 7.

Due to auto-ignition of hydrocarbons, low-temperature reactions start in the combustion chamber, which produce intermediate species (CH_2O, HO_2, and O radicals) [39]. After some time, thermal flame reactions start, followed by the main heat release. Domination of high-temperature oxidation (HTO) reactions of hydrocarbon fuel can be clearly seen in HRR of main combustion (Fig. 7). High concentrations of CH, H, and OH radicals are indicative of the dominance of high-temperature chemistry during bulk combustion [40]. These intermediate species are produced by thermal decomposition reactions, including the chain breaking of C–C bonds in the fuel. The other important concern of HCCI combustion is the steep rate of pressure rise produced by auto-ignition of nearly homogeneous fuel–air mixture. The in-cylinder gas temperatures, at which the low and high-temperature oxidation reactions start, are significantly different. The LTO and HTO start at ~790 and ~970 K, respectively, for n-pentane, which is independent

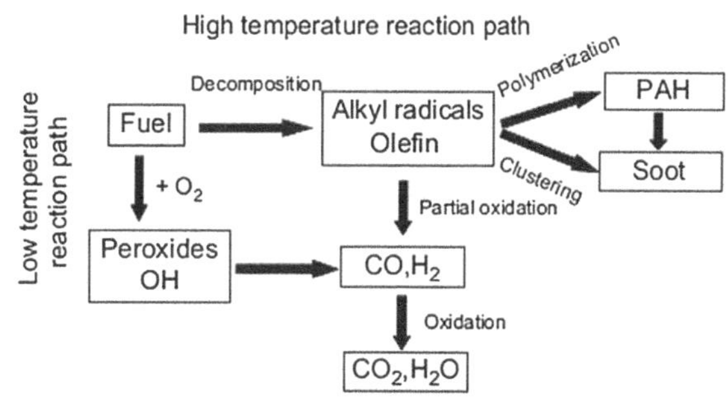

Fig. 7 Reaction pathways of hydrocarbon fuels in low- and high-temperature reactions [38]

of equivalence ratio, EGR rate and intake air temperature (T_i) [41]. Najt and Foster [42] reported that the auto-ignition of homogeneous fuel–air mixture was controlled by low-temperature (<1000 K) chemistry, and the bulk energy release was controlled by the high-temperature (>1000 K) chemistry dominated by CO oxidation.

4 Emission Characteristics of LTC

This section is divided into three subsections, namely regulated emissions, unregulated emissions, and particulate emissions.

4.1 Regulated Emissions from LTC

The basic motivation behind the development of this novel technology is its potential for significant reduction in exhaust emission as compared to conventional CI engines. In a conventional CI engine, there is a trade-off between NOx and PM. Figure 8 shows the advanced technology options, which can be used for simultaneous reduction of NOx and PM. In the CI engine, NOx is formed in the high-temperature zones under close to stoichiometric conditions and soot is formed in the fuel-rich spray core region. When engine runs at lower loads, its peak cylinder temperature is low, which reduces NOx formation but lower oxidation levels enhance soot formation. On the other hand, when engine runs at higher loads, its peak cylinder temperature is higher. This increases the NOx level but reduces the PM level due to improved oxidation at higher in-cylinder temperature. This proves that CI engine must use exhaust gas after treatment for NOx or/and PM. Although the in-cylinder average equivalence ratio is always lean, the combustion process

DEVELOPMENT SCENARIO (Heavy Duty Diesel)

	Euro 3 to Euro 4				Euro 4 to Euro 5				Euro 5 to Euro 6			
	NOx	PM	HC	CO	NOx	PM	HC	CO	NOx	PM	HC	CO
Emission reduction	30%	80%	30%	29%	43%	0	0	0	80%	50%	72%	0
Base Technology	High pressure CRDI, Fuel metering, Timing retard, EGR (2017)				-- (2019)				-- (2021)			
Engine-out emission control technology	Improved engine combustion, Improved engine calibration (PM), TCIC, cooled EGR				Further improved combustion & calibration, Multi-Phased injection, VGT, NOx control – cooled EGR				VGT, Advance combustion – PCCI, LTC			
After treatment Technology	NOx control - SCR (open loop), PM control – DPF				NOx control - SCR (closed loop), PM control - DPF				NOx control - SCR (closed loop), PM control - DPF			

Fig. 8 Advanced technology options for emissions control [43]

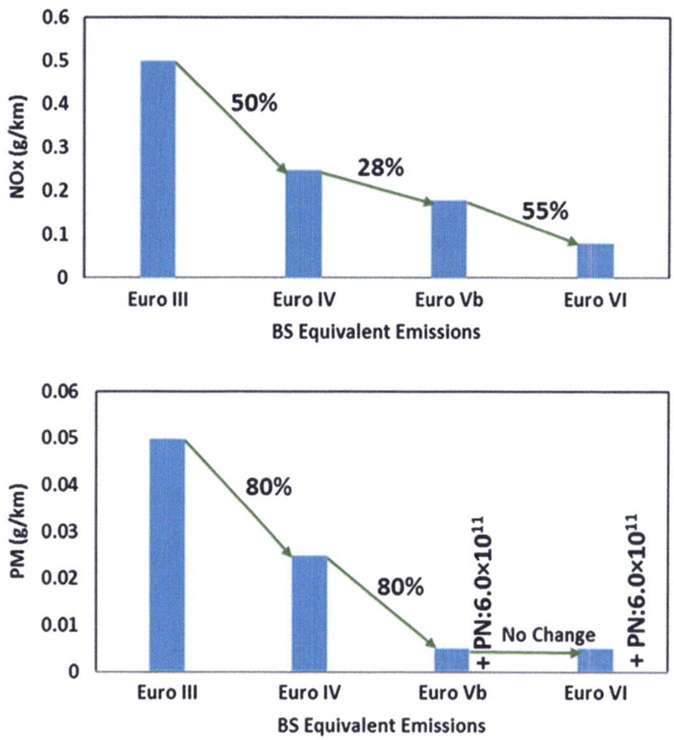

Fig. 9 Emissions from diesel passenger car [43]

does not complete in overall lean conditions. These characteristics show large potential for LTC to reduce NOx and PM levels simultaneously.

At present, a large fraction of NOx and PM concentration are originating from the heavy-duty vehicles, which use CI combustion. Figure 9 shows the emissions from light-duty vehicles and heavy-duty vehicles.

It becomes very challenging to develop a system, which can be used as an alternative to conventional CI engines. LTC technology is a potentially strong alternative to the CI combustion; however, till now, it can be used only at light engine loads.

Control of NOx levels is the most attractive aspect of LTC, and quantitatively, it reduces 90–98% NOx as compared to conventional diesel combustion [44, 45]. The basic reason of NOx formation in a standard CIDI engine is the presence of localized high-temperature regions. When the fuel is sprayed in the combustion chamber, oxygen interacts at the boundary surface between the air and the fuel droplets. This is the boundary where combustion takes place and generates high temperature locally. This provides favorable conditions for NOx formation because of availability of excess oxygen along with high temperatures in the localized regions. The underlying mechanism responsible for low NOx levels in LTC is the absence of high-temperature regions in the combustion chamber. The availability of excess oxygen is also reduced by introduction of EGR, which further reduces the NOx level during LTC. The combustion products such as CO_2 and H_2O present in EGR have a higher specific heat per unit mass compared to the fresh air, which absorbs the thermal energy and results in lower combustion temperatures, which slows down the chemical kinetics of fuel–air mixtures.

Takeda et al. [46] performed experiments in a direct injection four-stroke naturally aspirated single-cylinder engine. Three injectors (one at the center, two on the sides) were used. A dramatic reduction in NOx by early fuel injection timing due to lean fuel/air mixture was reported. Reduction in NOx emissions was accompanied by an increase in THC and CO levels due to incomplete combustion of over-lean fuel/air mixture. Gray et al. [47] reported 90–98% reduction in BSNOx under all operating conditions. Iwabuchi et al. [48] performed their experiments on PCCI combustion and reported ultra-low NOx emissions at a cost of slightly higher specific fuel consumption. Kimura et al. [49] developed a new combustion concept "Modulated Kinetics" (MK) and reported 90% reduction in NOx levels as compared to CIDI engines.

Hydrocarbon emissions from the LTC indicated incomplete combustion. For LTC, the whole cylinder volume was full of homogeneous fuel–air mixture and the combustion temperature was relatively lower, which promoted formation of HC emissions. Increasing EGR resulted in higher HC emissions primarily due to reduction of in-cylinder temperatures. The mixture trapped in crevices will be too cold to ignite during LTC. Fraction of HC emissions originating from crevice volume increases with increasing compression ratio. Due to increasing compression ratio, the amount of charge trapped in crevice volume increases; hence, HC emissions also increase.

It has been found that LTC engines usually have elevated CO emission levels, particularly at low loads. This was due to low in-cylinder temperature of LTC; hence, fuel near the cylinder walls does not burn. In the conventional CI combustion mode, a large fraction of CO is oxidized into CO_2 due to high combustion temperature. Low combustion temperature in case of LTC mode also reduces the oxidation of CO into CO_2.

4.2 Unregulated Emissions from LTC

Major pollutants emitted in HCCI combustion studies for different engine operating conditions and combustion chamber configurations are regulated emissions. The formation mechanism, sources, and harmful effects of unregulated emissions from HCCI combustion have not been fully explored and understood yet. Polycyclic aromatic hydrocarbons (PAHs) are potentially carcinogenic, while oxygenated hydrocarbons (OHC) such as aldehydes or ketones act as ozone precursors.

Many researchers have investigated the health effects and environmental effects of unregulated gaseous emission species (Table 1). Ravindra et al. [51] suggested

Table 1 Health and environmental effects of unregulated emission species [50]

Unregulated emission species	Possible health and environmental effects
Methane (CH_4)	• GHG with greenhouse index 21 times that of CO_2. • Simple asphyxiant, when inhaled • Leads to headache, dizziness, weakness, nausea, vomiting, and loss of consciousness
Normal-butane (n-C_4H_{10})	• Inhalation causes euphoria, drowsiness, narcosis, asphyxia, cardiac arrhythmia, and fluctuations in blood pressure
Isobutane (iso-C_4H_{10})	• Simple asphyxiant, when inhaled • Causes fatigue, dizziness, headache, and nervous system damage
Normal pentane (n-C_5H_{12})	• Affects central nervous system • Causes erythema, hyperemia, swelling, pigmentation, and anoxia • Burning sensation accompanied by itching and blisters
Normal octane (n-C_8H_{18})	• Giddiness, vertigo, skin redness and rashes, brain irritation, or apneic anoxia • Throat and lungs problems and headache
Ethylene (C_2H_4)	• Causes headache, drowsiness, dizziness, nausea, weakness, and unconsciousness • Causes irritation to respiratory system, alters carbohydrate metabolism • Acts as ozone formation agent
Acetylene (C_2H_2)	• Causes suffocation, dizziness, headache, unconsciousness, and nausea • Inhalation results in high blood pressure, fits and abnormal heart rhythm
Benzene (C_6H_6)	• Drowsiness, dizziness, rapid or irregular heartbeats, headache, tremors, confusion, unconsciousness, and carcinogenic • Chromosomal aberrations in human peripheral lymphocytes
Formaldehyde (HCHO)	• Irritation in eyes, nose, and throat, coughing, and skin irritation • Considered as human carcinogen, causes asthma-like respiratory problems • Affects pregnancy and reproductive system
Acetaldehyde (CH_3CHO)	• Irritation of skin, eyes, mucous membrane, throat, respiratory tract, nausea, vomiting and headache • Probable carcinogen
Formic acid (HCOOH)	• Causes teary eyes, running nose, coughing, sore throat, bronchitis, shortness of breath, pulmonary edema, liver and kidney damage • Burns tissues and membrane of the skin, and respiratory tract
Ethanol (C_2H_5OH)	• Causes unconsciousness • Affects formation of antidiuretic hormones, leading to brain disability • Affects nervous system of developing embryo and fetus
Sulfur dioxide (SO_2)	• Higher concentration (>100 ppm) causes danger to life and health • Burning sensation in nose and throat, breathing difficulties, and severe airway obstructions

that there should be appropriate regulations for PAHs. PAHs, carbonyl compounds, benzene, toluene, ethylbenzene and xylene (BTEX) are harmful for human health. These are categorized as "possible carcinogens."

Merritt et al. [52] carried out experiments in a light-duty high-speed diesel engine using different combustion modes and compared the different regulated and unregulated exhaust species emitted from PCCI and LTC combustion modes. They observed significantly higher carbonyl compounds from all LTC modes and PCCI combustion at lean conditions compared to baseline CI combustion. However, PCCI combustion at rich conditions produced much lower carbonyl emissions than diesel operations. In PCCI combustion, PAHs were found to be substantially higher compared to baseline diesel operation. Total NPAH emissions were much higher at rich and lean LTC operation compared to baseline diesel operation. Sluder et al. [53] reported that different organic species in engine exhaust including formaldehyde, 1,3-cyclo-pentadiene, and benzene increased at different rates, when combustion mode was changed from conventional CI to a PCCI, indicating a change in fundamental combustion process. Natti et al. [54] studied the effect of operating parameters such as swirl ratios, injection pressures, injection timings, and EGR rate on regulated and unregulated emissions and their sources in a high-speed direct injection (HSDI) diesel engine operating in LTC regime using low sulfur diesel. They reported high levels of volatile organic compounds (VOCs) and PAHs together with UHC and CO emissions during the LTC regime. Similar results were also reported by Bohac et al. [55] during their investigations on speciated hydrocarbon emissions from a diesel engine and diesel oxidation catalyst (DOC) using conventional and PCCI combustion.

Ogawa and Li [56] investigated the unregulated harmful emissions from a diesel engine under different operating conditions. They reported the significance of VOCs and aldehyde compounds at low-load or idling conditions compared to total hydrocarbons (THC). Also, there was notable increase in formaldehyde and acetaldehyde emissions with low coolant temperatures at light loads. Further, they observed a drastic increase in VOCs and some low-molecular HCs with intake oxygen content lower than 14% due to EGR. They finally concluded that oxidation catalysts were effective in reducing VOC emissions including aldehydes and some unsaturated HCs. However, at overall rich fuel–air mixture condition, use of catalysts showed no significant reduction in aromatics and methane generation from ultra-high EGR LTC.

4.3 Particulate Emission Characteristics of LTC

PM emissions, especially small-sized particles emitted from engines, have harmful impact on urban air quality and human health. Studies have already shown the correlation between human health and PM emissions; therefore, environmental protection agencies are more concerned about PM emitted from IC engines [57].

Depending on size, particulates are categorized into three distinct types, namely (i) nucleation mode (<50 nm), (ii) accumulation mode (50–1000 nm), and (iii) coarse mode (>1000 nm) [58–60] (Fig. 10). Area corresponding to the curve in any size range showed the concentration of particles in that size range (Fig. 10).

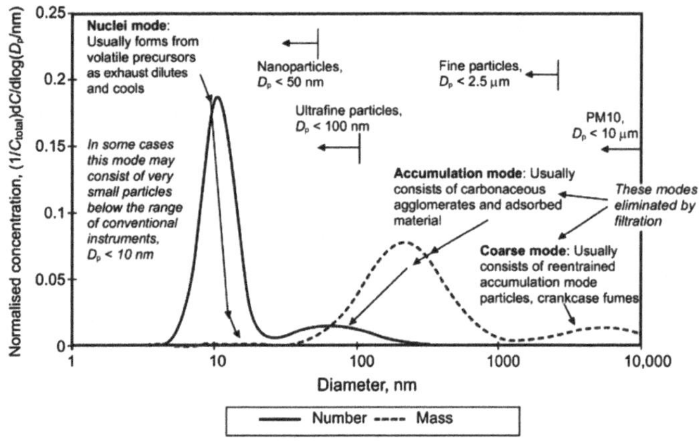

Fig. 10 Typical particle size distribution from IC engines [58]

The environmental impact of particles also depends on particle size since it influences residence time of the particulate in the atmosphere [61], optical properties of the particulate [62], particle surface area available for adsorption [63], its ability to participate in atmospheric chemistry [58], and its health impact [64].

LTC concept is known for its potential for very low NOx and PM emissions, which are usually close to negligible. However, recent investigations have shown that though the total mass of PM emitted from LTC is negligible; however, number concentration of relatively smaller particles (in the size range <100 nm mobility diameter) cannot be neglected [65–68]. This size range of particulate can be measured by modern nanoparticle sizing instruments, which are suitable for studies of LTC particulate. Singh and Agarwal [69] performed LTC experiments using different test fuels (mineral diesel, dieseline, diesohol, and diesosene) to investigate the effect of fuel volatility on particulate emission characteristics. The experiments were carried out at constant EGR and intake charge temperature at different engine loads. They observed that diesel particle number concentration was higher in ultra-fine range; however, it was slightly lower in nanoparticle range. Lower in-cylinder temperature with extremely lean operation prevented full oxidation of the boundary layer and crevice bound hydrocarbons, thereby increasing concentration of HC precursors, which enabled higher nucleation. Presence of large accumulation mode particles for all the test fuels was explained by the existence of at least some degree of diffusion burning.

Misztal et al. [66] carried out LTC experiments to investigate detailed characteristics of particulate emitted from DI-HCCI system using unleaded gasoline and NVO to capture the residuals. The main focus of this work was to explain the consequences of intake air heating on PM emissions in a LTC engine. This study suggested that by preheating the inlet air, PM emissions from LTC engines could be reduced. The effect was mainly due to higher in-cylinder temperature during the compression stroke, leading to improved fuel evaporation. Since evaporation and wall wetting phenomenon are unique to DI fueling strategies, this trend may not be expected as characteristic of all LTC engines. Another investigation by Misztal et al. [67] examined the role of injection timing in PM formation in the same engine.

Franklin [70] reported that HCCI combustion resulted in negligible accumulation mode (soot) particles; however, significant nucleation mode particles were present in the exhaust of a fully premixed HCCI engine. They suggested that volatile species present in the lubricating oil acted as precursors for these nucleation mode particles. Effect of intake air pressure on particle size–number distribution was studied by Desantes et al. [71]. They investigated the effect of engine parameters such as intake air oxygen content, intake air pressure on the engine out emissions. They observed that a slight increase in intake air pressure caused a significant reduction in CO, HC, PM mass and number emissions. Maurya et al. [68] reported the effect of port fuel injection timings on particulate emissions from a methanol-fueled HCCI engine. Effect of SoI timings on particulate emissions at different T_i for constant fuel quantity injected (25 mg/cycle) is shown in Fig. 11. They reported that the peak concentration of particles increased with increasing T_i up to 170 °C and further increase in T_i did not increase the peak concentration of particles.

Maurya et al. [68] also investigated the effect of SoI on particulate as a function of different fueling at constant T_i (Fig. 12).

Above studies suggested that in HCCI engines, significant number of particles remained in the size range <100 nm mobility diameter which cannot be neglected. Various researchers have already shown the health effects of particulate [72]; therefore, further research is needed to characterize PM formation in HCCI engines with different control strategies and new fuels.

Agarwal et al. [73] performed diesel HCCI experiments and investigated particulate emission characteristics at different λ and EGR rates. They observed that most diesel HCCI particles were ultra-fine particles. Increasing EGR rate resulted in higher particle number concentration. They also suggested that increasing EGR rate

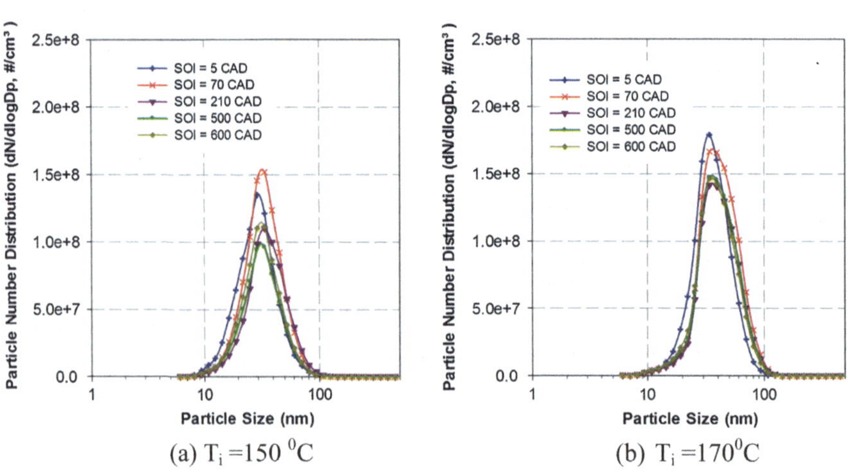

Fig. 11 Particle size–number distribution at 1500 rpm with varying SoI for different inlet air temperatures [68]

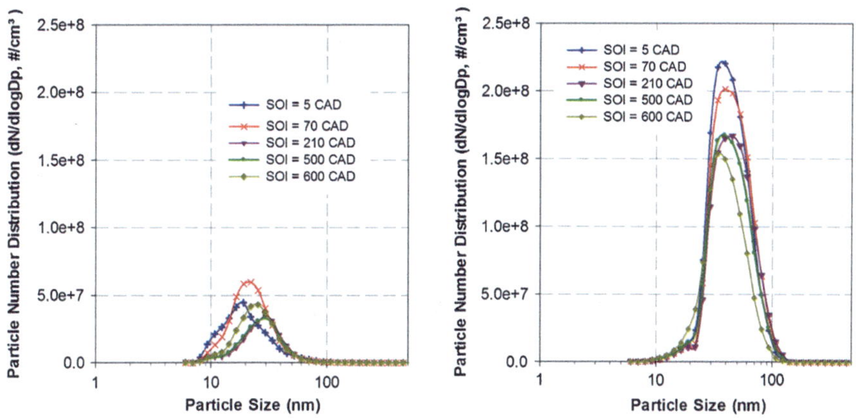

(a) 21 mg/stroke fuel (b) 29 mg/stroke fuel

Fig. 12 Particle size–number distribution at 1500 rpm for varying SoI for different fuel injection quantities [68]

and λ resulted in higher number of accumulation mode particles, which was mainly due to higher BSOF of the PM. According to Kittleson [58], approximately 10% (w/w) of PM is inorganic, which primarily includes trace metals and ash. In a study carried out by Agarwal et al. [74], gasoline-fueled HCCI combustion was investigated. They reported that PM emissions from HCCI engine largely depended on EGR rate, λ, and inlet air temperature (T_i). In their experiment, total particle number concentration increased with increasing T_i; however, it reduced for leaner mixtures. Increasing T_i led to lower surface area of the PM, which showed lesser toxic substance carrying capacity of PM and this was further reduced by decreasing λ. Many researchers experimentally investigated the particulate emission characteristics of biodiesel-fueled engines and reported that biodiesel showed sharp reduction in particle mass emissions compared to baseline mineral diesel. Park et al. [75] investigated the nanoparticle size distribution of particulate emitted by diesel engine fueled by 20% biodiesel blend and the application of EGR reduced particle number and mass emissions from both B20 and mineral diesel by $\sim$43%.

Several researchers studied the effect of fuel injection parameters and strategies on particulate number–size distribution using a single-cylinder research engine and reported that particulate number–size distribution reduced with increasing FIP [76, 77]. Total particulate numbers emitted by Karanja biodiesel blends were lower than mineral diesel [78]. Agarwal et al. reported the lowest particulate numbers emitted by 10% Karanja biodiesel blend [78]. Dhar and Agarwal [79] reported that total particulate number concentration of B20 and B50 were lower than baseline mineral diesel. Increasing FIP resulted in relatively lesser particulate number concentration. Fuel injection timing also affected the particulate number–size distribution, and it increased at retarded main injection timings. Di et al. [80] reported that with increasing oxygen content of diglyme (DGM) blended with ULSD, smoke opacity, particulate mass emission, and geometric mean diameter of particles reduced.

Previous studies reported that bioavailable, soluble, and particle-bound trace metals can harm the human health [81, 82]. Fine particle-associated trace metals (such as vanadium and aluminum) can affect iron (Fe) status of alveolar macrophage and alter the pulmonary immune competence of exposed hosts [83]. Nanoparticles (Dp < 50 nm) as carrier of heavy metals (such as cobalt and manganese) enhance the formation of reactive oxygen species (ROS) by a factor of eight compared to pure aqueous solution of the same metals [84]. Certain trace metals in the exhaust are of interest for their relevance to potential health effects, such as lead (Pb), manganese (Mn), nickel (Ni), arsenic (As), cadmium (Cd), and chromium (Cr) and have been classified by USEPA as mobile source air toxics (MSAT). Hu et al. [85] performed particulate bound trace metal analysis from a heavy-duty diesel engine equipped with emission reduction devices. They divided the trace metals into several categories, namely oil additives metals (Ca and Zn), Fe, Cu, platinum group metals (PGE), (V and Ti), MSAT metals (As, Cr, Pb, Mn, and Ni), and other metals. Valavanidis et al. [86] reported that deposition of these trace metals, which are originally from diesel particulate, in the lower airway of the human respiratory system could generate hydrogen radicals and then trigger production of oxygen free radicals, which can potentially cause both acute and chronic lung injuries. Agarwal et al. [73, 74] performed HCCI experiments to investigate particulate bound trace metal analysis from mineral diesel and gasoline-fueled HCCI engines. They reported that the trace metals detected were comparatively lower in PM emitted from diesel-fueled HCCI engine. Trace metal concentration and BSOF increased with increasing EGR rate. They reported that BSOF from gasoline HCCI was negligible; therefore, gasoline could be mixed with mineral diesel to reduce BSOF.

5 LTC Control Strategies

Stable and efficient operation of LTC engines need precisely controlled combustion timings. One of the main challenges in LTC engines is the combustion control since onset of combustion depends on in-cylinder temperature, pressure, and fuel–air mixing inside the combustion chamber, and there is no direct control for initiating the combustion. When combustion control is not fast enough, too advanced or too retarded combustion can take place in the engine. Too advanced combustion can yield unacceptable RoPR or unacceptable peak cylinder pressure, thus causing excessive noise, which may potentially damage the engine. Additionally, NOx emissions from the engine tend to increase with ignition advance [87]. Another driver for gaining an effective closed-loop combustion control is the fact that late combustion timing leads to incomplete combustion and increasing emissions of CO and HC. The worst case of "too late combustion" leads to a complete misfire, which if repeated, can cause the engine to stall.

Several means to actuate combustion phasing for LTC engine control have been suggested by various researchers [90–92] such as dual fuel, variable valve timing

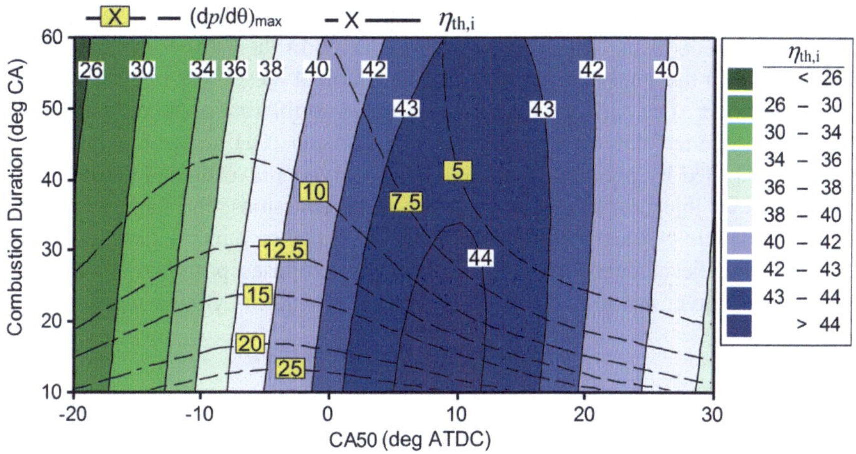

Fig. 13 Thermal efficiency variation with combustion phasing [18]

(VVT), variable compression ratio (VCR), and active thermal management. Riseberg et al. [93] investigated influence of NO on combustion phasing in a single-cylinder LTC engine. Isooctane/*n*-heptane blend (primary reference fuels; PRF), toluene/*n*-heptane mixture (Tertiary reference fuels; TRF), and full boiling range gasoline were tested at two different intake charge temperatures. They reported that NO concentration in the air significantly affected combustion phasing of the LTC engine using different test fuels. They explained this behavior using chemical kinetics theory. Asad et al. [18] showed a relationship between combustion duration, combustion phasing and indicated thermal efficiency (ITE). They stated that combustion phasing for the highest ITE lies in a small crank angle window between 7° and 12° aTDC. Any deviation (advanced or retarded combustion) resulted in a rapid drop in thermal efficiency (Fig. 13).

Following subsections describe different control strategies and their status in LTC engines.

5.1 Dual-Fuel Control Strategy

Dual-fuel combustion is commonly referred to as RCCI combustion and has been demonstrated extensively for mineral diesel–gasoline [27] and mineral diesel–ethanol blends [94]. In dual-fuel-type controls, two fuels with different auto-ignition properties are used. The system would have a main fuel with a high octane number and a secondary fuel with low octane number. Different auto-ignition properties of dual-fuel systems are used to control the combustion phasing in LTC as blending two fuels in different proportions changes their auto-ignition properties. Use of commercial fuels or mixtures of single-component and commercial fuels; PRFs have been investigated in several studies [95–97]. For better understanding of the dual-fuel mode operation of CI engines, Mancaruso and Vaglieco [98] performed PCCI experiments using ethanol and mineral diesel. In these experiments, ethanol was injected in the intake manifold; however, mineral diesel was injected directly into the cylinder. The experiments were performed at different premixed fuel ratios, and it was reported that the LTC governing parameter (OH^- radicals) was significantly controlled by premixed fuel ratio. Ma et al. [97] investigated the effects of different diesel injection strategies on combustion, emissions, and fuel economy of a modified single-cylinder diesel engine fueled by gasoline/mineral diesel dual fuel. This gasoline/diesel dual-fuel combustion mode proposes port fuel injection of gasoline and direct injection of mineral diesel with rapid in-cylinder fuel blending.

5.2 *Variable Compression Ratio (VCR) Strategy*

VCR can be used to control combustion phasing by increasing the compression ratio and charge temperature after the compression. VCR can be achieved by several different methods. Lack of control over an individual cylinder, which is necessary to obtain good combustion phasing control, is the main drawback of the VCR system. Cost and complexity of VCR systems are other major obstacles for their application in LTC engines. Asad et al. [18] carried out LTC experiments at different CR and reported that a lower CR improves combustion phasing, reduces maximum RoPR and peak cylinder pressure; therefore, LTC operating range can be increased (Fig. 14). A CR of 14–16 with advanced turbo-charging, VVT, and improved cylinder charge cooling may provide a realistic compromise between high efficiency CI combustion and clean diesel LTC.

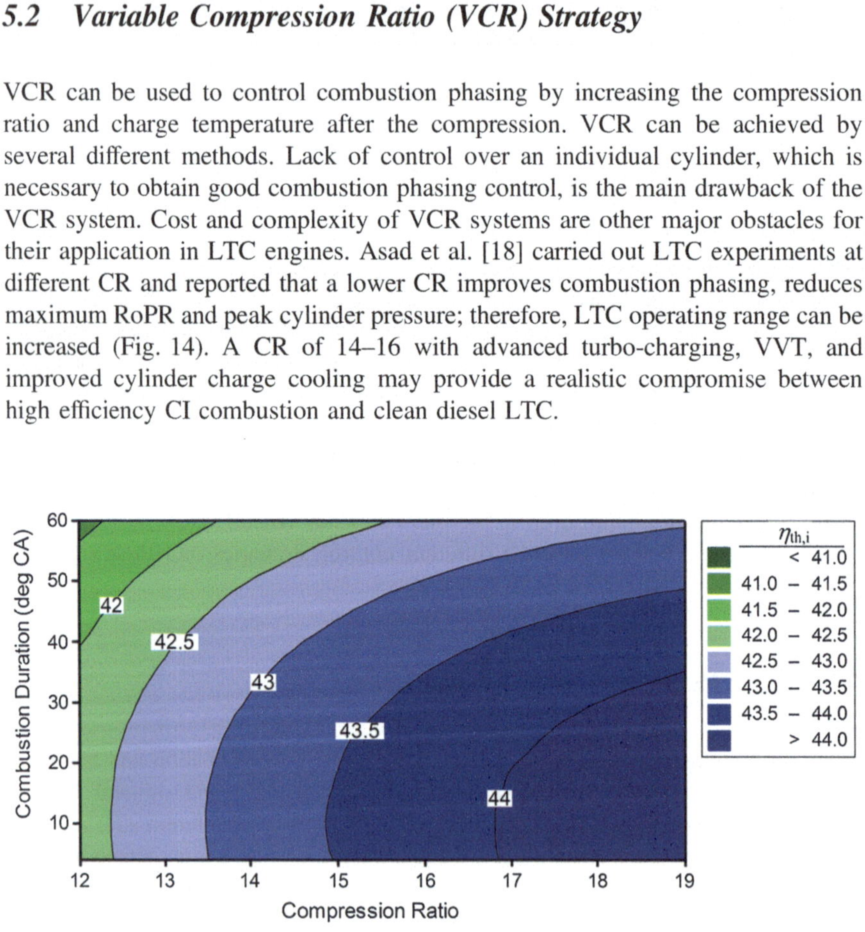

Fig. 14 ITE variations with the compression ratio [18]

5.3 *Exhaust Gas Recirculation Strategy*

EGR is essential to achieve simultaneous reduction in soot and NOx emissions from LTC without prohibitively high fuel consumption penalties due to poor combustion phasing. Tuning the amount of EGR is the most commonly used technique to adjust the in-cylinder temperature, which controls the SoC. Initially, Thring [99] achieved LTC by varying the inlet air temperature and EGR fraction over a range of equivalence ratios. High heat capacity constituents of EGR are CO_2, H_2O, N_2, O_2, CO, PM, HC, NOx, and other intermediate species of combustion reactions which control the LTC [100–102]. These constituents showed following four effects on the combustion and emissions. First was the preheating effect, in which the inlet charge temperature increased when hot EGR was mixed with the fuel–air mixture. Second was the dilution effect, where by the introduction of EGR led to substantial reduction in oxygen concentration. Third was the heat capacity effect, in which total heat capacity of the mixture of EGR, air, and fuel would be higher owing to higher heat capacity of CO_2 and water vapor. This would lead to a reduction in gas temperature at the end of the compression stroke. Fourth and final was the chemical effect, where unburnt combustion products in the EGR would take part in chemical reactions. HC, CO, CO_2, NO, H_2O, etc., in the EGR take part in chemical reactions and lead to a moderate effect on the reaction rates. Integrating all these effects, SoC during the LTC and combustion duration of overall HCCI-DI combustion can be controlled by regulating the EGR quantity. Therefore, EGR makes it possible to suppress the excessively advanced SoC by low-temperature reactions in the LTC phase. This excessively advanced and rapid combustion causes knocking, which limits the operating range of LTC [103]. EGR can be classified as internal EGR and external EGR. Internal EGR rate can be controlled by changing the valve overlap period and the external EGR rate can be adjusted by the combined effect of the exhaust backpressure and EGR. For high-octane fuels such as gasoline, NVO is recognized as one of the possible implementation strategies for LTC. The effect of inlet air temperature via NVO is insignificant when the engine runs well within the LTC operating range [104]. Cooled external EGR reduces the fuel–air mixture temperature in the compression stroke and hence delays the SoC of high cetane fuels such as biodiesel.

EGR stratification was a novel technique demonstrated by André et al. [105], which could be used for the controlling rapid HRR and combustion noise in the LTC engines thus extending the operating range of PCCI combustion engines. The exhaust gas was inducted into the combustion chamber via a helical port and induction of fresh air was through the tangential port. Thus, a delay in combustion, in the stratified case, was caused by the stratified exhaust gas present close to the TDC. This exhaust gas was present in the fuel-rich zone and caused delay in combustion, hence gaining a superior control over the combustion phasing. Even though this technique offered benefits of combustion control, it still needs to be explored in greater detail before its practical implementation because of its high dependence on various fuel injection parameters. Charge dilution using either EGR or non-reacting species such as CO_2 and N_2 in order to achieve LTC has been extensively analyzed. Kook et al. [106] and Kanda et al. [107] carried out experiments using various dilution rates to attain PCCI combustion. They observed a direct correlation of NOx and soot luminosity with adiabatic flame temperature and reported that adiabatic flame temperature decreased with addition of EGR. The high oxidation rates at higher peak temperatures led to reduction in CO emissions. However, the researchers also pointed out the need to control dilution ratios because maximum fuel conversion efficiency was obtained at moderate charge dilution levels. Over-dilution leads to poor trade-off between work conversion efficiency and combustion efficiency, thus reducing fuel conversion efficiency. Another advantage of charge dilution is increased ignition delay, which improves IMEP and COV. Asad et al. [18] performed LTC experiments using different EGR rates to postpone SoC by increasing the ignition delay and to reduce the severity of high RoPR. They stated that EGR was effective in delaying combustion phasing toward higher thermal efficiency window by withholding the cylinder charge from early ignition. Figure 15 shows that increasing EGR rate results in stable combustion. It was concluded that combustion efficiency reduced with increasing EGR, which was

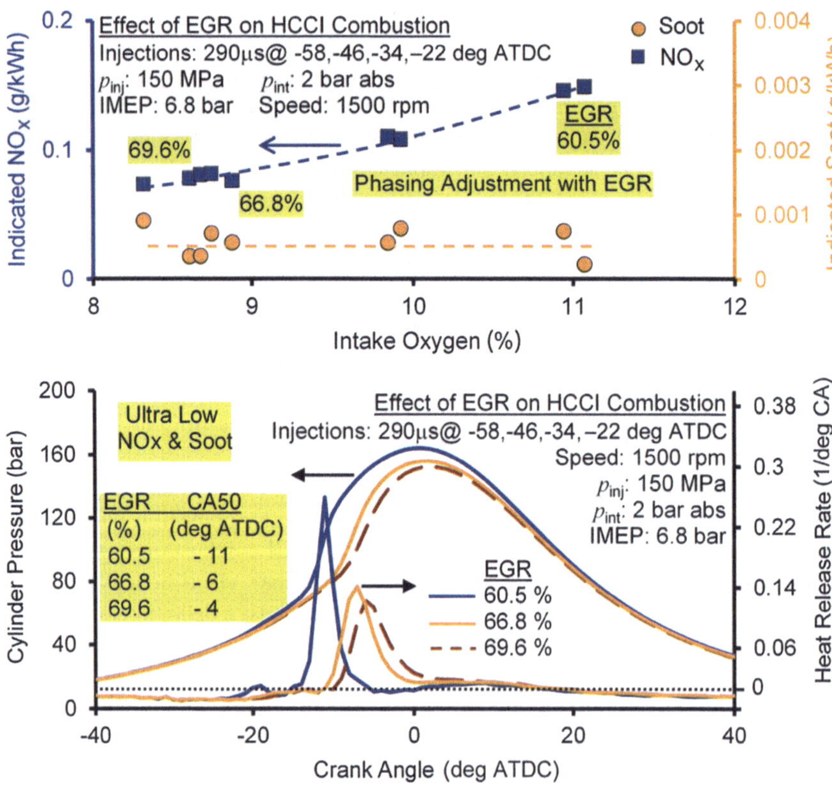

Fig. 15 Effect of EGR on HCCI combustion and emissions [18]

compensated by improvement in combustion phasing. This trade-off between combustion efficiency and emissions showed the importance of EGR in LTC engines.

5.4 Combustion Control by Fuel Additives

Fuel composition has little influence on reaction rate and reaction duration once reactions have been initiated; however, it defines the auto-ignition temperature, and hence can control SoC of LTC engines. Fuel composition mainly affects LTR, which, in turn, affects the start of main reactions. Main fuel parameters influencing the physical delay part of the ignition delay (ID) are density, heating value, and latent heat of vaporization. The chemical delay part of ID is influenced by auto-ignition and distillation properties of the fuel. An increase of fuel density, volatilization difficulties, and higher latent heat of vaporization (due to decreasing fuel–air mixture temperature), all result in delayed combustion start/ignition. Common ways to classify fuels are according to the ease of auto-ignition, which is defined as CN and resistance to auto-ignition, which is defined by ON. A high CN represents less resistance to auto-ignition, which includes straight chain paraffin (iso-cetane defines CN = 100). A high octane number represents the resistance against auto-ignition, which includes branched chain paraffins (isooctane defines ON = 100). Gasoline has a high octane number and therefore has little or no low-temperature reactions; and the combustion initiation takes place at around 950 K. Diesel-like fuels show significant low-temperature reactions and have initiation temperatures in the range of 750 K [108].

Starck et al. [109] carried out experiments in quest of impact of fuel properties on LTC and found lower CN fuels to be superior LTC fuels. They reported that a lower CN fuel and an optimum combustion speed could improve the LTC operation range because lower combustion speed results in more time for homogenization, resulting in superior LTC. Tanaka et al. [110] used a Rapid Compression Machine (RCM) to study the effects of fuel structure and additives on homogeneous charge compression ignition (HCCI) of pure hydrocarbon fuels and mixtures. They investigated the effect of fuel additives on LTC performance and found that fuels with saturated compounds lead to two-stage combustion and those with unsaturated compounds led to single-stage combustion. Higher octane number results in higher ignition delay and a lower burn rate.

5.5 LTC Control by Fuel Injection Strategy

Early injection led to impingement of mineral diesel on the cylinder walls, particularly at high BMEPs [111–113]. Single-injection LTC is applicable to limiting conditions, where tighter controls on operating conditions are required compared to conventional CI combustion. Issues of LTC such as high HRR and uncontrolled combustion can be resolved by using split fuel injection strategy. Kook and Bae [114] used split injection in order to promote low-temperature ignition and to gain better combustion control. Major quantity of fuel was injected early in the compression stroke (100° bTDC) to prepare premixed charge, and a pilot quantity was injected near TDC to control SoC. This resulted in better combustion characteristics, precise control of SoC, and higher BMEP. Results clearly showed that NOx emissions from PCCI combustion were much lower compared to CI combustion and advanced pilot injection (from 100° BTDC to 150° BTDC) showed further reduction in NOx.

The effect of split injection strategy on emission formation in a PCCI diesel engine was investigated by Horibe et al. [115] and Torregrosa et al. [116]. The researchers realized that higher thermal efficiency and lower NOx emissions at moderate loads could be achieved by single-injection strategy; however, they experienced problem of very high RoPR. In case of split injection strategy, a small quantity of pilot injection helped in suppressing higher RoPR. The application of early pilot injection resulted in a significant improvement in thermal efficiency and reduction in engine noise and emissions by optimizing SoI timings and EGR rate. They later dealt mainly with effect of pilot injection on engine noise and performance. Engine noise remained one of the major factors governing combustion and performance characteristics. They reported a reduction in BMEP with an increase in pilot quantity above 40%. Further, relatively higher engine noise was observed for almost all early single-injection timings, mainly due to very high RoPR. The inclusion of pilot injection proved to be very effective in reducing noise, but it also led to reduction in BMEP. Neely et al. [117] investigated the effect of number of pilot injections (up to 3) to achieve PCCI combustion in order to reduce NOx in light-duty and heavy-duty vehicles. They stated that an early single pilot injection was effective for ~14% NOx reduction, but it was at the expense of higher CO and BSFC. However, CO and BSFC penalty significantly reduced along with NOx by employing multiple pilot injections at lower loads. This was mainly due to superior fuel–air mixing. For heavy-duty vehicles, multiple pilot injections proved to be ineffective in reducing NOx compared to single pilot injection. In addition, multiple pilot injections also led to higher HC and CO emissions. Asad et al. [18] performed HCCI experiments and suggested that optimized SoI of the multiple injections could assist in preparation of a near-homogeneous charge that resulted in near-zero soot emissions. Yehliu [118] observed slightly higher NOx emissions from B100 compared to mineral diesel in single-injection mode. Split injection reduced NOx emissions from both fuels, but higher reduction was observed in case of B100 fueled engine, which showed 18% lower NOx emission compared to mineral diesel.

6 Path Forward for LTC

Recent research activities related to LTC have significantly influenced the global perspective about it because of development of different control mechanisms and new strategies. Hence, prior to commercial production of LTC engines for heavy-duty and light-duty applications, tremendous R&D effort is required. Some of the possible areas for future research for improving this LTC concept are as follows:

(i) Development of control methods for maintaining ignition timing at different engine loads and speeds is the first challenge, which needs to be resolved. Optimization of ignition timing is quite important for LTC engines compared to conventional engines because LTC engines have no direct control over ignition timing such as by using spark or fuel injection.
(ii) Methodology development to slowdown the rate of combustion in LTC engines at high engine loads is the next challenge, which needs to be resolved to prevent excessive noise and engine damage.
(iii) LTC engines emit very low NOx at low-to-medium engine loads; therefore, no emission control is required; however, at higher engine loads, NOx emission become too high, which required emission control equipment. Therefore, low-cost emission control equipment needs to be developed to control NOx emissions at higher engine loads.
(iv) For the development of production grade LTC engines, closed-loop control of different operating parameters needs to developed. Control mechanisms, sensors, and appropriate control algorithms are key enabling technologies for practical LTC engines.
(v) In conventional SI or CI engines, small differences in intake and exhaust flow between cylinders do not affect engine combustion and performance characteristics significantly. However, LTC engines are very sensitive to small changes in compressed-charge temperature, which leads to significant cylinder-to-cylinder variations in combustion timings. Therefore, research is required for development of intake and exhaust manifold designs for multi-cylinder engines so that difference in the inlet and exhaust flows between different cylinders can be minimized.

7 Conclusions

LTC is a combustion concept, which has evolved over decades in response to the need for improved thermal efficiency of gasoline-fueled engines and ultra-low NOx and soot emissions of diesel-fueled engines. Although remarkable progress has been made in LTC technology, large-scale production of LTC engines for commercial applications has encountered several challenges. Limited operating range, lack of direct control on SoC, homogeneous fuel–air mixture preparation and higher HC and CO emissions are the main obstacles faced by LTC technology's adaptation commercially. Several techniques have been developed for low-load application of LTC technology in heavy-duty engines; however, full load application even in light-duty engines has not been demonstrated till now. To control the SoC, some advanced concepts such as spark assisted LTC and laser assisted LTC have also been investigated, in which combustion events were precisely controlled using a spark plug or a laser. Different derivatives of LTC such as PCCI, HCCI were thoroughly investigated, and suitability of each derivative has been defined for a particular operating range. These derivatives of LTC can be achieved by varying engine operating and control parameters. Fuel flexibility is the other important feature of LTC. Fuel properties significantly affect chemical kinetics, which has a dominating role in LTC. In LTC, auto-ignition was controlled by modifications of the fuel properties in order to make it more chemically reactive by adding an ignition promoter or inhibitor. In all derivatives of LTC, control is the most critical issue, which was resolved using various techniques such as dual-fuel injection, VCR, VVT, EGR. However, these techniques have their own merits and challenges. Based on these research directions, several researchers proposed use of dual combustion system, which seems to be an effective solution for commercializing LTC technology.

References

1. Association IE (2011) World Energy Outlook 2011 Factsheet → International energy agency, World Energy Outlook 2011 Factsheet
2. Statistics I. CO_2 emissions from fuel combustion-highlights. IEA, Paris http://www.iea.org/co2highlights/co2highlights.pdf. Cited July 2011
3. Maricq MM (2007) Chemical characterization of particulate emissions from diesel engines: a review. J Aerosol Sci 38:1079–1118
4. Birol F (2010) World energy outlook 2010. In: International Energy Agency, vol 1
5. Knothe G (2010) Biodiesel and renewable diesel: a comparison. Prog Energy Combust Sci 36:364–373
6. Web Source: https://www.e-education.psu.edu/egee439/node/684. Accessed on 7th Mar 2016
7. Agarwal AK (1998) Vegetable oils versus diesel fuel: development and use of biodiesel in a compression ignition engine. TIDE. 8:191–204
8. Agarwal AK, Das LM (2001) Biodiesel development and characterization for use as a fuel in compression ignition engines. J Eng Gas Turbines Power 123:440–447
9. Urja A (2013) Ministry of new and renewable energy, vol 7(1). Government of India, New Delhi
10. Cozzi L, Head EMU (2012) World Energy Outlook 2012
11. Pundir BP (2007) Engine emissions: pollutant formation and advances in control technology: Alpha Science International, Limited
12. Fischer M, Werber M, Schwartz PV (2009) Batteries: Higher energy density than gasoline? Energy Policy 37:2639–2641
13. Heywood JB (1988) Internal combustion engine fundamentals. Mcgraw-hill, New York

14. Wåhlin F (2007) Experimental investigation of impinging diesel sprays for HCCI combustion. Doctoral thesis, Royal Institute of Technology, ISRN/KTHMMK/R-06/17-SE
15. Flynn PF, Hunter GL, Durrett RP, Farrell LA, Akinyemi WC (2000) Minimum engine flame temperature impacts on diesel and spark-ignition engine NOx production. SAE Technical paper 2000-01-1177
16. Van Basshuysen R, Schäfer F (2004) Internal combustion engine handbook-basics, components, systems and perspectives. SAE 2016-03-07
17. Dec JE (1997) A conceptual model of DI diesel combustion based on laser-sheet imaging. SAE Technical paper 970873
18. Asad U, Zheng M, Ting DSK, Tjong J (2015) Implementation challenges and solutions for homogeneous charge compression ignition combustion in diesel engines. J Eng Gas Turbines Power 137:101505
19. Asad U, Zheng M, Han X, Reader GT, Wang M (2008) Fuel injection strategies to improve emissions and efficiency of high compression ratio diesel engines. SAE Int J Eng 1:1220–1233
20. Kodama Y, Nishizawa I, Sugihara T, Sato N, Iijima T, Yoshida T (2007) Full-load HCCI operation with variable valve actuation system in a heavy-duty diesel engine. SAE Technical paper 2007-01-0215
21. Zhao H, Xie H, Peng Z (2005) Effect of recycled burned gases on homogeneous charge compression ignition combustion. Combust Sci Technol 177:1863–1882
22. Shi L, Cui Y, Deng K, Peng H, Chen Y (2006) Study of low emission homogeneous charge compression ignition (HCCI) engine using combined internal and external exhaust gas recirculation (EGR). Energy 31:2665–2676
23. Asad U, Divekar P, Zheng M, Tjong J (2013) Low temperature combustion strategies for compression ignition engines: operability limits and challenges. SAE Technical paper 2013-01-0283
24. Asad U, Zheng M (2009) Efficacy of EGR and boost in single-injection enabled low temperature combustion. SAE Int J Eng 2:1085–1097
25. Kimura S, Ogawa H, Matsui Y, Enomoto Y (2002) An experimental analysis of low-temperature and premixed combustion for simultaneous reduction of NOx and particulate emissions in direct injection diesel engines. Int J Engine Res 3:249–259
26. Akihama K, Takatori Y, Inagaki K, Sasaki S, Dean AM (2001) Mechanism of the smokeless rich diesel combustion by reducing temperature. SAE Technical paper 2001-01-0655
27. Kokjohn S, Hanson R, Splitter D, Reitz R (2011) Fuel reactivity controlled compression ignition (RCCI): a pathway to controlled high efficiency clean combustion. Int J Engine Res 12:209–226
28. Agarwal AK, Singh AP, Maurya RK (2017) Evolution, challenges and path forward for low temperature combustion engines. Prog Energy Combust Sci 61:1–56
29. Elkotb M (1982) Fuel atomization for spray modelling. Prog Energy Combust Sci 8:61–91
30. Lee CS, Park SW (2002) An experimental and numerical study on fuel atomization characteristics of high-pressure diesel injection sprays. Fuel 81:2417–2423
31. Zhao H (2007) HCCI and CAI engines for the automotive industry. Elsevier
32. Fish A (1968) The cool flames of hydrocarbons. Angew Chem Int Ed Engl 7:45–60
33. Singh AP, Agarwal AK (2012) Combustion characteristics of diesel HCCI engine: an experimental investigation using external mixture formation technique. Appl Energy 99:116–125
34. Curran HJ, Gaffuri P, Pitz WJ, Westbrook CK (1998) A comprehensive modeling study of n-heptane oxidation. Combust Flame 114:149–177
35. Westbrook CK, Warnatz J, Pitz WJ (1989) A detailed chemical kinetic reaction mechanism for the oxidation of iso-octane and n-heptane over an extended temperature range and its application to analysis of engine knock. Symposium (international) on combustion. Elsevier, pp 893–901

36. Curran HJ, Pitz W, Westbrook C, Callahan G, Dryer F (1998) Oxidation of automotive primary reference fuels at elevated pressures. Symposium (International) on combustion. Elsevier, pp 379–387
37. Chevalier C, Pitz W, Warnatz J, Westbrook C, Melenk H (1992) Hydrocarbon ignition: automatic generation of reaction mechanisms and applications to modeling of engine knock. Symposium (International) on combustion. Elsevier, pp 93–101
38. Machrafi H (2010) HCCI combustion chemistry reduced kinetic mechanisms and controlling strategies. In: Handbook of combustion. Wiley-VCH Verlag GmbH & Co
39. Li H, Prabhu SK, Miller DL, Cernansky NP (1994) Autoignition chemistry studies on primary reference fuels in a motored engine. SAE Technical paper 942062
40. Ranzi E, Faravelli T, Gaffuri P, Sogaro A (1995) Low-temperature combustion: automatic generation of primary oxidation reactions and lumping procedures. Combust Flame 102:179–192
41. Machrafi H (2008) Experimental validation of a kinetic multi-component mechanism in a wide HCCI engine operating range for mixtures of n-heptane, iso-octane and toluene: Influence of EGR parameters. Energy Convers Manag 49:2956–2965
42. Najt PM, Foster DE (1983) Compression-ignited homogeneous charge combustion. SAE Technical paper 830264
43. Olsson J (2001) Closed-loop control of an HCCI engine. Society of Automotive Engineers. SAE Technical paper 2001-01-1031. Report on "Mass Emission Standards for automobiles—Overview and Technical Details of BS IV, V and VI". Ministry of Road Transport & Highways. Accessed on 17/11/2017. http://pib.nic.in/newsite/backgrounders.aspx?relid=131993
44. Shawn MM (2004) Diesel HCCI with external mixture preparation. DEER 2004, Ohio State University
45. Shawn MM, Guezennec Y, Rizzoni G (2003) Mixed mode diesel HCCI with external mixture preparation: preliminary results. DEER 2003, Ohio State University
46. Takeda Y, Keiichi N, Keiichi N (1996) Emission characteristics of premixed lean diesel combustion with extremely early staged fuel injection. SAE Technical paper 961163
47. Gray III AW, Ryan III TW (1997) Homogeneous charge compression ignition of diesel fuel. SAE Technical paper 971676
48. Iwabuchi Y, Kawai K, Takeda Y (1999) Trial of new concept diesel combustion system—premixed compression—ignited combustion. SAE Technical paper 1999-01-185
49. Kimura S, Aoki O, Ogawa H, Muranaka S, Enomoto Y (1999) New combustion concept for ultra-clean and high-efficiency small DI diesel engines. SAE Technical paper 1999-01-3681
50. Agarwal AK, Shukla PC, Patel C, Gupta JG, Sharma N, Prasad RK et al (2016) Unregulated emissions and health risk potential from biodiesel (KB5, KB20) and methanol blend (M5) fuelled transportation diesel engines. Renew Energy 98:283–291
51. Ravindra K, Sokhi R, Van Grieken R (2008) Atmospheric polycyclic aromatic hydrocarbons: source attribution, emission factors and regulation. Atmos Environ 42:2895–2921
52. Merritt P, Huang Y, Khair M, Pan J (2006) Unregulated exhaust emissions from alternate diesel combustion modes. SAE Technical paper 2006-01-3307
53. Sluder C, Wagner R (2006) An estimate of diesel high-efficiency clean combustion impacts on FTP-75 after treatment requirements. SAE Technical paper 2006-01-3311
54. Natti KC, Bhattacharyya A, Kastury A, Henein NA, Bryzik W (2007) An analysis of regulated and unregulated emissions in a HSDI diesel engine under the LTC regime. SAE Technical paper 2007-01-0905
55. Bohac S, Han M, Jacobs T, Lopez A, Assanis D, Szymkowicz P (2006) Speciated Hydrocarbon emissions from an automotive diesel engine and DOC using conventional and PCI combustion. SAE Technical paper 2006-01-0201
56. Ogawa H, Li T (2010) Volatile organic compounds in exhaust gas from diesel engines under various operating conditions. Int J Engine Res 12. https://doi.org/10.1243/14680874JER595

57. Hall D, King D, Morgan T, Baverstock S, Heinze P, Simpson B (1998) A review of recent literature investigating the measurement of automotive particulate; the relationship with environmental aerosol, air quality and health effects. SAE Technical paper 982602
58. Kittleson DB (1998) Engines and nano-particles: a review. J Aerosol Sci 29(5–6):575–588
59. Abdul-Khalek I, Kittelson D, Brear F (1999) The influence of dilution conditions on diesel exhaust particle size distribution measurements. SAE Technical paper 1999-01-1142
60. Price P, Stone R, Misztal J, Xu H, Wyszynski M, Wilson T et al (2007) Particulate emissions from a gasoline homogeneous charge compression ignition engine. SAE Technical paper 2007-01-0209
61. Harrison R, Brimblecombe P, Derwent R, Dollard G, Eggleston S, Hamilton R et al (1996) Airborne particulate matter in the United Kingdom. Third Report of the Quality of Urban Air Review Group
62. Scherrer H, Kittelson D (1981) Light absorption cross-sections of diesel particles. SAE Technical paper 810181
63. Jakab GJ, Risby TH, Hemenway DR (1992) Use of physical chemistry and in vivo exposure to investigate the toxicity of formaldehyde bound to carbonaceous particles in the murine lung. Research report (Health Effects Institute), pp 1–39, discussion 41-9
64. Donaldson K, Beswick PH, Gilmour PS (1996) Free radical activity associated with the surface of particles: a unifying factor in determining biological activity? Toxicol Lett 88:293–298
65. Eastwood P (2008) Particulate emissions from vehicles. Wiley
66. Misztal J, Xu H, Tsolakis A, Wyszynski ML, Constandinides G, Price P et al (2009) Influence of inlet air temperature on gasoline HCCI particulate emissions. Combust Sci Technol 181:695–709
67. Misztal J, Xu H, Wyszynski M, Price P, Stone R, Qiao J (2009) Effect of injection timing on gasoline homogeneous charge compression ignition particulate emissions. Int J Engine Res 10:419–430
68. Maurya R, Srivastava D, Agarwal A (2011) Experimental investigations of particulate emitted by an alcohol-fuelled HCCI/CAI combustion engine. Int Energy J 12:29–38
69. Singh AP, Agarwal AK (2015) Diesoline, diesohol and diesosene fuelled HCCI engine development. J Energy Res Technol 138(5)
70. Franklin L (2010) Effects of homogeneous charge compression ignition (HCCI) control strategies on particulate emissions of ethanol fuel. University Of Minnesota
71. Desantes JM, López JJ, Redon P, Arregle J (2012) Evaluation of the Thermal NO formation mechanism under low-temperature diesel combustion conditions. Int J Engine Res. 1468087411429638
72. Shah SD, Cocker DR, Miller JW, Norbeck JM (2004) Emission rates of particulate matter and elemental and organic carbon from in-use diesel engines. Environ Sci Technol 38:2544–2550
73. Agarwal AK, Singh AP, Lukose J, Gupta T (2013) Characterization of exhaust particulates from diesel fueled homogenous charge compression ignition combustion engine. J Aerosol Sci 58:71–85
74. Agarwal AK, Gupta T, Lukose J, Singh AP (2015) Particulate characterization and size distribution in the exhaust of a gasoline homogeneous charge compression ignition engine. Aerosol Air Qual Res 15:504–516
75. Park K, Cao F, Kittelson DB, McMurry PH (2003) Relationship between particle mass and mobility for diesel exhaust particles. Environ Sci Technol 37:577–583
76. Agarwal AK, Srivastava DK, Dhar A, Maurya RK, Shukla PC, Singh AP (2013) Effect of fuel injection timing and pressure on combustion, emissions and performance characteristics of a single cylinder diesel engine. Fuel 111:374–383
77. Agarwal AK, Dhar A, Srivastava DK, Maurya RK, Singh AP (2013) Effect of fuel injection pressure on diesel particulate size and number distribution in a CRDI single cylinder research engine. Fuel 107:84–89

78. Agarwal AK, Dhar A, Gupta JG, Kim WI, Choi K, Lee CS et al (2015) Effect of fuel injection pressure and injection timing of Karanja biodiesel blends on fuel spray, engine performance, emissions and combustion characteristics. Energy Convers Manag 91:302–314
79. Dhar A, Agarwal AK (2015) Effect of Karanja biodiesel blends on particulate emissions from a transportation engine. Fuel 141:154–163
80. Di Y, Cheung CS, Huang Z (2009) Experimental investigation on regulated and unregulated emissions of a diesel engine fueled with ultra-low sulfur diesel fuel blended with biodiesel from waste cooking oil. Sci Total Environ 407(2):835–846
81. Hopke PK, Ito K, Mar T, Christensen WF, Eatough DJ, Henry RC et al (2006) PM source apportionment and health effects: 1. Intercomparison of source apportionment results. J Exposure Sci Environ Epidemiol 16:275–286
82. Mauderly JL (1994) Toxicological and epidemiological evidence for health risks from inhaled engine emissions. Environ Health Perspect 102:165
83. Minai L, Yeheskely-Hayon D, Yelin D (2013) High levels of reactive oxygen species in gold nanoparticle-targeted cancer cells following femtosecond pulse irradiation. Scientific reports 3:2146
84. Tao F, Goazalez F, Kobzik L (2003) Reactive oxygen species in pulmonary inflammation by ambient particulates. Free Radical Biol Med 35(4):327–340
85. Hu X, Ding Z, Zhang Y, Sun Y, Wu J, Chen Y et al (2013) Size distribution and source apportionment of airborne metallic elements in Nanjing, China. Aerosol Air Qual Res 13:1796–1806
86. Valavanidis A, Fiotakis K, Vlahogianni T, Bakeas EB, Triantafillaki S, Paraskevopoulou V et al (2006) Characterization of atmospheric particulates, particle-bound transition metals and polycyclic aromatic hydrocarbons of urban air in the centre of Athens (Greece). Chemosphere 65:760–768
87. Olsson J-O, Tunestål P, Johansson B, Fiveland S, Agama R, Willi M et al (2002) Compression ratio influence on maximum load of a natural gas fueled HCCI engine. SAE Technical paper 2002-01-0111
88. Tunestal P, Johansson B (2007) HCCI control. In: Zhao H (ed) HCCI and CAI engines for the automotive industry. Woodhead Publishing Limited, England
89. Bengtsson J, Strandh P, Johansson R, Tunestål P, Johansson B (2004) Closed-loop combustion control of homogeneous charge compression ignition (HCCI) engine dynamics. Int J Adapt Control Signal Process 18:167–179
90. Bengtsson J (2004) Closed-loop control of HCCI engine dynamics. Lund University
91. Agrell F, Ångström H-E, Eriksson B, Wikander J, Linderyd J (2003) Transient control of HCCI through combined intake and exhaust valve actuation. SAE Technical paper 2003-01-3172
92. Martinez-Frias J, Aceves SM, Flowers D, Smith JR, Dibble R (2000) HCCI engine control by thermal management. SAE Technical paper 2000-01-2869
93. Risberg P, Johansson D, Andrae J, Kalghatgi G, Björnbom P, Ångström H-E (2006) The influence of NO on the combustion phasing in an HCCI engine. SAE Technical paper 2006-01-0416
94. Shimasaki Y, Kobayashi M, Sakamoto H, Ueno M, Hasegawa M, Yamaguchi S et al (2004) Study on engine management system using in-cylinder pressure sensor integrated with spark plug. SAE Technical paper 2004-01-0519
95. Aldawood A, Mosbach S, Kraft M (2012) HCCI combustion control using dual-fuel approach: Experimental and modeling investigations. SAE Technical paper 2012-01-1117
96. Ma J, Lü X, Ji L, Huang Z (2008) An experimental study of HCCI-DI combustion and emissions in a diesel engine with dual fuel. Int J Therm Sci 47:1235–1242
97. Ma S, Zheng Z, Liu H, Zhang Q, Yao M (2013) Experimental investigation of the effects of diesel injection strategy on gasoline/diesel dual-fuel combustion. Appl Energy 109:202–212
98. Mancaruso E, Vaglieco B (2010) Optical investigation of the combustion behaviour inside the engine operating in HCCI mode and using alternative diesel fuel. Exp Thermal Fluid Sci 34:346–351

99. Thring RH (1989) Homogeneous-charge compression-ignition (HCCI) engines. SAE Technical paper 892068
100. Ladommatos N, Abdelhalim S, Zhao H, Hu Z (1996) The dilution, chemical, and thermal effects of exhaust gas recirculation on diesel engine emissions-part 1: effect of reducing inlet charge oxygen. SAE Technical paper 961165
101. Ladommatos N, Abdelhalim SM, Zhao H, Hu Z (1998) Effects of EGR on heat release in diesel combustion. SAE Technical paper 980184
102. Ladommatos N, Abdelhalim SM, Zhao H, Hu Z (1996) The dilution, chemical, and thermal effects of exhaust gas recirculation on diesel engine emissions—part 2: effect of carbon dioxide. SAE Technical paper 961167
103. Ying W, Li H, Jie Z, Longbao Z (2009) Study of HCCI-DI combustion and emissions in a DME engine. Fuel 88:2255–2261
104. Shi Y, Ge HW, Brakora JL, Reitz RD (2010) Automatic chemistry mechanism reduction of hydrocarbon fuels for HCCI engines based on DRGEP and PCA methods with error control. Energy Fuels 24(3):1646–1654
105. André M, Walter B, Bruneaux G, Foucher F, Mounaïm-Rousselle C (2012) Exhaust gas recirculation stratification to control diesel homogeneous charge compression ignition combustion. Int J Engine Res. 1468087412438338
106. Kook S, Bae C, Kim J (2007) Diesel-fuelled homogeneous charge compression ignition engine with optimized premixing strategies. Int J Engine Res 8:127–137
107. Kanda T, Hakozaki T, Uchimoto T, Hatano J, Kitayama N, Sono H (2005) PCCI operation with early injection of conventional diesel fuel. SAE Technical paper 2005-01-3837
108. Lu X, Han D, Huang Z (2011) Fuel design and management for the control of advanced compression-ignition combustion modes. Prog Energy Combust Sci 37:741–783
109. Starck L, Lecointe B, Forti L, Jeuland N (2010) Impact of fuel characteristics on HCCI combustion: performances and emissions. Fuel 89:3069–3077
110. Tanaka S, Ayala F, Keck JC, Heywood JB (2003) Two-stage ignition in HCCI combustion and HCCI control by fuels and additives. Combust Flame 132:219–239
111. Nathan SS, Mallikarjuna J, Ramesh A (2007) Effect of mixture preparation in a diesel HCCI engine using early in-cylinder injection during the suction stroke. Int J Automot Technol 8:543–553
112. Iwabuchi Y, Kawai K, Shoji T, Takeda Y (1999) Trial of new concept diesel combustion system-premixed compression-ignited combustion. SAE Technical paper 1999-01-0185
113. Hashizume T, Miyamoto T, Akagawa H, Tsujimura K (1998) Combustion and emission characteristics of multiple stage diesel combustion. SAE Technical paper 980505
114. Kook S, Bae C (2004) Combustion control using two-stage diesel fuel injection in a single cylinder PCCI engine. SAE Technical paper 2004-01-0938
115. Horibe N, Harada S, Ishiyama T, Shioji M (2009) Improvement of premixed charge compression ignition-based combustion by two-stage injection. Int J Engine Res 10:71–80
116. Torregrosa A, Broatch A, García A, Mónico L (2013) Sensitivity of combustion noise and NOx and soot emissions to pilot injection in PCCI Diesel engines. Appl Energy 104:149–157
117. Neely GD, Sasaki S, Leet JA (2004) Experimental investigation of PCCI-DI combustion on emissions in a light-duty diesel engine. SAE Technical paper 2004-01-0121
118. Yehliu PC (2007) NOx emissions from heavy duty engine equipped with advanced CRDI system. Report on Advanced Combustion Technology. Accessed on 12th Mar 2009

Characterization of Ringing Operation in Ethanol-Fueled HCCI Engine Using Chemical Kinetics and Artificial Neural Network

Rakesh Kumar Maurya and Mohit Raj Saxena

Abstract The homogeneous charge compression ignition (HCCI) strategy is an advanced engine combustion concept having higher thermal efficiency while maintaining the NO_x and soot emission to an ultra-low level. Intense ringing operation in HCCI engine is one of the major challenges at high engine load conditions, which limit the HCCI engine operation range and can also damage engine parts. Ethanol is a promising alternative to conventional fuel, especially for utilization in advanced engine combustion modes such as HCCI. This chapter presents the overview of HCCI combustion along with its numerical simulation using stochastic reactor model. This chapter also presents detailed characterization of ringing operation, and HCCI operating range of ethanol-fueled HCCI engine. Ringing operation is typically characterized by either ringing intensity or peak pressure rise rate (PPRR). Characterization of PPRR and its prediction using artificial neural network (ANN) in ethanol-fueled HCCI engine is also presented. The ANN model is of utility to identify engine operating limits to avoid the ringing operation.

Keywords Combustion engine · Ringing intensity · HCCI · Artificial neural network · ANN · Ethanol · Ignition · Stochastic reactor model

1 Introduction

An increase in the petroleum prices, depletion of fossil fuel resources, and stringent emission norms demands to develop an efficient engine either using alternative fuel or by developing an advance combustion strategy. In past few decades, several alternative fuels as well as advanced combustion strategies in low-temperature

R. K. Maurya (✉) · M. R. Saxena
Advanced Engine and Fuel Research Laboratory, Department of Mechanical Engineering, Indian Institute of Technology Ropar, Rupnagar 140001, Punjab, India
e-mail: rakesh.maurya@iitrpr.ac.in

D. K. Srivastava et al. (eds.), *Advances in Internal Combustion Engine Research*, Energy, Environment, and Sustainability,
https://doi.org/10.1007/978-981-10-7575-9_3

combustion (LTC) regime are proposed for internal combustion (IC) engines. Among all the LTC strategies, homogeneous charge compression ignition (HCCI) is intensively investigated by the researchers of engine combustion community [1–5].

HCCI engine has an ability for achieving higher thermal efficiency while maintaining nitrogen oxides (NO_x) and particulate matter (PM) emissions below the current stringent emission mandates [1]. However, HCCI engine emits higher hydrocarbons (HCs) and carbon monoxide (CO) due to lower combustion temperature in the combustion chamber [1, 6]. In HCCI combustion strategy, autoignition of homogeneous fuel–air mixture occurs in the combustion chamber and whole charge autoignite simultaneously, which results in very high heat release rate. Thus, peak pressure rise rate (PPRR) is also very high due to very short combustion time period in comparison to conventional combustion modes. Since whole premixed charge autoignites in the cylinder, HCCI combustion strategy has no direct control of ignition timing similar to spark timing in spark-ignition (SI) engines and fuel injection timing in compression ignition (CI) engines. Combustion phasing in HCCI engines is typically controlled by indirect methods such as exhaust gas recirculation (EGR), intake air pressure and temperature, variable valve timing (amount of residuals) etc. [7–9]. Additionally, HCCI engine has lower operating range because of intense ringing operation or very high PPRR during rich air–fuel mixture operation. Rich mixture operation at high engine load is limited by high combustion noise level due to rapid increase in the PPRR. A more complete discussion on pressure oscillations and combustion noise can be found in the reference [5].

To tackle the challenges of higher IC-engine emissions and rapid depletion of fossil resources, biofuels can be a better replacement of conventional fossil fuels. The HCCI combustion strategy has fuel flexibility of using gasoline-like as well as diesel-like fuels, however, appropriate combustion control strategy needs to be selected for autoignition of charge depending on the fuel quality. Several studies demonstrated the HCCI engine operation using various fuels such as ethanol, butanol, methanol, hydrogen [2–5, 7–9]. Typically, alcohols have higher octane number and oxygen content and other characteristics are similar to gasoline. Higher oxygen content in alcohols leads in reduction of exhaust emissions species [3, 8–10]. In HCCI engine operation, alcohol fuels can be used either neat or blended with conventional fuel. Ethanol is intensively used in HCCI combustion engines by the researchers [11–15]. Ethanol is a renewable fuel, which is generally produced by the fermentation of sugarcane or by biomass. Utilization of ethanol in HCCI combustion engines combines the advantage of biofuel and advanced engine combustion modes. A study reported that combustion of alcohol fuels in a controlled autoignition (CAI) combustion engine has higher thermal efficiency as compared to hydrocarbon fuel [16]. Alcohol fuels (i.e., ethanol and methanol) in CAI combustion strategy can also be operated with wider range of air–fuel ratio and exhaust gas recirculation (EGR) as compared to gasoline fuel. A study showed that 100% methanol HCCI operation has more advanced combustion timing and reduced combustion duration as compared to ethanol and gasoline [17]. Additionally, alcohol-fueled HCCI operation has more capability for leaner combustion as compared to gasoline within achievable HCCI operating range. With 100% methanol and 100% ethanol fuel, HCCI engine can be operate up to 1.5 and 1.4 relative air–fuel ratio (λ), respectively, while in case of 100% gasoline operation, it can be operated up to 1.3 relative air–fuel ratio (λ). Alcohol-fueled HCCI engine has potential to operate at high speed and lower load range efficiently, while neat gasoline operation is limited by misfire [17]. In HCCI combustion engine, air–fuel ratio significantly affects the IMEP, while combustion phasing (CA_{50}) is affected by inlet air temperature [18]. At higher engine speed, higher intake air temperature is required for autoignition for gasoline fuel [18]. Study also reported that for a constant relative air–fuel ratio (λ), ethanol- and methanol-fueled HCCI engine can be operated with higher IMEP for CA_{50} position corresponding to the best thermal efficiency [18]. It is also found that methanol-fueled HCCI operation has lower indicated thermal efficiency as compared to gasoline and ethanol. However, higher combustion efficiency obtained with methanol and ethanol-fueled HCCI operation as compared to gasoline HCCI operation. Moreover, higher ringing intensity was observed with methanol and ethanol HCCI operation for all the tested operating load condition for optimum CA_{50} position. Studies demonstrated that HCCI operation has lower NO_x emissions as compared to conventional engines and has potential to reduces PM emissions simultaneously [1, 2, 8, 18]. In another study, PM emission characteristics are investigated for methanol, ethanol, and gasoline-fueled HCCI engine [19]. Study showed that particle number increases, when engine was operated with richer air–fuel mixture for all the tested air intake temperature conditions and fuels. Methanol HCCI operation was found to have lower total particle number concentration as compared to ethanol and gasoline.

Various studies demonstrated the benefits of ethanol-fueled HCCI engine, however high load operation is limited by ringing operation. To understand the ringing operation in HCCI engine, several experimental and numerical strategies are used by researchers. This chapter presents the characterization of ringing operation using chemical kinetics mode and corresponding HCCI operating range of ethanol HCCI engine. For real-time estimation of PPRR at high load, artificial neural network (ANN) model is proposed and discussed in this chapter. Before discussing the ringing characteristics and operating range, brief methodology of model is presented in the next section.

2 Modeling Methodology

2.1 Chemical Kinetic Model

To understand the HCCI combustion process and determine the HCCI operating range, chemical kinetic model is used. Different models are used for simulation of HCCI combustion including single zone, multi-zone model, 3D CFD models. In this chapter, numerical HCCI simulation is performed using stochastic reactor

Table 1 Specifications of HCCI engine

Parameter	Specifications
Bore/Stroke	120.65/140 mm
Connecting rod length	260 mm
Displacement volume	1600 cm^3
Compression ratio	21:1
Intake valve closing (IVC)	−167° ATDC
Exhaust valve opening (EVO)	141° ATDC
Wall temperature	450 K

model (SRM). The chemical kinetic simulation is performed on Kinetic SRM software [20]. A reduced ethanol mechanism developed by Maurya and Akhil [11], consisting of 47 number of species and 272 reactions is used for simulation of ethanol HCCI combustion using SRM. The details of the development of ethanol-reduced mechanism can be found in original study [11]. All the simulations are conducted during closed valve conditions. In this study, the simulated results are validated by comparing the experimental in-cylinder combustion pressure trace in the previously published study [21]. The detailed specification of the four-stroke engine used for numerical simulation is provided in Table 1. Homogeneous fuel–air mixture is prepared using port fuel injection strategy, and combustion phasing control is achieved using intake air preheating.

A more complete detail of the governing equations of SRM model can be found in the studies [20, 22, 23]. This model is based on probability density function which predicts better results close to 3D-CFD analysis. In SRM, statistical homogeneity is assumed throughout the combustion chamber which means that probability density function is not changed throughout the combustion chamber. SRM determines the progression of N_s chemical species, their temperature and mass fraction as a function of time. The 'N_{s+1}' random scaler variables are combined into the vector $\psi = (\psi_1, \psi_2, \psi_3, \cdots, \psi_{N_s}, \psi_{N_{s+1}})$ whose probability density function is 'f' [20]. In IC engines, the density of charge (air–fuel mixture) is varying in combustion cycles. Thus, the mass density function was used in place of probability density function in SRM [24]. The mass density function (F) is related to probability density function (f) and represented as

$$F(\psi, t) = \rho(\psi) f(\psi; t) \tag{1}$$

where ρ denotes the mass density.

The variation in mass density function with time is represented as follows [24]

$$\frac{\partial}{\partial t}F(\psi;t) = -\sum_{j=1}^{N_s+1}\frac{\partial}{\partial\psi_j}\left[G_j(\psi)F(\psi;t)\right] + \sum_{j=1}^{N_s+1}\frac{\partial}{\partial\psi_j}\left[A(\psi)F(\psi;t)\right] - \frac{1}{V}\frac{\mathrm{d}V}{\mathrm{d}t}F(\psi;t) - \frac{\partial}{\partial\psi_{N_s+1}}\left[U\left(\psi_{N+1}\right)F(\psi;t)\right] + \frac{F_c(\psi;t)}{\tau_{\mathrm{crev}}} - \frac{F(\psi;t)}{\tau_{\mathrm{cycle}}} + \frac{F_f(\psi;t)}{\tau_f} \tag{2}$$

where

the term $-\sum_{j=1}^{N_s+1}\frac{\partial}{\partial\psi_j}\left[G_j(\psi)F(\psi;t)\right]$ describes the in-cylinder chemistry process
the term $\sum_{j=1}^{N_s+1}\frac{\partial}{\partial\psi_j}\left[A(\psi)F(\psi;t)\right]$ describes the turbulent mixing
the term $\frac{1}{V}\frac{\mathrm{d}v}{\mathrm{d}t}F(\psi;t)$ describes the piston movement
the term $-\frac{\partial}{\partial\psi_{N_s+1}}\left[U\left(\psi_{N+1}\right)F(\psi;t)\right]$ describes the convective heat transfer
the term $\frac{F_c(\psi;t)}{\tau_{\mathrm{crev}}} - \frac{F(\psi;t)}{\tau_{\mathrm{cycle}}}$ describes the crevice flow
the term $\frac{F_f(\psi;t)}{\tau_f}$ describes the fuel injection [20]

This multi-dimensional mass density function is then computed by using Monte Carlo particle method with a second-order operator splitting algorithm [23]. The chemical reactions have significant effect on the temperature and composition of the species in the combustion chamber. The effect is described by a function $G(\psi)$ and represented as

$$G_j(\psi) = \frac{M_j\dot{\omega}_j}{\rho} \quad j = 1, 2, 3, \ldots, N_s \tag{3}$$

$$G_{N_s+1}(\psi) = -\frac{1}{c_v\rho}\sum_{j=1}^{N_s} e_j m_j \dot{\omega}_j - \frac{p}{c_v\rho}\frac{\mathrm{d}V}{\mathrm{d}t} \tag{4}$$

where

$\dot{\omega}_j$	denotes the production rate
M_j	is the molecular mass
e_j	is the specific internal energy of species j
c_v	is the specific heat capacity at constant volume
m and V	are the total mass and volume of cylinder

Using this model, HCCI combustion is simulated with a reduced reaction mechanism (47 species and 272 reactions) of ethanol. To validate chemical kinetic model, in-cylinder pressure trace of simulation and engine experiments are compared. Figure 1 shows the comparison of numerically simulated and experimental pressure trace. The figures depict that numerical simulation is in good agreement with experimental data as both cylinder pressure traces are in close proximity. After validation of simulation with experiments, numerical simulation is conducted at

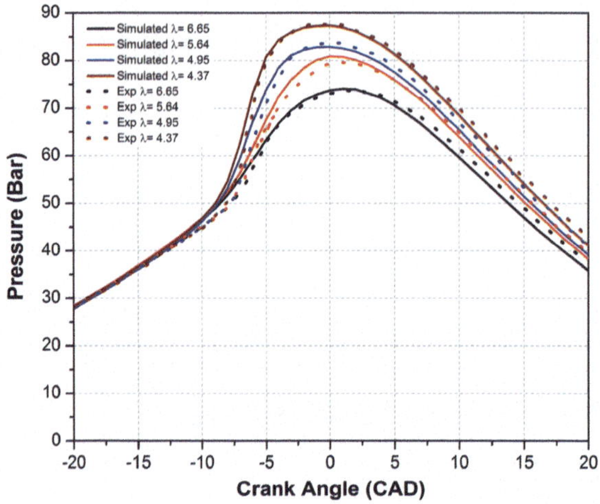

Fig. 1 Comparison of simulated (using SRM) and experimental in-cylinder pressure data of ethanol HCCI combustion [11]

Table 2 Operating conditions simulated for determination of HCCI engine operating range using ethanol

Variables	Values
Engine speed, N (rpm)	1000, 1500, 2000, 2500 and 3000
Intake manifold temperature, T_{in} (°C)	80–160
Relative air–fuel equivalence ratio ( )	1.5–8
Exhaust gas recirculation, EGR (%)	0

different engine operating conditions to determine the HCCI operating range. The operating test conditions used for simulation are given in Table 2.

2.2 *Artificial Neural Network*

Artificial neural network (ANN) model can be used for real-time control of ringing or misfire operation of HCCI combustion engine. ANN is a computational model, which consists of input layer, hidden layer (either one, two, or three), and output layers of neurons. Typical architecture of ANN is illustrated in Fig. 10. In ANN, connection of the one neuron to another neuron is associated with numeric values known as weight. The output 'h_i' of any neuron 'i' in the hidden layer is represented as [25]

$$h_i = \sigma\left(\sum_{j=1}^{N} V_{ij}x_j + T_i^{\mathrm{hid}}\right) \tag{5}$$

where

$\sigma(\,)$	represents the activation function
‘N’	is the total number of input neurons
‘V_{ij}’	denotes the weights
‘x_j’	is the inputs to input neurons
‘T_i^{hid}’	denotes the threshold terms of the hidden neuron

The architecture structure of ANN is typical feed-forward type with tangent sigmoid for the activation function for the hidden and output layers [26]. The activation function is a typically a sigmoid function and is represented as [25].

$$\sigma(u) = \frac{1}{1+\exp(-u)} \tag{6}$$

After deciding the input, hidden, and output layer, a neural network is trained by presenting to it a set of input data which are known as training set. The purpose of training is to minimize the error function by adjusting the weights between the neurons to get the desired output data. The error function is defined as sum of squared differences between the ANN output and desired output. Physical model for combustion analysis of internal combustion engine is very complex because of highly nonlinear thermodynamics reactions and variations in the pressure waves.

In past few years, several studies used ANN for internal combustion engines applications [26–32]. Studies developed an ANN model to predict the performance and emissions characteristics of SI engine fueled with ethanol/gasoline-blended fuel [26, 27]. Another study developed an ANN model for the prediction of performance and emission characteristics of diesel engine fueled with biodiesel/diesel blends at different injection timings [28]. ANN is also used in advanced combustion modes such as HCCI, for the prediction of misfire and ringing operations [31, 32]. ANN computational time is also very less as compared to experiments or other simulation analysis. Therefore, ANN model can be an important tool for real-time control of advanced combustion engines.

3 Ringing Characteristics and HCCI Operating Range

This section discusses the ringing characteristics and HCCI operating range of ethanol-fueled HCCI engine. Figure 2 illustrates the effect of relative fuel ratio (λ) on in-cylinder pressure trace and rate of heat release (ROHR) at constant intake temperature. It is well known that in HCCI combustion, autoignition of nearly homogeneous mixture takes place in the combustion chamber and charges

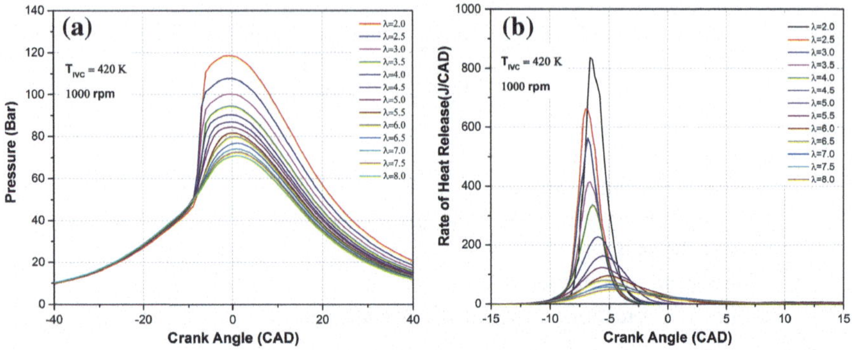

Fig. 2 Variation of **a** in-cylinder pressure trace and **b** rate of heat release with relative fuel ratio (λ) in ethanol HCCI combustion at 1000 rpm [12]

autoignite at multiple sites spontaneously. This phenomenon leads to very high heat release rate, particularly at richer fuel–air mixture, which is clearly illustrated in Fig. 2b. Magnitude of ROHR is typically higher than conventional compression ignition engines. Figure 2a depicts that richest mixture (λ = 2) leads to the highest peak cylinder pressure, and it decreases as mixture becomes leaner (λ increases). At rich mixture operating conditions, higher amount of charge is autoignited that leads to very high peak pressure. Richer mixture has very high combustion rate resulting into higher ROHR due to high reactivity of richer fuel–air mixture (Fig. 2b). Position of peak ROHR advances with richer fuel–air mixture because oxidation reaction rate depends on fuel concentrations, and exponentially on combustion temperature. HCCI combustion rate is mainly affected by engine operating parameters that affect species concentration and combustion temperature. Figure 2 also depicts that rich mixture operation leads to very fast combustion and very high pressure rise rate. The very high pressure rise rate leads to higher combustion noise and ringing operation.

Since there is no direct control on HCCI combustion rate (kinetically controlled combustion), a criterion is needed to define stable HCCI engine operating range. The HCCI combustion is limited by several possible operational limits such as combustion noise limits, emission limits, peak cylinder pressure limit, and combustion instability limit. Typically, HCCI combustion is mainly affected by high- and low-load limits, which occur by engine knock (high peak pressure rise rate) and combustion instability (misfire/partial burn or high cyclic variations), respectively. Higher operating load is generally limited by ringing operation (high combustion noise) or severity of engine knock. To characterize the severity of knock, a threshold value of peak pressure rise rate (PPRR) is used in HCCI combustion [5]. Acceptable limit of PPRR is less than 5 MPa/ms.

At lower engine loads, engine is operated on leaner fuel–air mixtures, which lead to lower temperature and retarded combustion phasing. Autoignition of charge highly depends on cylinder pressure and temperature. Therefore, combustion efficiency decreases at lower temperature and late combustion phasing. Poor combustion phasing results into partial burn of fuel, and too late combustion phasing can have possibility of misfire. Poor combustion efficiency (η_{com}) resulting from lower temperature leads to higher CO and THC emissions from HCCI engine. To define the misfire limit, a study used combustion efficiency less than 85% as misfire [12] and limits the lower engine load limit. Higher and lower engine load limits can be defined using the limits of PPRR and combustion efficiency. Figure 3 illustrates the ethanol HCCI operating range at different engine speed and compression ratios using the limits PPRR <5 MPa/ms and combustion efficiency >85%. For every λ, inlet temperature is selected which corresponds to maximum engine efficiency. The figure shows that higher compression ratio has higher operating range. Higher engine speed has relatively lower range of λ for all the three compression ratios.

Engine operation at particular value of λ can be achieved at a range of intake temperatures, where PPRR and combustion efficiency is within the specified limits. Results shown in this chapter (Figs. 3, 4 and 5), is presented for the intake temperature at which engine efficiency is highest. Figure 4 shows the temperature at intake valve closing (T_{IVC}) for ethanol HCCI operation at different compression ratios. The T_{IVC} is directly related to inlet charge temperature. Since simulations are conducted during valve-closed conditions, data of T_{IVC} is presented. Figure depicts that lower IMEP conditions need higher T_{IVC} for each engine speeds due to leaner HCCI engine operation. Additionally, T_{IVC} requirement is lower for higher compression ratio at higher engine loads.

One of the main advantages of HCCI engine is the higher thermal efficiency because of various factors such as leaner air–fuel mixture operation, higher compression ratios, shorter combustion duration, and lower heat transfer. At a particular compression ratio, HCCI engine efficiency is typically determined by combustion efficiency and ignition timing for any operating condition. Both, combustion

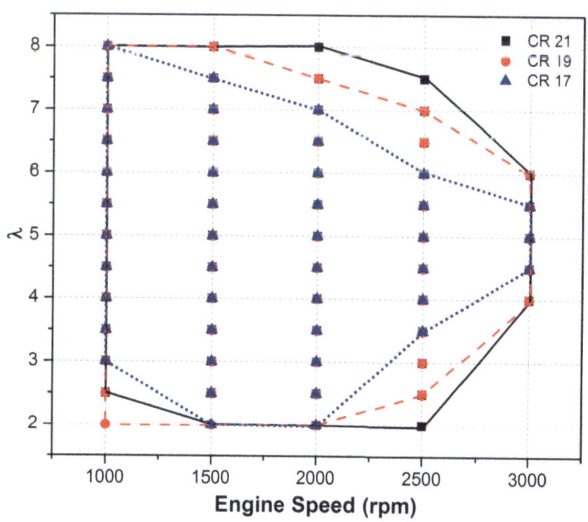

Fig. 3 Ethanol HCCI operating range for compression ratios 17, 19, and 21 [12]

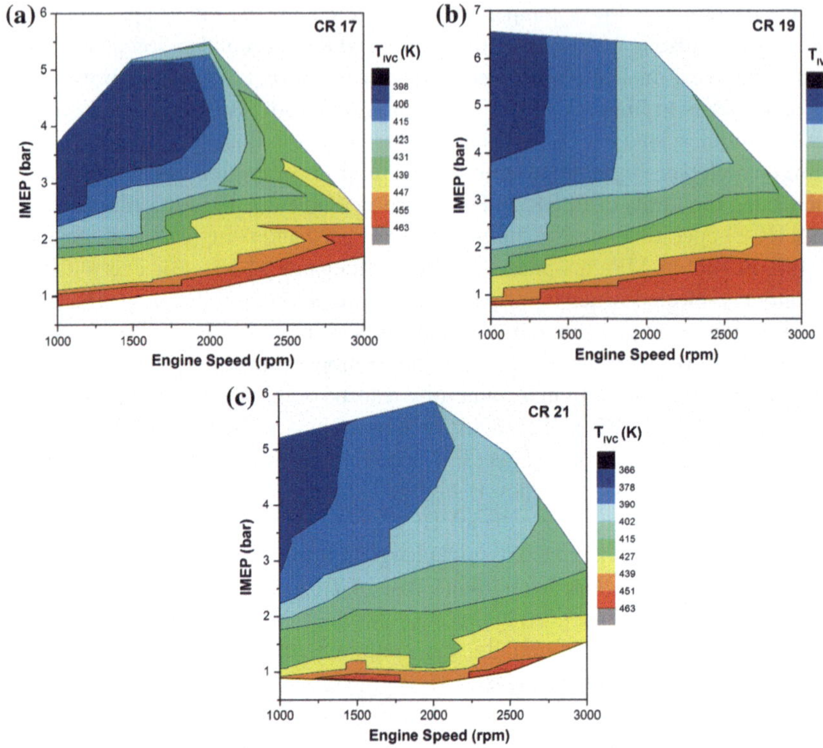

Fig. 4 Intake valve closing temperature variations in ethanol HCCI engine for compression ratios 17, 19, and 21 [12]

efficiency and ignition timings are affected by inlet temperature. On the optimal inlet temperature (shown in Fig. 4), corresponding thermal efficiency map is also presented in Fig. 5 for different compression ratios. The figure depicts that maximum indicated efficiency achieved is around 50% for all the compression ratios, which affected by engine speed and load conditions. Vertical contour lines illustrate that engine efficiency is mainly governed by engine speed at optimal combustion phasing. The HCCI engine load has relatively lower influence on thermal efficiency for optimum combustion phasing. At each test compression ratio, thermal efficiency decreases at higher engine speeds due to lower combustion efficiency and heat transfer [12].

Engine operation at richer fuel–air mixture above the high load limit can lead to severe cylinder pressure oscillations in HCCI combustion engine. These pressure oscillations (termed as 'ringing') are very similar to knock in a spark-ignition engine. Pressure oscillations in HCCI cylinder produce a large amount of combustion noise in the engine structure. For evaluation of the combustion noise 'ringing intensity' (RI) is calculated using the following equation in HCCI engine [33]:

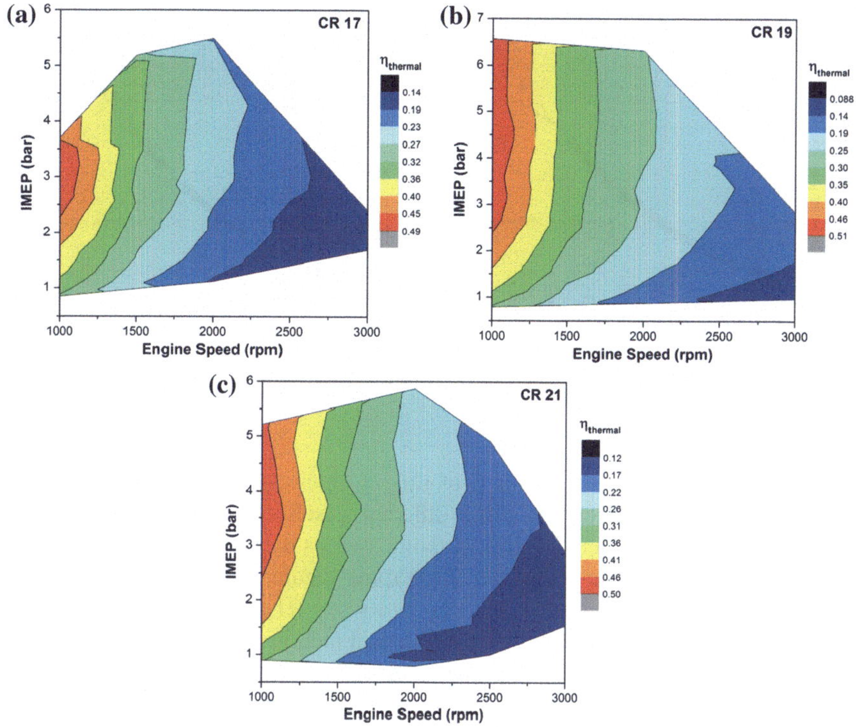

Fig. 5 Thermal efficiency variation in ethanol HCCI engine for compression ratios 17, 19, and 21 [12]

$$\mathrm{RI} = \frac{\sqrt{\gamma R T_{\max}}}{2\gamma P_{\max}} \left[\beta \left(\frac{\mathrm{d}P}{\mathrm{d}t}\right)_{\max}\right]^2 \tag{7}$$

where $P_{\max}$ is the peak in-cylinder pressure, $(\mathrm{d}P/\mathrm{d}t)_{\max}$ is the maximum pressure rise rate, and $T_{\max}$ is the maximum of mass averaged in-cylinder temperature (calculated using ideal gas law). γ is the ratio of specific heats (C_p/C_v), and R is the gas constant. β is a tuning parameter, which relates the amplitude of pressure pulsations to the maximum pressure rise rate and is set to 0.05 here.

Other method used to define high load limit for HCCI engine is by using the pressure rise rate (PRR). Excessive PRR leads to not only noisy operation but also structural damage to the engine. Rate of pressure rise is calculated using fourth order central difference method:

$$\frac{\mathrm{d}P}{\mathrm{d}\theta} = \frac{-P(\theta + 2h) + 8P(\theta + h) - 8P(\theta - h) + P(\theta - 2h)}{12h} \tag{8}$$

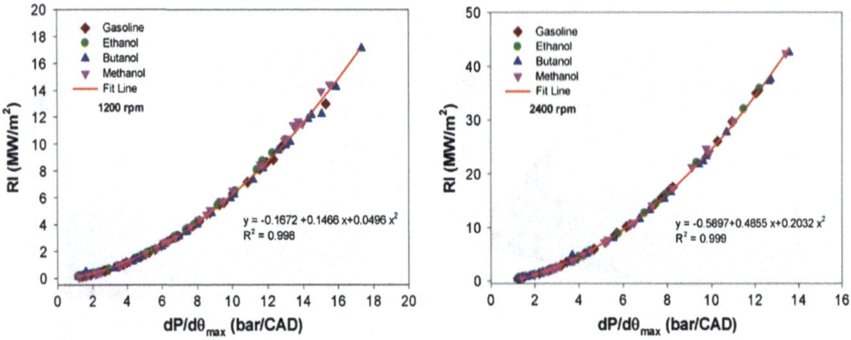

Fig. 6 Illustration of relationship between PPRR and ringing intensity at 1200 and 2400 rpm in HCCI engine using different gasoline-like fuels

Figure 6 illustrates the correlation of ringing intensity (RI) with PPRR at two engine speeds in HCCI engine using different gasoline-like fuels. RI is plotted against PPRR for the λ, and inlet temperature sweeps data using gasoline, ethanol, methanol, and butanol at 1200 and 2400 rpm. Figure depicts that RI strongly correlates with PPRR. Regression of data shows that RI dependence with PPRR is quadratic. Therefore, PPRR can also be used as a measure for emitted engine noise for HCCI knock in a naturally aspirated engine.

Figure 7 shows the variation of ringing intensity at different inlet temperatures with HCCI combustion phasing (CA50) using gasoline and ethanol fuel. For both fuels, ringing intensity increases with advancement of combustion phasing (CA50). Figure 7 also illustrates that CA50 is a major factor that affects the ringing operation in HCCI engine. To reduce the RI in HCCI combustion, the CA50 needs to be retarded. Regression of the data shows that third-order polynomial curve captures well the trend of RI with combustion phasing for both fuels. Therefore, RI can be estimated by CA50 at any inlet temperature.

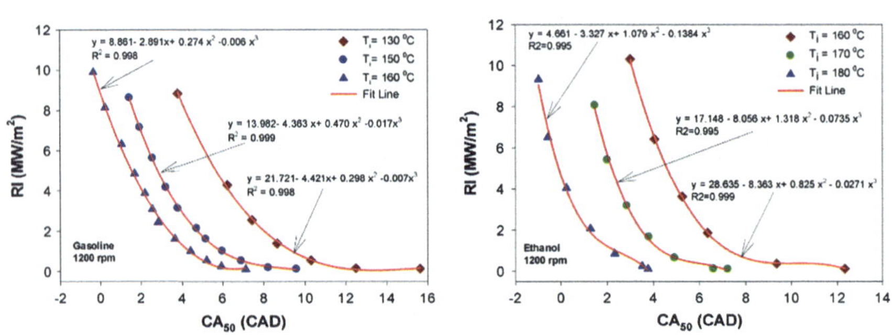

Fig. 7 Effect of combustion phasing on RI in HCCI combustion using gasoline and ethanol at 1200 rpm [18]

To determine the general trend of RI with different combustion and performance parameters, experiments are conducted at different engine speed, equivalence ratio, and different intake temperatures in an ethanol-fueled HCCI engine [32]. Figure 8 presents the RI as function of start of combustion (SOC), crank angle position of 50% heat release (CA50), crank angle position of 90% heat release (CA90), and burn duration (BD) at 100 steady-state engine operating conditions. A moderate second-order polynomial correlation is present between RI and combustion phasing parameters (R^2 ranges from 0.38 to 0.70). More delayed and longer combustion duration occurs in the test conditions with lower RI value (Fig. 8a, b). Highest correlation coefficient of 0.70 is found between RI and BD, which indicate higher ringing operation at lower combustion duration due to very high combustion rate. Figure 8 depicts that the high RI value test conditions have 1 < SOC < 4 CAD aTDC, 4 < CA50 < 7 CAD aTDC, 9 < CA90 < 14 CAD aTDC and 8 < BD < 10 CAD conditions.

Figure 9 shows the variation of exhaust temperature (T_{exh}) and adiabatic temperature (T_{ad}), unburned hydrocarbon (uHC), and carbon monoxide (CO) emission as function of RI.

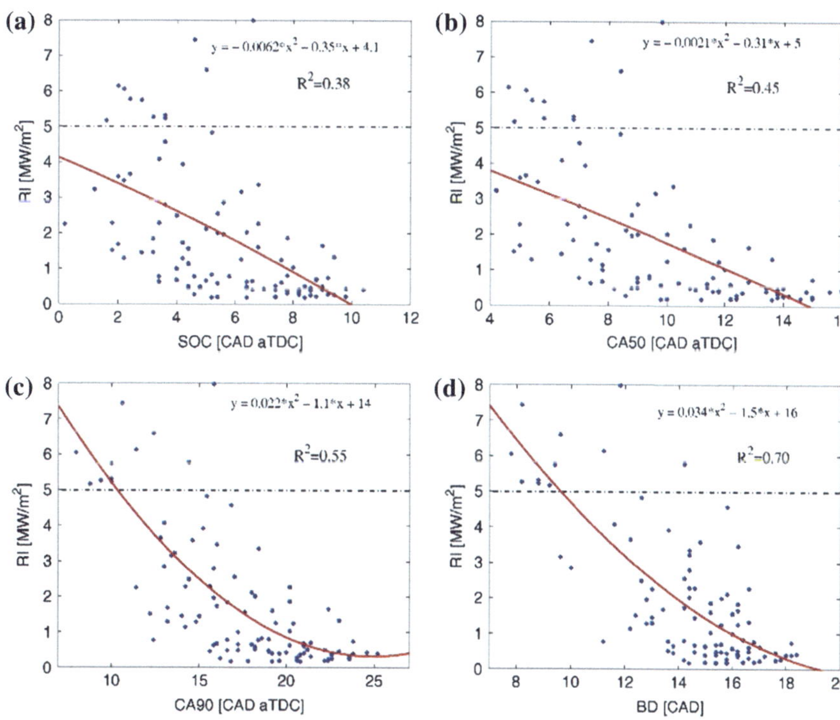

Fig. 8 Ringing intensity as a function of SOC, CA50, CA90, and BD for 100 steady-state HCCI operating conditions [32]

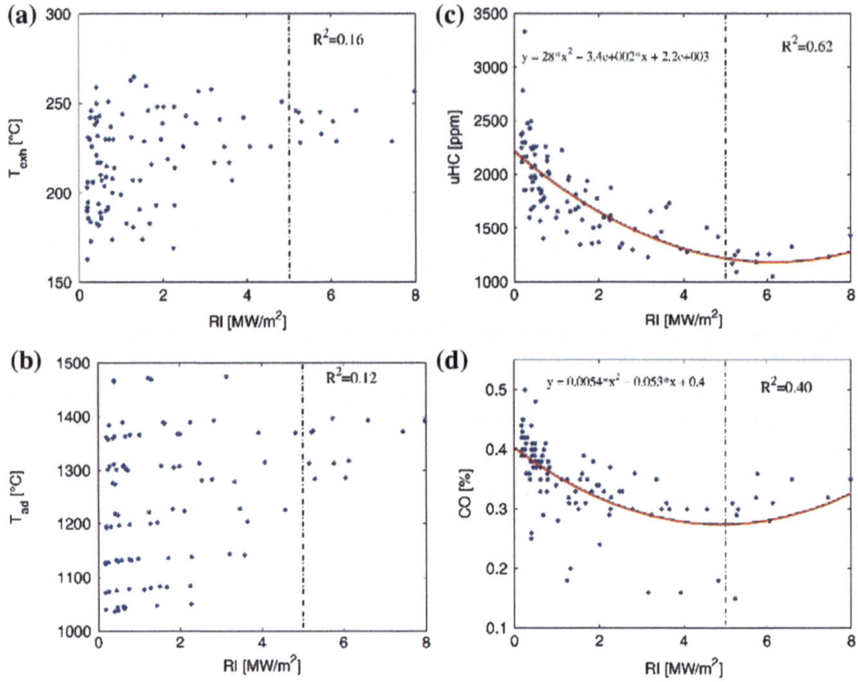

Fig. 9 Variations in uHC and CO emissions, exhaust temperature (T_{exh}), and adiabatic temperature (T_{ad}) as function of RI for 100 steady-state HCCI operating conditions [32]

Figure 9a, b depicts that there is no apparent correlation between exhaust and adiabatic temperature with RI. Exhaust temperature is an important parameter for effective conversion on HC and CO emissions by oxidation catalyst. The HC and CO emissions are relatively higher in HCCI engine, and thus, it is necessary to mitigate. At some of the engine operating conditions, exhaust temperature is very low in comparison to light-off temperature of oxidation catalyst. Very weak correlation occurs between CO emission and RI (Fig. 6d). At higher RI test conditions, HC emissions are lower (Fig. 6c). Additionally, significant correlation exists between HC emission and RI.

It is observed from regression analysis that correlation between combustion/performance parameters and RI are not strong. Thus, for characterization and control of RI operation some other model is required, which is also computationally less time consuming. The artificial neural network (ANN) model is one possible option and is discussed in next section.

4 Artificial Neural Network Model for Ringing Operation

This section provides the details of ANN model used for prediction of PPRR and RI. The ANN model is developed by using input parameters of operating conditions (i.e., RPM, λ, T_{ivc}) and combustion parameters (i.e., P_{max}, CA_{50}, $CA_{90\text{-}10}$). ANN model for PPRR in ethanol-fueled HCCI engine is illustrated in Fig. 10. In this analysis, Levenberg–Marquardt algorithm, training function is used to make an ANN model. To design an accurate ANN model, neurons are varying from 5 to 40 (in case of operating parameters) and 5 to 30 (in case of combustion parameter). With further increases in number of neurons, the correlation starts decreasing. The results of ANN model with number of neurons are presented in Table 3a, b. The best result in case of PPRR prediction from performance parameters is obtained

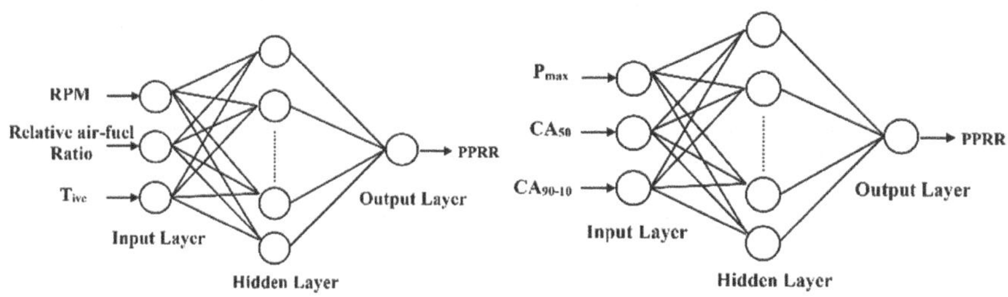

Fig. 10 Architecture of ANN model for operating parameters (RPM, λ, T_{ivc}) and combustion parameters (P_{max}, CA_{50}, $CA_{90\text{-}10}$)

Table 3 **a** Comparison of RMSE and R^2 between predicted and kinetically simulated PPRR for various numbers of neurons for engine operating parameters. **b** Comparison of RMSE and R^2 between predicted and kinetically simulated PPRR for various number of neurons for engine combustion parameters

Number of neurons	RMSE (MPa/ms)	R^2 (−)
a		
5	1.28	0.96
10	1.09	0.97
15	0.95	0.98
20	0.87	0.98
25	1.05	0.97
30	0.89	0.98
35	**0.53**	**0.99**
40	1.04	0.97
b		
5	2.15	0.86
10	2.15	0.89
15	2.58	0.85
20	**2.01**	**0.91**
25	3.43	0.77
30	4.27	0.70

with 35 neurons, while the best result in case of PPRR prediction from combustion parameters is obtained with 20 neurons (as indicated with bold values in Table 3).

The ANN model is developed using MATLAB toolbox. Figure 11 demonstrates the comparison of kinetically simulated and ANN predicted values for 179 steady-state operating points for operating and combustion parameters. ANN model results depicted that there is strong correlation between the combustion parameters and the PPRR (as shown in Fig. 11). Comparison of simulated results and ANN results indicates that ANN model predicts PPRR well with strong correlation coefficient. In addition, ANN model indicates strong correlation between the combustion parameters and PPRR, which were not present in the regression analysis. Moreover, ANN predicts PPRR in ethanol-fueled HCCI engine with an average error of less than 5%. The computational time of ANN prediction is also very less as compared to kinetically simulation or experimental time and has very low cost.

A study used in-cylinder pressure at 5, 10, 15 CAD aTDC and peak cylinder pressure to determine the correlation with RI [32]. Study demonstrated the strong correlation between cylinder pressure at specified points and ringing intensity. Figure 12 shows the comparison of predicted (Sim.) and experimental (Exp.) values

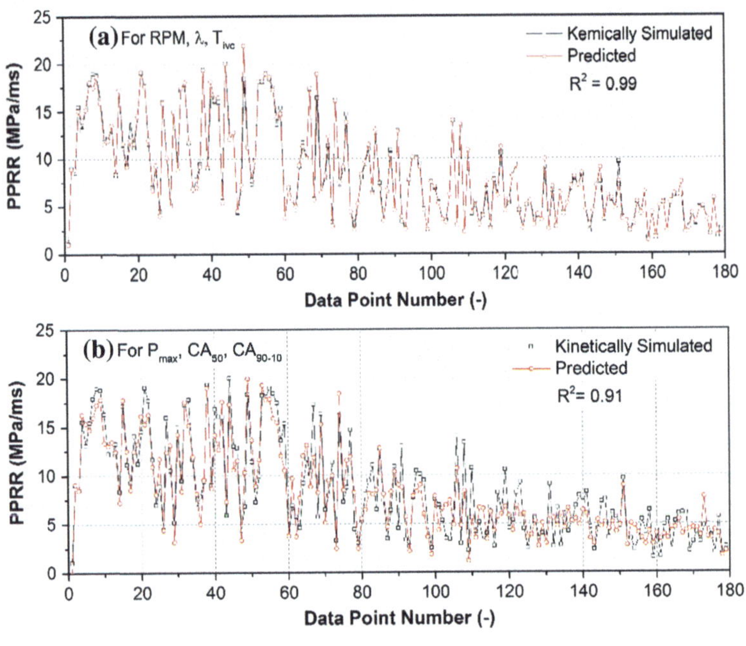

Fig. 11 Comparison of chemical kinetically simulated and ANN predicted values for 179 steady-state operating points **a** for RPM, λ, T_{ivc}; **b** for P_{max}, CA_{50}, $CA_{90\text{-}10}$

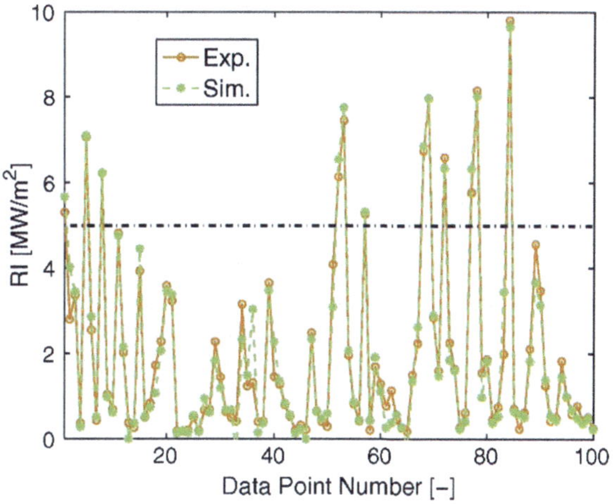

Fig. 12 Comparison of predicted (Sim.) and experimental (Exp.) values of RI for different test points [32]

of RI for different test points. It can be observed from Fig. 12 that ANN model is able to predict RI well with good accuracy (less than 4.2% error).

It can be summarized that ANN model can be used to characterize the ringing operation in HCCI engine for combustion control applications.

5 Summary

This chapter presents the characterization of ringing operation using chemical kinetics and artificial neural network (ANN) in ethanol-fueled HCCI engine. Chemical kinetic simulation is performed using stochastic reactor model. The chemical kinetic model is able to accurately determine the cylinder pressure trace and peak pressure rise rate. The HCCI operating range using ethanol is defined using PPRR and combustion efficiency. Ethanol HCCI operating range at compression ratios 17, 19 and 21 is estimated and presented in this chapter. HCCI operating load range decreases with engine speed. Thermal efficiency up to 50% is achieved for each compression ratio depending on the engine operating condition. Thermal efficiency map at optimal combustion phasing is presented in this chapter. Regression analysis indicated weak correlation between ringing intensity and combustion parameters. Artificial neural network model is developed for identification of PPRR and RI. The ANN model is able to predict PPRR well with strong correlation coefficient of 0.99 (in case of operating parameters) and 0.91 in case of combustion parameters. Ringing intensity is also predicted with good accuracy by ANN model. It is concluded that ANN model can be used to characterize the ringing operation in HCCI engine for combustion control applications.

Acknowledgements Authors gratefully acknowledge the help received from Mr. Nekkanti Akhil in SRM simulations. Help and assistance of Mr. Yogendra Kumar Vishwakarma in generating data from ANN module of MATLAB is highly appreciated and acknowledged.

References

1. Agarwal AK, Singh AP, Maurya RK (2017) Evolution, challenges and path forward for low temperature combustion engines. Prog Energy Combust Sci 61:1–56
2. Maurya RK, Akhil N (2017) Comparative study of the simulation ability of various recent hydrogen combustion mechanisms in HCCI engines using stochastic reactor model. Int J Hydrogen Energy 42(16):11911–11925
3. Maurya RK, Agarwal AK (2015) Combustion and emission characterization of n-butanol fueled HCCI engine. J Energy Res Technol 137(1):011101
4. Singh AP, Agarwal AK (2017) Partially homogenous charge compression ignition engine development for low volatility fuels. Energy Fuels 31(3):3164–3181
5. Maurya RK (2018) Characteristics and control of low temperature combustion engines: employing gasoline, ethanol and methanol. Springer. ISBN 978-3-319-68507-6
6. Khandal SV, Banapurmath NR, Gaitonde VN, Hiremath SS (2017) Paradigm shift from mechanical direct injection diesel engines to advanced injection strategies of diesel homogeneous charge compression ignition (HCCI) engines-A comprehensive review. Renew Sustain Energy Rev 70:369–384
7. Jafarmadar S, Nemati P, Khodaie R (2015) Multidimensional modeling of the effect of Exhaust Gas Recirculation (EGR) on exergy terms in an HCCI engine fueled with a mixture of natural gas and diesel. Energy Convers Manag 105:498–508
8. Maurya RK, Agarwal AK (2014) Effect of intake air temperature and air–fuel ratio on particulates in gasoline and n-butanol fueled homogeneous charge compression ignition engine. Int J Engine Res 15(7):789–804
9. He BQ, Yuan J, Liu MB, Zhao H (2014) Combustion and emission characteristics of a n-butanol HCCI engine. Fuel 115:758–764
10. Bhaskar PB, Srihari S (2017) A study of emission and performance characteristics on HCCI engine using methanol blends (No. 2017-28-1963). SAE Technical paper
11. Maurya RK, Akhil N (2016) Numerical investigation of ethanol fuelled HCCI engine using stochastic reactor model. Part 1: development of a new reduced ethanol oxidation mechanism. Energy Convers Manag 118:44–54
12. Maurya RK, Akhil N (2016) Numerical investigation of ethanol fuelled HCCI engine using stochastic reactor model. Part 2: parametric study of performance and emissions characteristics using new reduced ethanol oxidation mechanism. Energy Convers Manag 121:55–70
13. Bendu H, Deepak BBVL, Murugan S (2017) Multi-objective optimization of ethanol fuelled HCCI engine performance using hybrid GRNN–PSO. Appl Energy 187:601–611
14. Swanson JJ, Franklin LM, Bika AS, Kittelson DB (2017) Size and volatility of particle emissions from an ethanol-fueled HCCI engine. Aerosol Sci Technol 51(5):614–625
15. Singh E, Waqas M, Johansson B, Sarathy M (2017) Simulating HCCI blending octane number of primary reference fuel with ethanol (No. 2017-01-0734). SAE Technical paper
16. Oakley A, Zhao H, Ma T, Ladommatos N (2001) Dilution effects on the controlled auto-ignition (CAI) combustion of hydrocarbon and alcohol fuels
17. Xie H, Wei Z, He B, Zhao H (2006) Comparison of HCCI combustion respectively fueled with gasoline, ethanol and methanol through the trapped residual gas strategy (No. 2006-01-0635). SAE Technical paper
18. Maurya RK, Agarwal AK (2014) Experimental investigations of performance, combustion and emission characteristics of ethanol and methanol fueled HCCI engine. Fuel Process Technol 126:30–48

19. Maurya RK, Agarwal AK (2015) Experimental investigations of particulate size and number distribution in an ethanol and methanol fueled HCCI engine. J Energy Res Technol 137 (1):012201
20. Srm suite user manual v8.5.0. cmcl innovations, 27 Nov 2015
21. Christensen M, Johansson B, Einewall P (1997) Homogeneous charge compression ignition (HCCI) using isooctane, ethanol and natural gas-a comparison with spark ignition operation (No. 972874). SAE Technical paper
22. Bhave A, Balthasar M, Kraft M, Mauss F (2004) Analysis of a natural gas fuelled homogeneous charge compression ignition engine with exhaust gas recirculation using a stochastic reactor model. Int J Engine Res 5(1):93–104
23. Bhave A, Kraft M (2004) Partially stirred reactor model: analytical solutions and numerical convergence study of a PDF/Monte Carlo method. SIAM J Sci Comput 25(1):1798–1823
24. Bernard G, Scaife M, Bhave A, Ooi D, Dizy J (2016) Application of the SRM engine suite over the entire load-speed operation of a US EPA Tier 4 capable IC engine (No. 2016-01-0571). SAE Technical paper
25. Wang SC (2003) Artificial neural network. In: Interdisciplinary computing in java programming. Springer, US, pp 81–100
26. Najafi G, Ghobadian B, Tavakoli T, Buttsworth DR, Yusaf TF, Faizollahnejad M (2009) Performance and exhaust emissions of a gasoline engine with ethanol blended gasoline fuels using artificial neural network. Appl Energy 86(5):630–639
27. Kiani MKD, Ghobadian B, Tavakoli T, Nikbakht AM, Najafi G (2010) Application of artificial neural networks for the prediction of performance and exhaust emissions in SI engine using ethanol-gasoline blends. Energy 35(1):65–69
28. Pai PS, Rao BS (2011) Artificial neural network based prediction of performance and emission characteristics of a variable compression ratio CI engine using WCO as a biodiesel at different injection timings. Appl Energy 88(7):2344–2354
29. Choi Y, Chen JY (2005) Fast prediction of start-of-combustion in HCCI with combined artificial neural networks and ignition delay model. Proc Combust Inst 30(2):2711–2718
30. Rezaei J, Shahbakhti M, Bahri B, Aziz AA (2015) Performance prediction of HCCI engines with oxygenated fuels using artificial neural networks. Appl Energy 138:460–473
31. Bahri B, Aziz AA, Shahbakhti M, Said MFM (2013) Understanding and detecting misfire in an HCCI engine fuelled with ethanol. Appl Energy 108:24–33
32. Bahri B, Shahbakhti M, Kannan K, Aziz AA (2016) Identification of ringing operation for low temperature combustion engines. Appl Energy 171:142–152
33. Eng JA (2002) Characterization of pressure waves in HCCI combustion (No. 2002 01 2859). SAE Technical paper

Variable Valve Actuation Systems

Dhananjay Kumar Srivastava, Abhimanyu Das and Nitish Kumar Singh

Abstract Internal combustion (IC) engines today represent a class of heat engines marked by their high power-to-weight ratio, making them the suitable choice for portable power solutions. Being reliable and robust, their widespread use in commercial vehicles is, therefore, implicitly justified. Being a heat engine, the efficiency and performance of an internal combustion engine are limited by the temperature of heat addition and rejection. Moreover, with the inherent irreversible and non-ideal nature of the various processes of the power cycle, a fraction of the ideal thermodynamic efficiency is realised accounting for the low overall thermal efficiency. The text that follows is centred around the gas exchange process in an IC engine. The working of conventional camshaft-driven valve train systems, which have been in use for quite a long time, has been discussed followed by its limitations and their repercussions on the performance and efficiency of an IC engine. The origins of the unavoidable pumping losses accompanying load control using a throttle valve have been explained. An overview of the various strategies and methods used in commercial vehicles to mitigate such losses (variable valve timing and variable valve lift) has been given while providing some insight into the working of some experimental variable valve actuation systems. The discussion then shifts to fully flexible camless valve actuation systems explaining the working of some popular actuation systems, highlighting their advantages and limitations. The basic control logic of such systems is then discussed followed by a list of some unique attributes and advantages of the same. Few experimental results from the literature have also been cited to substantiate the utility of variable valve actuation systems.

Keywords Valve train · Camshaft · Throttling · Pumping losses
Valve lift · Valve timing · Valve event duration · Intake valve opening (IVO)
Intake valve closing (IVC) · Exhaust valve opening (EVO) · Exhaust valve closing (EVC) · Variable valve actuation (VVA) · Variable valve lift (VVL)
Variable valve timing (VVT) · Camless system

D. K. Srivastava (✉) · A. Das · N. K. Singh
Department of Mechanical Engineering, Indian Institute of Technology Kharagpur, Kharagpur 721302, West Bengal, India
e-mail: srivastava@mech.iitkgp.ernet.in

D. K. Srivastava et al. (eds.), *Advances in Internal Combustion Engine Research*, Energy, Environment, and Sustainability,
https://doi.org/10.1007/978-981-10-7575-9_4

Abbreviations and Acronyms

3D	Three-dimensional
BDC	Bottom dead centre
CI	Compression ignition
ECU	Engine control unit
EEVC	Early exhaust valve closing
EEVO	Early exhaust valve opening
EGR	Exhaust gas recirculation
EIVC	Early intake valve closing
EIVO	Early intake valve opening
EPVA	Electro-pneumatic valve actuation
EVC	Exhaust valve closing
EVO	Exhaust valve opening
FFVA	Fully flexible valve actuator
HC	Hydrocarbon
IC	Internal combustion
IVC	Intake valve closing
IVO	Intake valve opening
LEVC	Late exhaust valve closing
LEVO	Late exhaust valve opening
LIVC	Late intake valve closing
LIVO	Late intake valve opening
PMEP	Pumping mean effective pressure
RPM	Revolutions per minute
SI	Spark ignition
TDC	Top dead centre
VCP	Variable cam phaser
VCT	Variable camshaft timing
VTEC	Variable valve timing and lift electronic control
VVA	Variable valve actuation
VVEL	Variable valve event and lift
VVL	Variable valve lift
VVT	Variable valve timing
VVTi	Variable valve timing with intelligence

1 Introduction

Right from the inception of the four-stroke IC engine, valves and their design have played an important role in the gas exchange process. The gas exchange process in any IC engine deals with two fundamental functions. The first is the induction of fresh air–fuel mixture, as in a port injection spark ignition (SI) engine, or air, as in a compression ignition (CI) engine or direct injection into the engine cylinder(s)

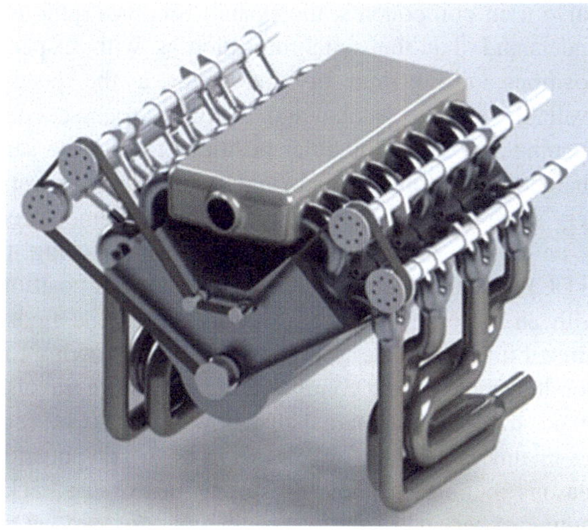

Fig. 1 Conventional valvetrain

during the intake stroke (here onwards, we will be referring both air and air–fuel mixture as charge). The second function is the ejection of combustion products from the engine cylinder(s) after the power stroke. Additionally, an important function of the gas exchange process interlinks intake and exhaust events to produce a mixed bag of effects, which can either improve or deteriorate the performance of an IC engine. Valve events, as a result, strongly dictate the overall efficiency and performance of an internal combustion engine.

In order to control the opening and closing of the intake and exhaust valves, IC engines, for a long time, have used mechanical couplings to coordinate timing of the valve events with crankshaft rotation. In order to increase the power output for a given cylinder displacement, IC engines need to operate at high rotational speeds. This, in turn, implies that the valves have a very small time period for their opening and closing. The situation is self-explanatory and why a mechanically linked valve train, marked by its robustness and reliability, was the obvious choice for the early designers of the IC engine. Most IC engines today use a cam and follower mechanism for actuating the intake and exhaust poppet valves. Depending on the number of the cylinders and valves, a number of cam lobes are fastened on to a shaft which is called as the camshaft. A chain drive mechanism connects the crankshaft and the camshaft(s) with a gearing ratio to rotate the camshaft(s) once for every two rotations of the crankshaft as shown in Fig. 1.

1.1 Limitations of Conventional Valve Actuation Systems

With a mechanical coupling in place between the crankshaft and the camshaft(s), one no longer needs to worry about the synchronisation between valve event timing and engine speed. This makes the system reliable, and with proper design of the valve train components, the system becomes quite robust. However, one needs to understand that the synchronisation is with respect to crank and cam angular positions and not time. In other words, as the speed of the engine increases, time available for the intake and exhaust processes decrease although the relative opening and closing angular positions remain the same. Though the mass of fresh charge that the engine needs to draw in is speed invariant, the time available for the very same process keeps on reducing as engine speed increases. The performance of an engine with the valves opening and closing at appropriate top dead centre (TDC) and bottom dead centre (BDC) angular positions of the crankshaft will suffer with an increase in engine speed. The engine finds it difficult to breathe in the limited time available at high engine speeds owing to increase frictional losses and possible choked flow conditions, as the charge travels at high velocities through the intake tract. Similarly, exhaust flow will also be restricted at high speeds due to lesser time available for expansion of the combustion products in addition to the reasons mentioned above, resulting in exhaust back pressure. Consequently, the engine experiences a low volumetric efficiency. One can reason that the valve opening and closing angular positions may be spread out, i.e. the intake valve(s) can open early before the TDC and close late after BDC, and similarly with the exhaust valve(s). This does turn out to be beneficial at high engine speeds with added inertia of the fast-moving intake and exhaust streams. The performance in such a situation, however, would suffer at low speeds due to loss of effective compression of the pre-combustion mixture. The compressed charge could now exit through late closing intake valve(s) during compression or combustion products might enter into the intake manifold due to late closing exhaust valve(s) during the exhaust stroke and dilute the incoming fresh charge. The latter is an issue associated with what is popularly known as valve overlap. Parvate-Patil et al. [1] have assessed the philosophies concerning the timing of different valve events and their effect on the pressure–volume (p-V) cycle of the engine. With fixed angular timings of the valves, most engines with camshaft-driven valve trains are designed with a compromise between high- and low-speed performances. The valve opening and closing positions are chosen so as to have a good performance in a narrow speed range at which the engine operates most of the time. This is the reason one finds racing engines with large valve duration camshafts and commercial vehicles with comparatively smaller duration camshafts. Hence, one finds engines whose valve trains are particularly tuned and optimised for a specific range of engine speed. Operating the engine at any speed outside this narrow band usually incurs performance and efficiency drops.

The lift profile of the valves in almost all camshaft-driven valve trains have the same nature evident from their lift curves. Alterations, no doubt, do exist but are mainly centred around the maximum lift and the duration of the lift event. This is due to the fact that cam lobes need to have a continuous shape or profile in order to have smooth operation. As the camshaft rotates at very high speeds, the associated accelerations and dynamic loads leave little room for aggressive lifts, necessary to have a good airflow. Moreover, one also has to ensure negligible valve seating velocities in order to maintain the structural integrity of the valve train. Assanis and

Polishak [2] have developed and used a computer simulation to study the effect of cam design and resulting valve events on engine performance. Another limitation of a camshaft-driven valve train is the load and speed invariant, fixed valve lift. Particularly at low engine loads or speeds or both, as in idling, when the required charge mass flow rate is small, a large (full) valve lift implies a large flow area for the incoming charge and consequently, a small charge intake velocity. This results in low in-cylinder turbulence which may lead to poor combustion efficiency and higher emissions particularly in SI engines. A direct consequence of this limitation is the requirement of a throttle valve in SI engines for controlling the mass flow rate of incoming charge.

1.2 Throttling and Pumping Losses

A throttle valve sits between the ambient atmosphere and the air intake system and operates by simply controlling the area available for the fresh air to flow into the intake tract, thereby controlling the mass flow rate of air into the engine. For wide open throttle conditions, when one needs maximum power output from the engine, the obstruction to airflow due to the valve is negligible. However, operating at part load conditions with reduced mass flow rates, the throttle valve creates restriction for the incoming airflow which manifests itself as a pressure drop. This pressure drop across the throttle valve implies that the intake system downstream of the throttle valve is now in partial vacuum. As such, there is a pressure differential across the piston with near atmospheric pressure on the crankcase side and sub-atmospheric pressure on the intake (or exhaust) valve side. The charge induction or cylinder filling process which should ideally take place at zero-pressure differential across the piston now requires additional work to draw charge into the cylinder. Hence, the air induction process which is theoretically represented as a straight line in the p-V diagram for an Otto, Diesel or dual power cycle becomes the lower curve representing the intake pumping work in the pumping work loop. The exhaust process similarly takes place against a pressure differential across the piston with near atmospheric pressure on the crankcase side and slightly higher pressure than atmospheric on the exhaust (or intake) valve side due to restricted exhaust flow or insufficient time for blow-down. The exhaust process is therefore represented by the upper curve of the pumping work loop in the p-V diagram rather than theoretical straight line. Usually, the pressure differential due to throttling is relatively more feasible to manage as compared to the one on the exhaust side and increases with a decrease in throttle angle or increase in flow restriction at the throttle. The size of the pumping loop is representative of the specific work that is lost due to throttling and exhaust back pressure (shaded region in Fig. 2). Thus, the throttling process while trying to reduce the mass flow rate of incoming air also reduces the net specific work output of the engine with part of the work produced by the engine lost to the pumping process. If it were possible to avoid the pumping work, the same net specific work output could have been

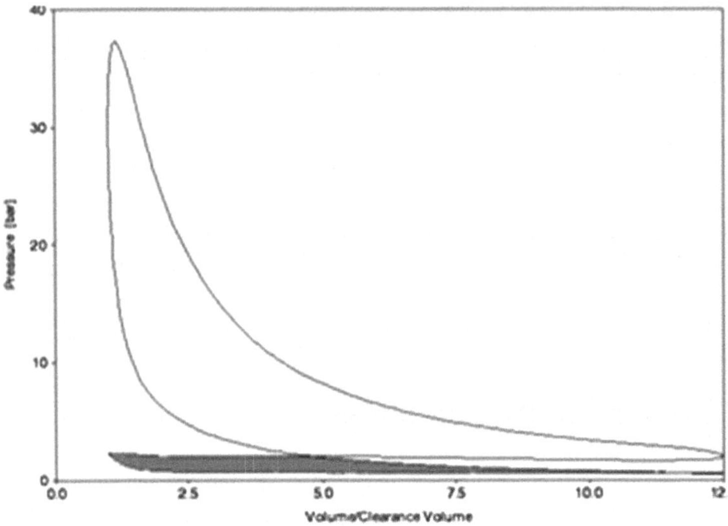

Fig. 2 p-V diagram for a conventional four-stroke SI engine

obtained at the expense of lesser fuel consumption. Thus, part load operation by the aid of a throttle valve ends up reducing the overall brake thermal efficiency of the engine in addition to reducing the mass flow rate of air being drawn into the engine.

2 Variable Valve Timing and Lift

An intuitive solution to the problem of pumping work in throttled engine operation would be to do away with the throttle valve itself and directly control the timing as well as lift of the intake valve(s) in order to regulate the airflow into the engine. The exhaust pumping work can be similarly minimised by suitable controlling the lift and timing of the exhaust valve(s) to reduce exhaust back pressure. In order to incorporate such flexibility into the valve train, one can make significant modifications to the camshaft. One can also introduce intermediate systems or mechanisms between the camshaft and the valves, or in the most extreme case, one can completely do away with the camshaft and use a variety of electronically assisted control systems to directly govern the lift and timing of the valves. The aforementioned methods to control the lift and timing of the intake and exhaust valves of an IC engine are collectively known as variable valve timing (VVT), variable valve lift (VVL) or variable valve actuation (VVA) systems. Modifications to the camshaft to have variable valve lift and timing control have been carried out by many automotive manufacturers, and such technologies are being used in many vehicles of everyday use. The last alternative of replacing the camshaft with a new control system is popularly known as the camless valve train system and is seeing a lot of research and development.

Researchers have investigated a number of ways to reduce the pumping work encountered during throttled engine operation. Shelby et al. [3] have outlined a new method for calculating the pumping work and indicated work in a four-stroke IC engine and analysed the role of VVT, intake charge dilution (stratified lean operation) and cylinder deactivation in reducing pumping work by its reflection on pumping mean effective pressure (PMEP). Schirm et al. [4] have studied the effect of late IVC on pumping losses of an unthrottled engine, in which the relative phase between the two intake cams of the same cylinder, can be varied and have reported a reduction in fuel consumption at part load conditions. Koo and Bae [5] have investigated the effects of an adaptive valve timing control system on volumetric efficiency, engine torque and pumping losses through computer simulation and have reported improved low-end torque (by about 6% at 1000 RPM and 10% at 5000 RPM) and reduced pumping losses. Fontana et al. [6] have numerically and experimentally analysed the combustion process of a small displacement, two valves, SI engine, with variable valve timing, and have reported up to 36% reduction in pumping work and 7.7% improvement in thermal efficiency. Moro et al. [7] have presented the thermodynamic analysis of an innovative engine load control based on electronic control of intake and exhaust valves and internal exhaust gas recirculation (EGR).

A variable valve actuation system as mentioned previously allows one to flexibly control the timing and lift of the engine valves according to the speed and load on the engine. VVA systems today not only target at reducing pumping losses but also at improving the overall efficiency of the engine by functioning as a better air management system as compared to fixed valve lift and timing systems. Schechter and Levin [8] have discussed the advantages that a variable valve actuation system, particularly a camless valve train can impart to a SI engine. A good volumetric efficiency across the entire speed range, in-cylinder charge motion control, efficient charge mixing, improved combustion, improved low-end torque, improved idle stability, selective cylinder deactivation and internal exhaust gas recirculation are to name a few of the functionalities that a VVA system can add to an engine apart from reducing pumping losses. As cited by Shelton [9], Ahmad and Theobald [10] have classified the different VVA systems into five different groups based on the level of sophistication and the extent of timing variation. Most of today's production, VVA systems from various automotive manufacturers and new systems being investigated by researchers easily fit into any one of these groups. The classification may be given as systems which have the following attributes:

i. Discretely variable lift and duration (interdependent)
ii. Continuously variable lift and duration (interdependent)
iii. Fixed lift and discretely variable phasing
iv. Fixed lift and continuously variable phasing
v. Fixed lift and discretely variable duration
vi. Fixed lift and continuously variable duration
vii. Discretely variable lift and discretely variable phasing (independent)

viii. Discretely variable lift and continuously variable phasing (independent)
ix. Discretely variable lift and discretely variable phasing (interdependent)
x. Continuously variable lift, duration and phasing (independent)

2.1 Current Variable Valve Actuation Technologies

2.1.1 Cam Switching

Honda's Variable Valve Timing and Lift Electronic Control (VTEC) system is a classic example of discrete cam switching VVA technology. The VTEC system as shown in Fig. 3 consists of two sets of different-sized cam lobes mounted on the same camshaft and two sets rocker arms hinged about the same point. The two identical cam lobes at the two ends of the assembly have their lift profile optimised for low engine speeds. As such, they have low lift and short duration. The larger cam lobe at the centre has its lift profile optimised for high engine speeds. As such, it has a higher lift and a longer duration as compared to the other cam lobes. The two identical rocker arms at the two ends directly act on the poppet valves to actuate them, while the central rocker arm does not have any direct linkage with the valves. The three rocker arms can be linked together to work as a single rocker arm by using a solenoid valve controlled, oil pressure actuated locking pin. At low engine speeds, the locking pin is unengaged and the poppet valves are actuated by the low lift and short duration cam lobes via the rocker arms at the two ends to ensure the required charge flow. At high engine speeds, the locking pin is engaged when the three rocker arms are following the base circles of their respective cam lobes. Once locked, the rocker arm assembly is now actuated by the high lift and long duration cam lobe while the other two cam lobes lose contact with their follower rocker arms as soon as the lift event on the central cam lobe initiates. Thus, the demand for increased charge flow at high engine speeds is met by a sudden increase in valve lift

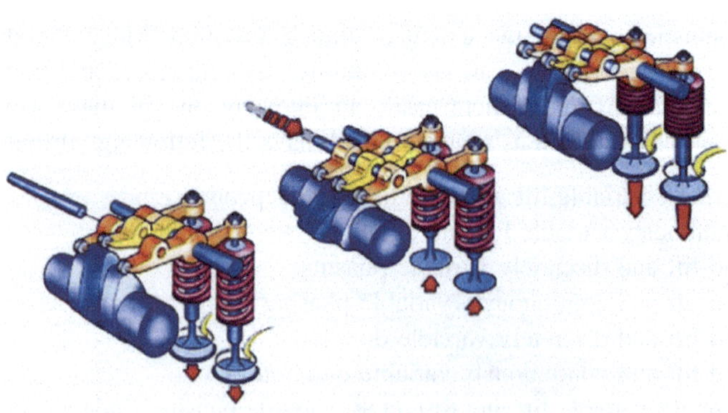

Fig. 3 Honda VTEC system, Honda Motor Company Website, 2004, reproduced from [11]

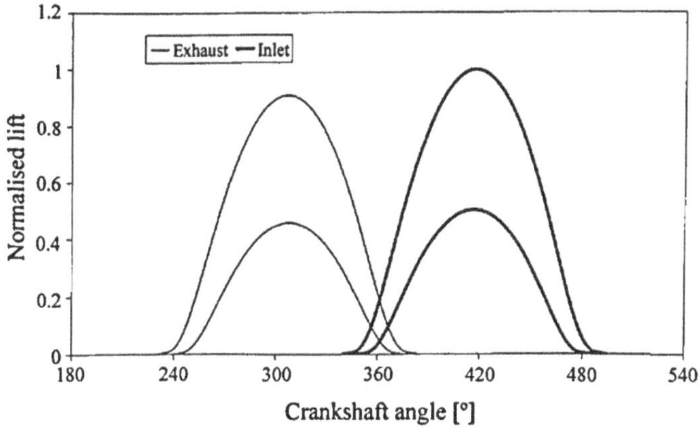

Fig. 4 Valve lift profiles for cam switching systems (lift normalised with maximum lift), reproduced from [11]

and duration which is marked by a step increase in engine torque. This can lead to driveability issues and is usually accompanied by changes in fuel injection and ignition timing to smoothen out the shift [11]. The system again switches back to low lift and short duration operation when engine speed falls. The operating engine speed at which the switch takes place lies in a narrow speed range and is controlled by the engine control unit (ECU). Audi's cam switching VVA system, Valve lift uses helical grooves machined on individual cam lobe assemblies to switch cams. Each cam lobe assembly comprises of two cam lobes having different lift and timing mounted next to each other and capable of sliding over the camshaft on sleeves. Whenever the ECU senses a need for changing the lift and timing, a solenoid controlled and engine oil pressure actuated switching pin engages on to the groove on the concerned cam lobe assembly. With the rotation of the camshaft, the pin traverses the helical groove, which in turn slides that particular cam assembly over the camshaft, switching to a new cam lobe with altered lift and timing. Mercedes Benz uses slightly altered variant of the same technology going by the brand name Camtronic [12] (Fig. 4).

2.1.2 Cam Phasing

Cam phasing is a form of VVT in which the relative phase difference between the crankshaft and the camshaft is varied by suitable means. Cam phasing in its standalone form does not change the valve lift or valve event duration. Hence, it is associated only with advancing or retarding valve opening time and consequently valve closing time. The phase difference between the crankshaft and the camshaft and between multiple camshafts can be introduced in a number of ways. Almost all methods of cam phasing use a camshaft that can rotate up to a certain extent with respect to its timing drive (sprocket, pulley or gear). A popular cam phasing mechanism uses a suitable flow control valve to controls the flow of hydraulic fluid (usually engine oil) into an internal vane mechanism mounted coaxially between the timing drive and the camshaft. The housing is made integral with the timing drive while the vanes rotate with the camshaft. The amount of oil held between the vanes and the housing, decides the relative phase difference between the camshaft and the timing drive, and consequently the crankshaft. Such a system has been used by Delphi in its variable cam phaser (VCP) system [13]. Depending on the construction of the control valve, both discrete as well as continuously variable cam phasing operations can be achieved. Much like Audi's valve lift, cam phasing can also be achieved by using helical splines as shown in Fig. 8 [11], where the camshaft slides and rotates along a helical path, relative to its timing drive to achieve discretely or continuously variable cam phasing. Another method to achieve variable cam phasing involves changing the path or the shape of the path of the timing chain or belt by using movable idler sprockets/pulleys or chain/belt guides. The one shown in Fig. 7 uses a hydraulic adjuster to change the path of the timing chain between two camshafts to change their relative timing [11]. BMW's VANOS, Ford's Variable Camshaft Timing (VCT), Dodge's Variable Valve Timing (VVT) and Toyota's Variable Valve Timing with intelligence (VVTi) are some production cam phasing VVT technologies. Porsche's Variocam Plus combines cam phasing and cam switching technologies for their VVA systems [12] (Figs. 5 and 6).

2.1.3 MultiCam Systems

Multicam VVA systems allow continuous adjustment of both valve lift and valve event duration [14]. Such systems use an intermediate cam lobe and follower between the camshaft and the poppet valves to incorporate another degree of freedom to control the valve motion. The motion of the intermediate cam lobe is controlled by the ECU, which in turn decides the motion of the intermediate follower acting on the poppet valves, deciding the final lift and duration. The intermediate cam lobe essentially changes the fulcrum of the intermediate rocker arm that actuates the valves, in turn changing the rocker ratio. The multicam system can be considered analogous to a proportional gain element in control theory having its value varying between zero and one, altering the maximum valve lift and event duration in the same ratio. The construction of the multicam system as such makes valve lift and event duration interdependent rather than providing independent control over both. The BMW Valvetronic [15], Nissan Variable Valve Event and Lift (VVEL) and Toyota Valvematic are some examples of multicam systems that have been used in production engines [12, 14] (Fig. 9).

Other VVA systems that have been investigated upon include the electrically driven cam, non-constant velocity drive, three-dimensional cam lobe and variable follower height system [11]. Kreuter et al. [16] have presented a mechanical continuously variable valve timing system which uses two camshafts together acting on the intake valve(s) to output a displacement equal to the sum of the effective

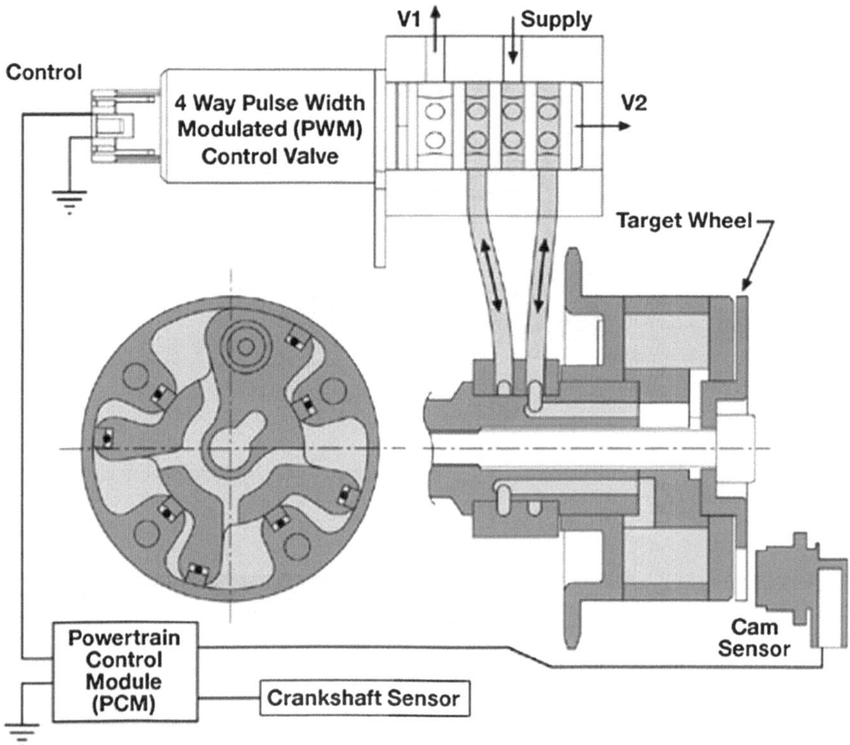

Fig. 5 Typical vane type VCP system, Delphi Corporation Website, 2004, reproduced from [11]

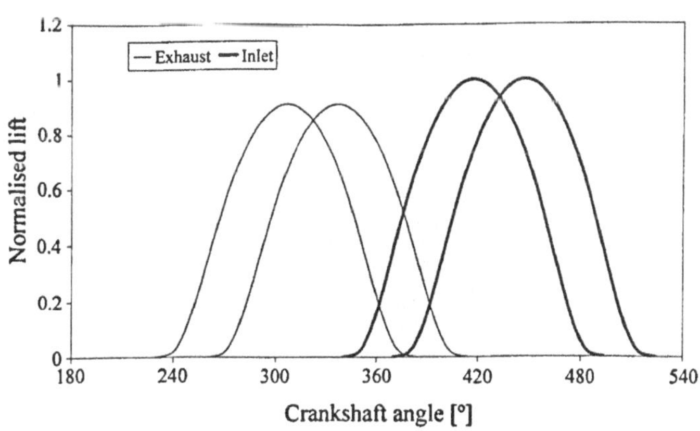

Fig. 6 Valve lift profiles for cam phasing systems (lift normalised with maximum lift), reproduced from [11]

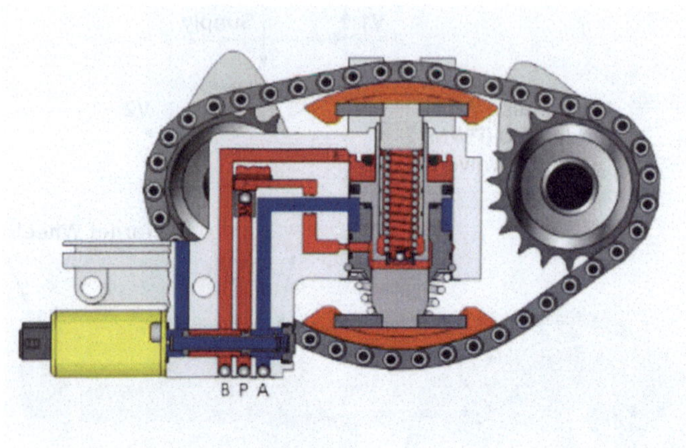

Fig. 7 Typical VCP system based on drive path change. Hilite International Website, 2004, reproduced from [11]

Fig. 8 Typical helical spline type VCP system, Auto Innovations Website, 2004, reproduced from [11]

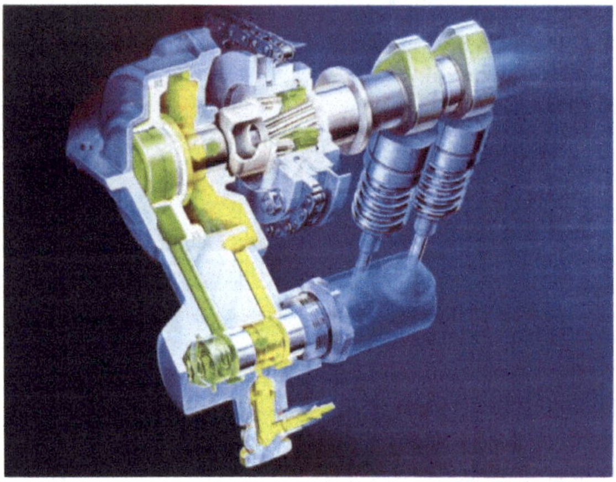

Fig. 9 Valvetronic system. BMW AG Website, 2004, reproduced from [11]

displacements of the two cams. Hara et al. [17] have developed an eccentric VVT control scheme in which valve timing is continuously controlled by varying the camshaft angular speed by means of an offset between the centre of the camshaft and that of the medium member that transfers crankshaft torque to the camshaft. Fallahzadeh et al. [18] have designed and developed VVA system that uses a sliding three-dimensional cam to continuously alter the valve lift and cam phasing via a differential gear train. Three-dimensional cams usually cannot independently control the lift and duration of a valve event, with the two being interrelated. Another kind of VVT technology, helical camshafts, allows a continuous variation of valve event duration. Characterised by an unique helical sliding between the two sections of the cam lobe about the camshaft, the sliding process leads to a decrease in the cam lobe base circle arc length while an increase in the nose arc length. As a result, the increased duration is obtained at full valve lift. Although promising in performance, helical camshafts have been limited to experimental systems mainly due to the economic reasons. Maas et al. [19] have discussed the development of a continuous VVL system, which in accordance with thermodynamic base studies show a potential of additional 2–5% savings in fuel consumption compared to an camless IC engine with variable intake and exhaust camshaft timing (Figs. 10 and 11).

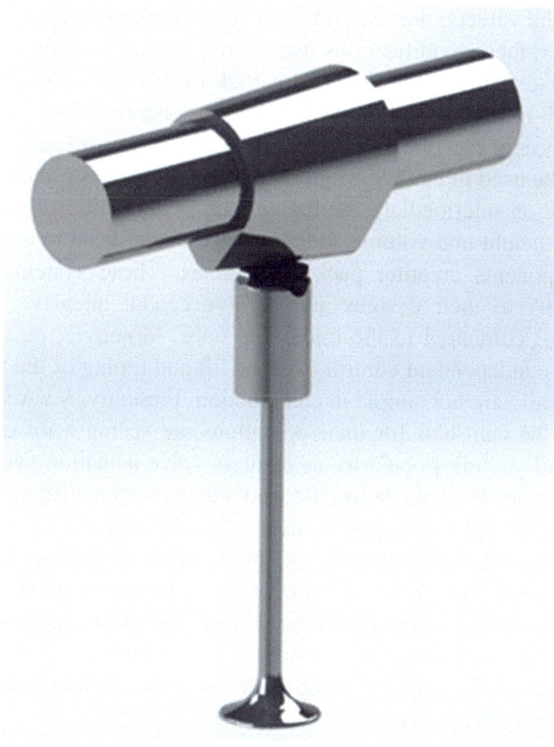

Fig. 10 Schematic of a three-dimensional cam lobe system

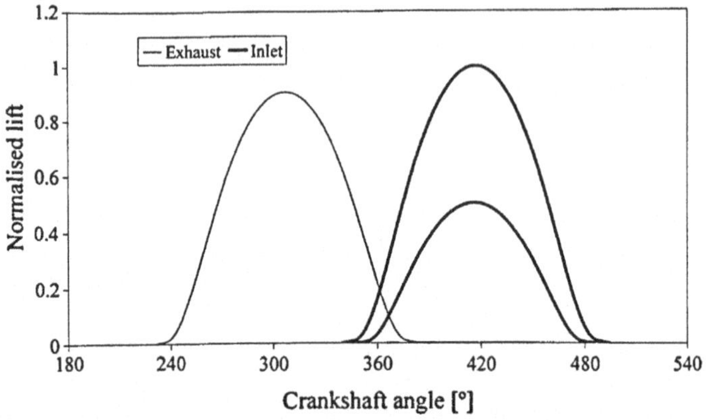

Fig. 11 Valve lift profiles for three-dimensional cam lobe systems (lift normalised with maximum lift), reproduced from [11]

2.2 *Camless Variable Valve Actuation Systems*

VVA systems that use modifications to the camshaft or use intermediate mechanisms between the valve(s) and the camshaft are no doubt reliable at high operating speeds. However, these modifications usually require the addition of extra moving mass to the valve train as compared to a traditional camshaft-driven valve train design. The high rotational speeds, at which these operate, bring in large dynamic forces that the system must be able to handle. This requires robust components and mechanisms to be used in the VVA system making them heavy and bulky. Usually, systems that use an intermediate oscillating cam or a significantly modified camshaft have extra height and volume added to the cylinder head to accommodate the additional components creating packaging issues. These systems usually have added complexity to their designs and involve capital intensive high precision manufacturing as compared to the traditional ones. Moreover, most of these systems do not have independent control over the lift and timing of the valves, and the ones that do usually are not simple in construction. Presently, VVA systems that do not depend on the camshaft for their operations are seeing a lot of research and development and gaining popularity as camless valve actuation systems. Camless engines use a variety of methods to independently control the lift and timing of the valves which include electromagnetic, hydraulic and pneumatic actuation systems with some even using a combination of these actuation methods. The camshaft, now done away with, the valve lift profile is no longer restricted the kinematic limitations of the design of the cam lobe. Hence, the theoretically best square lift profile can be practically realised as a trapezoidal or top hat lift profile, the shape being determined by the valve opening and closing speeds, allowable acceleration and seating velocity (Fig. 12).

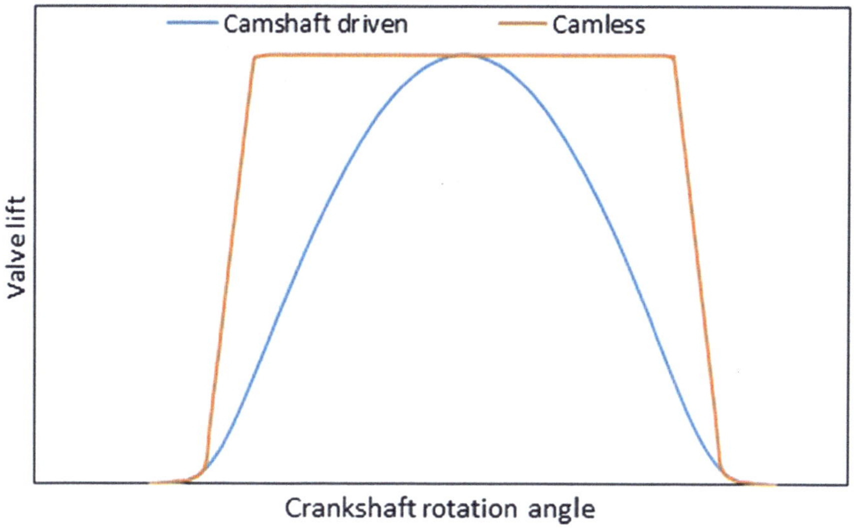

Fig. 12 Comparison between camless and conventional valve lift profile

2.2.1 Classification of Camless Variable Valve Actuation Systems

Current camless VVA systems can be broadly classified into the following three categories based on the type of actuation system used:

i. Electro-mechanical camless VVA system
ii. Electro-hydraulic camless VVA system
iii. Electro-pneumatic camless VVA system.

Electro mechanical Camless VVA System

Electro-mechanical camless systems use a combination of electromagnets and mechanical springs to achieve the desired valve motion in the required time. Sugimoto et al. [20] have studied, designed and tested an electro-mechanical VVA system. The functioning of the electro-pneumatc camless system studied by Mercorelli [21] is schematically shown in Fig. 13. The basic construction of the VVA system includes an armature disc mounted concentrically and rigidly with the valve stem and placed symmetrically placed between two electromagnets with the help of two springs (not necessarily identical). This configuration is referred to as the equilibrium state. Whenever the valve needs to open, the lower electromagnet is energised by applying a suitable voltage across its coil while the upper electromagnet is kept de-energised. This attracts the armature in the downward directions and the valve opens up. In the meantime, the upper spring gets elongated and the

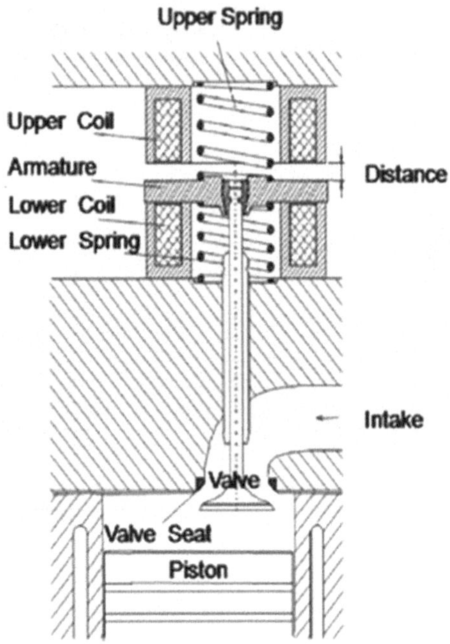

Fig. 13 Electro-mechanical valve actuator cross-section, reproduced from [21]

lower spring gets compressed. The energy stored in the springs is used to return the valve to its equilibrium position when the coils are de-energised. Similarly, for closing the valve, the lower electromagnet is de-energised while the upper electromagnet is energised, compressing the upper spring and elongating the lower. In order to hold the valve in open, closed or any desired lift position, the corresponding electromagnets have to be kept energised accordingly during the dwell period so that the spring force and electromagnetic force balance out. The desired actuation time decides the deceleration required at the end of the stroke, which in turn decides the stiffness of the springs used. The spring stiffness ultimately dictates the voltage applied across the coils. A simpler version of this design involves only one electromagnet and one mechanical spring, with the closed valve as the equilibrium position. An appropriate damping mechanism is usually used to control the velocity of the valve during closing. Sugimoto et al. [20] have used a small hydraulic tappet as a valve damping mechanism in thcir dcsign. A closed-loop control circuit is usually used in such applications tracking the valve motion using position sensors for enhanced reliability and reduced cycle to cycle variation. Cope and Wright [22] have designed, built and tested a novel electromagnetic fully flexible valve actuator (FFVA) which offers variable lift and timing combined with controllable seating velocity. The main advantages of using an electro-mechanical camless engine are its high valve opening and closing speeds, and its relatively easy system implementation. Montanari et al. [23] and Butzmann et al. [24] have presented control difficulties related to valve seating velocity and cost problems with expensive position sensors in electro-mechanical valve actuation systems [25]. Other problems include high power consumption, high noise, residual magnetic characteristics even after demagnetization and difficulty in case of high force applications [23]. The motion of the valves and the electro-mechanical actuation also suffers from parasitic losses, such as friction, ohmic resistance and eddy-current losses [26].

Electro-hydraulic Camless VVA System

A hydraulic camless engine converts the flow of hydraulic fluid into valve motion in response to an electrical signal. The basic construction of the arrangement involves a hydraulic cylinder whose piston is attached to a poppet valve. The flow of fluid into and out of the cylinder is controlled using high-frequency flow control valves. Figure 14 shows the schematic of the electro-hydraulic VVA system investigated by Zhao et al. [27]. In order to open the valve, hydraulic fluid from the reservoir is pressurised using a pump and is made to fill up the upper chamber of the

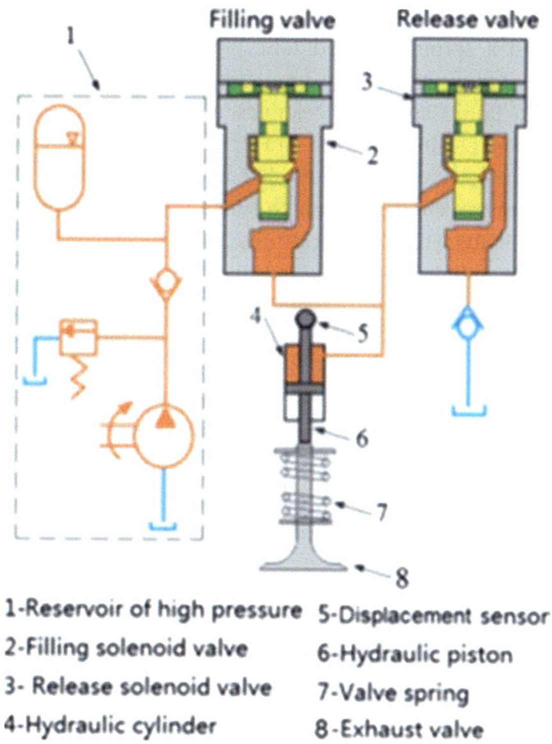

Fig. 14 Principle diagram of electro-hydraulic exhaust camless variable value actuation, reproduced from [27]

cylinder. At the same time, hydraulic fluid previously stored in the lower chamber is expelled to the reservoir. This makes the piston move downwards, while opening the poppet valve and compressing the valve spring. To hold the valve in open position or any desired lift position, the flow control valves restrict any fluid flow into or out of the cylinder. The fluid pressure forces and spring force then balance among themselves to determine the equilibrium hold position (dwell at full lift or partial lift). The lift of the valve can also be controlled by modulating the cylinder filling pressure, but such a technique usually is not adopted for real world on the go applications where the time available for such modulations is extremely small. To close the poppet valve, the flow control valves allow the high-pressure fluid to fill up the lower chamber, while expelling fluid from the upper chamber into the reservoir. This time the compressed spring also assists in closing the valve. A suitable damping mechanism is deployed just before the valve hits its seat to control the seating velocity. A simpler version of this design as shown in Fig. 14 involves fluid flow only into the upper chamber with valve closure being completely taken care of by the spring force. The performance and maximum operating speeds of such a system totally depend on precise control of fluid flow to ensure accurate positioning of the engine valves [25]. As such, it requires high performance flow control valves with very low response times and high frequency position sensors that cause increase in system production cost. Schechter and Levin [8] have developed a spring-less hydraulic pendulum for their electro-hydraulic VVA system that exploits the elastic properties of a compressed hydraulic fluid, which, acting as a liquid spring, accelerates and decelerates each engine valve during its opening and closing motions. Brader [28] has designed and manufactured an experimental piezo-electrically controlled hydraulic actuator to be used in an electro-hydraulic camless engine. Although suited for high force applications, some of the potential problems with basic electro-hydraulic actuation without closed-loop feedback systems include high energy consumption, valve seating velocity control and valve timing repeatability at different operating temperatures due to variation of oil viscosity with temperature [25]. Nam and Choi [25] have proposed an adaptive valve actuation system that is robust to oil temperature variations and can be realised by a simple valve timing controller based on valve opening and closing timing detection. Moreover, leakage in a hydraulic system can always be an issue.

Electro-pneumatic Camless VVA System

A pneumatic camless engine uses air pressure to actuate a piston–cylinder arrangement which in order actuates the poppet valve(s). Directional control valves are used to control the movement of pressurised air in and out of the pneumatic piston cylinder arrangement and consequently the timing of the poppet valves. Pneumatic camless VVA systems usually use additional hydraulic or electromagnetic systems to achieve variable lift and acceptable seating velocities. Figure 15 shows a schematic representation of an electro-pneumatic valve actuation (EPVA) system. The electro-pneumatic system, similar to an electro-hydraulic system,

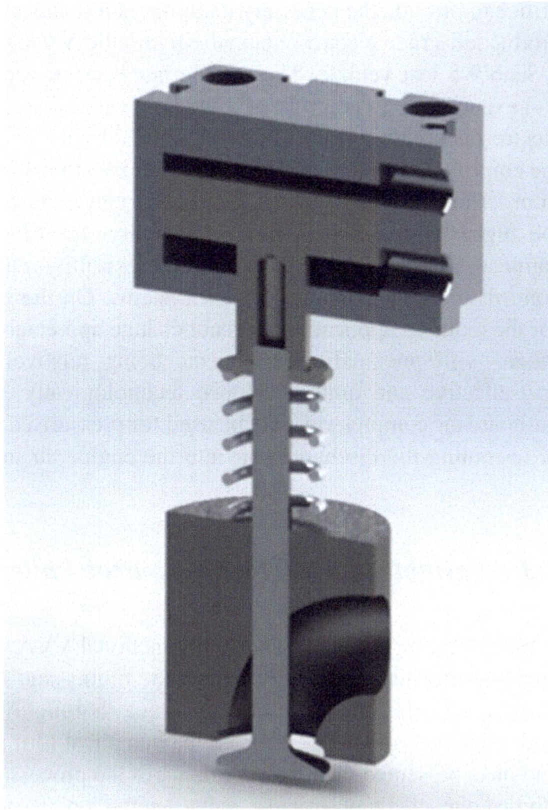

Fig. 15 Schematic sectional view of an electro-pneumatic valve actuation system

consists of a pneumatic cylinder whose piston is attached to the poppet valve. The cylinder can be either double acting, in which both opening and closing of the valve are realised using pressurised air, or single acting in which valve opening is realised using pressurised air and valve closure takes place with the help of a return spring. The EPVA system investigated by Ma et. al [29] uses a single acting cylinder. For opening the valve, pressurised air is allowed to fill the upper chamber of the cylinder while the lower chamber is open to the atmosphere. The pressure difference across pneumatic piston moves the poppet valve downwards while compressing the valve spring. At the same time, hydraulic chamber at the top of the assembly is filled with an incompressible fluid (oil). In order to hold the valve, open at the desired lift, the control valves for both air and oil close, trapping the air and the oil in their respective chambers. This is followed by transient force balancing due to expansion of the pressurised air, after which a steady lift is achieved. In order to close the valve, the air and oil chambers are depressurised by opening the flow control valves; air and oil are forced out of the chambers due to the spring force. Just before closure, a damping mechanism is engaged to control the seating velocity and the valve gently gets closed. In this case, the oil is made to flow through an orifice to provide the necessary damping before closure. Free valve has developed a production-grade electro-pneumatic-hydraulic VVA system that has been tested on a Saab 9-5 test vehicle. The vehicle has been driven 10,000 km with both intake and exhaust valves controlled by the Free valve system [30]. The functionality of an electro-pneumatic system is mainly limited by the response time and service life of the control valves. The quality of air goes into the system should also be maintained in order to avoid clogging of the control valves. Osborne et al. [31] have reported the highest noise and power consumption for EPVA systems among all other camless alternatives. Moreover, compressibility effects and leakage can make controls difficult without feedback systems. On the brighter side, behaviour of air for the required application does not change appreciably in the working temperature range, with pneumatic components being relatively simple, compact, reliable, cost-effective and quite-developed technologically. Compressed air cylinders or on-board air compressors may be used for pressurised air supply with the possibility of re-routing the discharged air into the engine air intake system [9].

2.3 Variable Valve Timing Control Philosophy

While designing a reliable and fully functional VVA system is crucial for improved engine performance, optimisation of the timing and lift of the valve events are an equally important task. Researchers have thoroughly investigated the same using both computer simulation assisted mathematical formulation as well as experiments and have presented their understanding of the process. Bohac and Assanis [32] have studied the effect of exhaust valve timing on gasoline engine performance and hydrocarbon emissions using both numerical simulation as well as experiments. Fontana et al. [6] have numerically analysed the influence of variable valve timing on the combustion process of a small SI engine. Fallahzadeh et al. [18] have used numerical simulations and experiments to study the effect of IVC time on engine performance, fuel consumption and emissions. Mudka et al. [33] have conducted a theoretical investigation on the effect of early intake valve closing (EIVC) on thermal efficiency of an IC engine. Ashhab et al. [34] have designed an adaptive nonlinear controller to coordinate intake valve lift and duration according to engine speed and load, by using high sampling rate intake manifold pressure and flow sensors. Trask et al. [35] have experimentally optimised the timing of all the valve events of a camless engine for a number of commonly encountered engine operation conditions. Wilson et al. [36] have completely mapped the tumble, swirl and flow characteristics of a four-valve engine, which can be studied to optimise the lift in a camless VVA. With a flexible VVA system, one can independently vary the timing of each valve event. However, the number of timing combinations possible with such a system is very large, making it necessary to understand the implication of each event on the performance and efficiency of the engine. The basic philosophies behind variable valve event timing as presented by Parvate-Patil [1] have been listed below.

2.3.1 Late Intake Valve Closing (LIVC)

During LIVC, at low engine speeds, a part of the compressed charge, at a pressure slightly higher than the atmospheric, is expelled back to the intake manifold during the early compression stroke. This reduces the partial vacuum created in the intake manifold due to throttling resulting in reduction of intake pumping work during the subsequent intake stroke. As such, LIVC can be used as a means of both throttled and throttle-less load control at low engine speeds. Asmus [37] has stated that the volumetric efficiency at high speeds and high loads can be improved by LIVC as the momentum of the fast moving charge during the intake stroke helps in cramming in more charge into the cylinder even during the upward movement of the piston. Going by the same reasoning, LIVC will lead to poor volumetric efficiency at low speeds and high load conditions.

2.3.2 Early Intake Valve Closing (EIVC)

EIVC can be used as a throttle-less load control method by closing the intake valve (s) as soon as the desired amount of charge is admitted into the cylinder during the intake stroke. This leads to a small amount of expansion work during the intake stroke. EIVC can also incur some pumping losses due to the low lift of the valves which can be taken care of with faster valve lifts [38]. Altogether, the pumping work during EIVC is much less as compared to a conventional throttled engine.

2.3.3 Late Intake Valve Opening (LIVO)

With LIVO, the piston needs to do expansion work during the early intake stroke [37] which leads to increase intake pumping work. However, the increase in pumping work due to LIVO is compensated for improved cylinder charging as the piston attains an appreciable velocity when the intake valve(s) open. A higher charging velocity leads to increase turbulence which results in better combustion efficiency and lower unburnt hydrocarbon (HC) emissions. Volumetric efficiency remains almost unaltered with LIVO at low engine speeds as the amount of charge drawn in during the early intake stroke is negligible due to low piston velocity.

2.3.4 Early Intake Valve Opening (EIVO)

EIVO during the late exhaust stroke increases the valve overlap which leads to the flow of exhaust gases into the intake manifold because of the cylinder-intake manifold pressure gradient [38]. At low loads, this is detrimental to engine performance as it dilutes the fresh charge in the intake manifold leading to lean and unstable combustion. At high loads, however, the same phenomenon helps in realising internal exhaust gas recirculation (EGR) which helps in bringing down peak cylinder temperatures and hence NOx emissions [39]. EIVO also helps in reducing exhaust pumping work as less burnt gases are expelled during the exhaust stroke. EIVO is particularly beneficial at high speeds as the piston attains an appreciably high velocity during the early intake stroke and more air can be effectively drawn into the cylinder.

2.3.5 Early Exhaust Valve Closing (EEVC)

EEVC reduces the possibility of valve overlap which is beneficial at low loads. EEVC also reduces exhaust gas scavenging, as less time is available for expelling the burnt gases out of the cylinder, leading to lower exhaust pumping work. However, the exhaust gases that remain in the cylinder dilute the fresh charge leading to reduce thermal efficiency, particularly at high loads and speeds, while at the same time reduced unburnt HC emissions due to repeated combustion cycles.

2.3.6 Late Exhaust Valve Closing (LEVC)

With LEVC, the valve overlap period increases which results in the flow of exhaust gases into the intake manifold [37]. This, in turn, reduces the quantity of fresh charge that is drawn into the cylinder while increasing the inlet manifold pressure. The reduced partial vacuum in the intake manifold results in the reduction of intake pumping work. At high speeds, a large valve overlap helps in scavenging to improve engine breathing, while at low loads and speeds, a small or negligible overlap helps in maintaining charge quality and stable combustion. LEVC also helps in reduction of unburnt HC emissions as the combustion products undergo repeated combustion cycles due charge dilution with the same. According to Siewert [38], LEVC is less effective in reducing HC emissions as compared to EEVC.

2.3.7 Early Exhaust Valve Opening (EEVO)

EEVO is particularly beneficial at high loads and speeds when the time available for the gas exchange process is very small. Opening the exhaust valve during the power stroke does reduce engine power, but if timed correctly, the loss in expansion work is well compensated for by a significant reduction in exhaust pumping work and enhanced scavenging. At low loads, the expansion work is significantly reduced and EEVO can drastically reduce the already low power output of the engine. Hence, by appropriately timing the EVO, one can control the effective expansion ratio.

2.3.8 Late Exhaust Valve Opening (LEVO)

LEVO results in increased expansion work during the power stroke. However, a large cylinder pressure at the end of the expansion stroke increases the exhaust pumping work particularly at high speeds and loads. LEVO can, however, improve the power output at low loads and speeds, when maximum work needs to be extracted from the expansion process and a sufficient amount of time is available for the cylinder pressure to reduce to a low value. LEVO also helps in reducing unburnt HC emissions as the exhaust gases have sufficient time for oxidation [40].

2.4 Attributes Unique to Camless Valve Actuation Systems

Fully flexible VVA systems which can independently control the lift and timing of the valve events over the operating speed range of an engine offer numerous advantages and possibilities over conventional engines. Some of them unique to camless VVA systems as documented by Schechter and Levin [8] have been listed below.

2.4.1 Efficient and Stable Idle

A camless system can essentially eliminate the need for throttling which is a major culprit behind low idle efficiency and stability in conventional engines. By suitably controlling the opening and closing times, and lift of the intake valves and corresponding timing of the exhaust valves to provide minimal or no overlap, the combustion efficiency during idling can be greatly improved. Altogether, a camless VVA system can allow an engine to have a stable idle with reduced fuel consumption and emissions.

2.4.2 Faster Burn Rate

For having good combustion efficiency, cylinder charging should be accompanied by turbulence for efficient charge mixing. However, at low engine speed and loads, with full valve lift, the charging velocity drops significantly due to reduced charge mass flow rate and increased flow area, leading to unstable and non-uniform combustion. Using a camless VVA system, one can reduce the duration of the intake stroke to draw in charge only when the piston velocity is sufficiently high or reduce the maximum valve lift to increase the charging velocity.

2.4.3 Variable Compression Ratio

By varying the timing of IVC, one can vary the effective compression ratio of an engine without a corresponding change in the expansion ratio. This can be particularly beneficial for forced induction engines where knocking becomes an issue at high load conditions. The compression ratio of such engines is usually reduced to avoid knocking, but this leads to lower thermal efficiency at low loads. With a flexible VVA system in place, EIVC or LIVC can be suitably used for efficient low load operation with reduced knocking probability at high loads. This eliminates the need for waste gating in turbocharged engines, leading to improved utilisation of exhaust waste heat.

2.4.4 Internal EGR

In conventional engines, EGR is usually realised by using a separate valve between the intake and the exhaust manifolds to regulate the amount of exhaust gases recirculating into the engine. With a camless VVA system, the timing of the EVC and IVO can be appropriately tuned to provide the necessary overlap and corresponding amount of internal EGR. This can help in reducing NOx emissions by bringing down the peak cylinder temperature.

2.4.5 Charge Motion Control

The ability to individually control the lift and timing of each valve in a camless engine allows one to control the in-cylinder charge motion. In a cylinder with two or more intake valves, unequal lifts and timing can be used to vary the swirl and tumble characteristics of the incoming charge during cylinder filling providing a variety of flow patterns. A particular valve can also be deactivated (held in closed position) to divert all the flow through a fewer number of valves at a higher velocity for increased turbulence.

2.4.6 Variable Engine Displacement

A camless system may be used to deactivate all the valves of a particular cylinder, effectively deactivating that cylinder when coupled with deactivation of fuel injection. As such, during part load operation, the total displacement of a multi-cylinder engine can be reduced, with only a few cylinders working efficiently at a higher load. This reduces part load fuel consumption and unburnt HC emissions. The valves are usually deactivated after the intake stroke in order to reduce the amount of residual gases trapped in the deactivated cylinders for low in-cylinder pressure.

2.4.7 Change in Firing Order

The selection of the firing order of a particular multi-cylinder engine is done with an aim to reduce vibrations. In conventional engines, the firing order decides the arrangement of the cam lobes on the camshaft and as such, once fixed, it is not possible to change the firing order. However, the nature of engine vibrations varies with engine speed and load and there arises a need for changing the firing order to reduce the same at different operating conditions. With a camless system, changes to the firing order and other necessary changes in valve timing can be easily made to provide smooth engine operation across the entire speed range.

2.4.8 Regenerative Braking

A camless VVA system can be used to realise dynamic braking by using the engine as a compressor during braking. During dynamic braking, the exhaust valves and ignition are deactivated, and the momentum of the vehicle via the transmission is used to compress the charge drawn in during the intake stroke. The compressed charge is then expelled around TDC back into the intake manifold or into a reservoir, which can be later drawn in for engine supercharging during acceleration. The compression work was done by the engine partly slows down the vehicle in conjunction with wheel braking and provides an efficient way to use the energy lost during braking.

2.5 *Experimental Results on Implementation of VVA Systems*

Çinar and Akgün [41] have reported a 5.1% increase in torque at low speeds and a 4.6% increase in torque at high speeds for their test engine operating with variable IVC timing camshaft. Specific fuel consumption has been reported to have decreased by 5.3 and 2.9% at low and high engine speeds, respectively, along with reduction in HC and CO emissions at high engine speeds. Fallahzadeh et al. [18] have reported a 15–20% improvement in brake thermal efficiency for their test engine due to EIVC at an engine speed of 1600 rpm accredited to lower residual gas fraction and reduced heat transfer. A 20–25% decrease in fuel consumption has also been reported due to reduced pumping loss results during part load operation at the same speed along with 4–5% reduction in exhaust gas temperature and 50–60% increase in intake manifold pressure. Bohac and Assanis [32] during their experiments to study the effect of exhaust valve timing on gasoline engine performance and HC emissions, using switchable camshafts, have reported a 27% decrease in start-up HC emissions by advancing EVO by 60° and advancing EVC by 40°. Sugimoto et al. [20] while investigating their electromagnetic VVA system have reported a 10% reduction in BSFC due to late IVC as compared to conventional cam driven valve trains again accredited to reduction in pumping losses. By fixing IVC timing to BDC, an approximate improvement of 20% in torque output at low- and mid-engine speeds has been recorded due to improvement in volumetric efficiency. The camshaft-driven eccentric VVA system developed by Hara et al. [17] has shown a 10–15% improvement in low- and mid-engine speed torque and 10% improvement in maximum power. A 5% reduction in fuel consumption, 19% reduction in HC emission and 29% reduction in NOx emissions have also been reported. Li et al. [42] have reported a 11.8% torque increase at 2400 RPM and 9.9% torque increase at 5400 RPM along with 1.2% increase in maximum torque at 3900 RPM for their variable cam phasing system after proper ECU tuning. A 6.9% and 23.3% improvement in BSFC have also been reported at 2800 RPM and 5400 RPM, respectively, but with a slight reduction in baseline BSFC of 1.4–4.4% between 1200 and 2400 RPM. A 59.4% reduction in CO emissions as compared to the baseline has also been recorded. Schirm et al. [4] have reported a mean reduction of 6.4% at 25% load, 3.7% at 50% load and 1.4% at 75% load in fuel consumption while running the test engine in unthrottled operation mode by late IVC using a variable cam phasing system.

3 Summary

An overview of the development of valve train systems has been presented in the text. The importance of valves in an IC engine and the functioning of a mechanically coupled, camshaft-driven valve train with focus on its simplicity and reliability has been discussed. The limitations of a camshaft-driven valve train with respect to varying engine speed and load control have been highlighted to substantiate the need for a variable valve actuation system. Pumping losses during the throttling process for load control have been explained along with a few strategies to reduce them. The working of few production level, camshaft-driven, variable valve timing, variable valve event duration and variable valve lift technologies has been described covering both discrete and continuous control systems. Drawbacks of camshaft operated variable valve actuation systems have been discussed to sketch the functional requirements of a camless valve train. Three types of camless valve trains along with their advantages and limitations have been explained. The basic variable valve timing control logic has been described and some unique attributes of camless engines have been listed. Some experimental results have also been cited from literature highlighting the improvement in performance, efficiency and emissions after implementation of variable valve actuation systems.

References

1. Parvate-Patil G, Hong H, Gordon B (2003) An assessment of intake and exhaust philosophies for variable valve timing. SAE Technical paper 2003-32-0078. https://doi.org/10.4271/2003-32-0078
2. Assanis DN, Polishak M (1990) Valve event optimization in a spark-ignition engine. J Eng Gas Turbines Power 112/341
3. Shelby M, Stein R, Warren C (2004) A new analysis method for accurate accounting of IC engine pumping work and indicated work. SAE Technical paper 2004-01-1262. https://doi.org/10.4271/2004-01-1262
4. Schirm E, De Jesus G, Valle R (2003) Performance controlling of spark ignition engine by the variations of the intake valve opening angle. SAE Technical paper 2003-01-3720. https://doi.org/10.4271/2003-01-3720
5. Koo JM, Bae CS (2000) Effect of breathing characteristics on the performance in spark-ignition engines. In: Proceedings of FISITA World Automotive Congress, 12–15 June 2000, Seoul, Korea, paper code F2000A027
6. Fontana G, Galloni E, Palmaccio R, Torella E (2006) The influence of variable valve timing on the combustion process of a small spark-ignition engine. SAE Technical paper 2006-01-0445. https://doi.org/10.4271/2006-01-0445
7. Moro D, Ponti F, Serra G (2001) Thermodynamic analysis of variable valve timing influence on SI engine efficiency. SAE Technical paper 2001-01-0667. https://doi.org/10.4271/2001-01-0667
8. Schechter M, Levin M (1996) Camless engine. SAE Technical paper 960581. https://doi.org/10.4271/960581
9. Shelton P (2008) A review of variable valve timing devices. Mechanical Engineering Undergraduate Honors Theses 17, http://scholarworks.uark.edu/meeguht/17
10. Ahmad T, Theobald M (1989) A survey of variable-valve-actuation technology. SAE Technical paper 891674. https://doi.org/10.4271/891674
11. Nouhov D (2004) Investigation of the effect of inlet valve timing on the gas exchange process in high-speed engines. Retrieved, 09:40, 10 July 2017 from https://dspace.lboro.ac.uk/2134/7682
12. Concepcion M (2013) Automotive variable valve timing & lift explained, 2 edn, 13 June 2013. ISBN/EAN13: 1490422463/ 9781490422466
13. Gasoline Engine Management Systems Article. Retrieved, 09:38, 10 July 2017 from http://www.delphi.com/manufacturers/auto/powertrain/gas/valvetrain/vcp
14. Autozine VVT Article. Retrieved 17 Jan 2012 from http://www.autozine.org/technical_school/engine/vvt_5.html
15. Autospeed Valvetronic Article. Retrieved 17 Jan 2012 from http://www.autospeed.com/cms/article.html?&title=BMWs-Valvetronic&A=111539
16. Kreuter P, Heuser P, Reinicke-Murmann J (1998) The meta VVH system—a continuously variable valve timing system. SAE Technical paper 980765. https://doi.org/10.4271/980765
17. Hara S, Hidaka A, Tomisawa N, Nakamura M et al (2000) Application of a variable valve event and timing system to automotive engines. SAE Technical paper 2000-01-1224. https://doi.org/10.4271/2000-01-1224
18. Fallahzadeh F, Subrahamanyam J, Sharma V, Gajendra Babu M (2005) Simulation and evaluation of a variable valve timing single cylinder spark ignition engine. SAE Technical paper 2005-01-0765. https://doi.org/10.4271/2005-01-0765
19. Maas G, Neukirchner H, Dingel O, Predelli O (2004) Potential of an innovative, fully variable valvetrain. SAE Technical paper 2004-01-1393. https://doi.org/10.4271/2004-01-1393
20. Sugimoto C, Sakai H, Umemoto A, Shimizu Y et al (2004) Study on variable valve timing system using electromagnetic mechanism. SAE Technical paper 2004-01-1869. https://doi.org/10.4271/2004-01-1869

21. Mercorelli P (2013) A Lyapunov-based adaptive control law for an electromagnetic Actuator. AASRI Procedia 4:96-103. https://doi.org/10.1016/j.aasri.2013.10.016
22. Cope D, Wright A (2006) Electromagnetic fully flexible valve actuator. SAE Technical paper 2006-01-0044. https://doi.org/10.4271/2006-01-0044
23. Montanari M, Ronchi F, Rossi C, Tonielli A (2004) Control of a camless engine electromechanical actuator: position reconstruction and dynamic performance analysis. IEEE Trans Ind Electron 51(2):299–311. https://doi.org/10.1109/TIE.2004.825230
24. Butzmann S, Melbert J, Koch A (2000) Sensorless control of electromagnetic actuators for variable valve train. SAE Technical paper 2000-01-1225. https://doi.org/10.4271/2000-01-1225
25. Nam K, Choi SB (2012) Development of a camless engine valve actuator system for robust engine valve timing control. Int J Veh Syst Model Test 7(4):372–389, 2012. https://doi.org/10.1504/IJVSMT.2012.049429
26. Schernus C, van der Staay F, Janssen H, Neumeister J et al (2002) Modeling of exhaust valve opening in a camless engine. SAE Technical paper 2002-01-0376. https://doi.org/10.4271/2002-01-0376
27. Zhang S, Zhao Z, Zhao C, Zhang F, Wang S (2016) Development and validation of electro-hydraulic camless free-piston engine. Appl Therm Eng 102:1197–1205. https://doi.org/10.1016/j.applthermaleng.2016.03.093
28. Brader JS (1995) Development of a piezoelectric controlled hydraulic actuator for a camless engine. Master Science Thesis, Boston University
29. Ma J, Schock H, Carlson U, Hoglund A et al (2006) Analysis and modeling of an electronically controlled pneumatic hydraulic valve for an automotive engine. SAE Technical paper 2006-01-0042. https://doi.org/10.4271/2006-01-0042
30. FreeValve Technology Article. Retrieved, 3:22, 9 July 2017 from http://www.freevalve.com/technology/freevalve-technology/
31. Osborne R, Stokes J, Lake T, Carden P et al (2005) Development of a two-stroke/four-stroke switching gasoline engine—the 2/4SIGHT concept. SAE Technical paper 2005-01-1137. https://doi.org/10.4271/2005-01-1137
32. Bohac S, Assanis D (2004) Effect of exhaust valve timing on gasoline engine performance and hydrocarbon emissions. SAE Technical paper 2004-01-3058. https://doi.org/10.4271/2004-01-3058
33. Żmudka Z, Postrzednik S, Przybyła G (2016) Throttleless control of SI engine load by fully flexible inlet valve actuation system. Combust Engines 164(1):44–48. ISSN 2300-9896
34. Ashhab M, Stefanopoulou A, Cook J, Levin M (1998) Camless engine control for a robust unthrottled operation. SAE Technical paper 981031. https://doi.org/10.4271/981031
35. Trask N, Hammoud M, Haghgooie M, Megli T et al (2003) Optimization techniques and results for the operating modes of a camless engine. SAE Technical paper 2003-01-0033. https://doi.org/10.4271/2003-01-0033
36. Wilson N, Watkins A, Dopson C (1993) Asymmetric valve strategies and their effect on combustion. SAE Technical paper 930821. https://doi.org/10.4271/930821
37. Asmus T (1982) Valve events and engine operation. SAE Technical paper 820749. https://doi.org/10.4271/820749
38. Siewert R (1971) How individual valve timing events affect exhaust emissions. SAE Technical paper 710609. https://doi.org/10.4271/710609
39. Hara S, Nakajima Y, Nagumo S (1985) Effects of intake-valve closing timing on spark-ignition engine combustion. SAE Technical paper 850074. https://doi.org/10.4271/850074
40. Stein R, Galietti K, Leone T (1995) Dual equal VCT-a variable camshaft timing strategy for improved fuel economy and emissions. SAE Technical paper 950975. https://doi.org/10.4271/950975

41. Çinar C, Akgün F (2007) Effect of intake valve closing time on engine performance and exhaust emissions in a spark ignition engine. J Polytech 10(4):371–375
42. Li L, Tao J, Wang Y, Su Y et al (2001) Effects of intake valve closing timing on gasoline engine performance and emissions. SAE Technical paper 2001-01-3564. https://doi.org/10.4271/2001-01-3564

Performance, Combustion, and Emissions Characteristics of Conventional Diesel Engine Using Butanol Blends

Mohit Raj Saxena and Rakesh Kumar Maurya

Abstract Energy security concern and stringent emission legislations norms demand a clean and high fuel conversion efficiency engines. Diesel compression ignition (CI) engines are more preferred over the spark-ignition (SI) engines in commercial applications due to their higher fuel conversion efficiency. Present chapter focuses on the effect of butanol addition in the diesel fuel on the combustion and emissions characteristics of a diesel engine. Butanol has inimitable properties, which makes it more suitable candidate fuel for diesel engine in comparison to other alcohol fuels such as ethanol and methanol. Combustion characteristics of the engine are analyzed from heat release analysis of measured in-cylinder pressure data at different engine operating conditions. Combustion stability is also discussed with respect to diesel engine operation with butanol blends. Carbon monoxide (CO), unburned hydrocarbons (HC), and nitrogen oxides (NO_x) emission characteristics of diesel engine using butanol blends are discussed in this chapter. Special emphasis is placed on the discussion of particulate emission (soot particle numbers) in diesel engine with butanol blends.

1 Introduction

In past few decades, alternative fuels are intensively investigated for internal combustion (IC) engine application to fulfill the energy demand across the world. With urbanization, the number of engines for automotive and off-road applications (such as tractors, trucks, agricultural equipment's, power generators) is increasing rapidly. Most of the commercial application engines are equipped with compression ignition (CI) engines due to their higher brake thermal efficiency and fuel economy. However, conventional compression ignition (diesel) engines typically emit oxides of nitrogen (NO_x) and particulate matter (PM) in relatively higher concentrations.

M. R. Saxena · R. K. Maurya (✉)
Advanced Engine and Fuel Research Laboratory, Department of Mechanical Engineering, Indian Institute of Technology Ropar, Rupnagar 140001, Punjab, India
e-mail: rakesh.maurya@iitrpr.ac.in

D. K. Srivastava et al. (eds.), *Advances in Internal Combustion Engine Research*, Energy, Environment, and Sustainability,
https://doi.org/10.1007/978-981-10-7575-9_5

Reduction in formation of these emissions during combustion itself is a bigger challenge for automotive industries as well as engine researchers. In addition to this, there exits a trade-off between the PM and NO_x in conventional diesel engine, which makes the problem more challenging. The NO_x emissions are higher mainly due to higher combustion temperature, and PM emission is higher due to the combustion of heterogeneous mixture in conventional diesel engine.

It is well known that the exhaust emissions from the combustion of fossil fuels significantly affecting the human health as well as the environment [1, 2]. To keep these emissions below or within the emission mandate limits, several after-treatment devices (such as diesel particulate filter, lean NO_x trap) are installed in the exhaust line of modern automotive engines. However, after-treatment devices have a penalty impact on the fuel economy of the engine. Utilization of renewable or biofuel is one of the potential ways to reduce these emissions from conventional diesel engines. Biodiesel was considered as a good alternative fuel for diesel engine and has a lower carbon monoxide (CO), hydrocarbon (HC), smoke and PM emissions. However, combustion of biodiesel in CI-engine emits higher NO_x emissions [3, 4]. In addition to this, it is considered to be presently impractical for diesel engine vehicles due to its limitations such as excessive carbon deposition, oxidation stability, and it may also cause corrosion in vehicle material, which leads to engine durability issues. Alcohols such as methanol, ethanol, butanol are renewable fuels with relatively lower cetane numbers, which are considered as good alternative fuels and has a potential to reduce the emissions from combustion engines [5–7]. Methanol and ethanol are considered as potential alternative fuels for spark-ignition engines. Various studies investigated the performance characteristics of methanol-/ethanol-fueled spark-ignition engines [8–11]. In Brazil, ethanol was used as an alternative fuel more than past 35 years [12]. A study reported that in European countries, blending of methanol with gasoline was used since 1970 [5]. Both these alcohols could be used in spark-ignition engine either in pure form or in blended form with gasoline. However, these alcohol fuels are typically not considered as good replacement of diesel fuel for CI-engines due to their higher auto-ignition temperature and lower cetane number. In CI-engines, methanol/ethanol could be used as either in blended form with conventional diesel or by separately injecting in the intake manifold and running the engine in dual fuel mode. Ethanol/methanol also has miscibility issues with diesel fuel, and thus, additives/solvents are required to avoid the fuel layer separation in the fuel tank. Butanol is an another alcohol fuel, which can be used as an alternative fuel for combustion engines and it has a better physical and chemical properties as compared to methanol and ethanol for CI-engine perspective. Benefits of butanol in comparison to ethanol and methanol are given below [5];

- Higher heating value as compared to methanol and ethanol.
- Lower volatility, which may reduce the tendency of vapor lock problem and cavitation.

- Less cold start problem in comparison to ethanol-/methanol-fueled engines (due to lower heat of vaporization, and ignition temperature of butanol is also lower as compared to ethanol and methanol).
- No miscibility issue with diesel as well as gasoline.
- Butanol has higher flash point temperature and lower vapor pressure point and thus, it is safer as compared to ethanol and methanol.
- Less corrosive in comparison to ethanol.
- Higher cetane number as compared to methanol and ethanol thus easier to auto-ignite in CI-engine.

Butanol is a renewable biofuel and can be easily produced by the fermentation process. A schematic diagram of common butanol production process is provided in the Fig. 1 and typical properties of the methanol, ethanol, butanol, and diesel are compared in Table 1.

In past few years, researchers intensively investigated the butanol as an alternative fuel for CI-engines [13–17]. A study optimizes the fuel injection parameters for the performance and emissions characteristics of butanol-/diesel-fueled

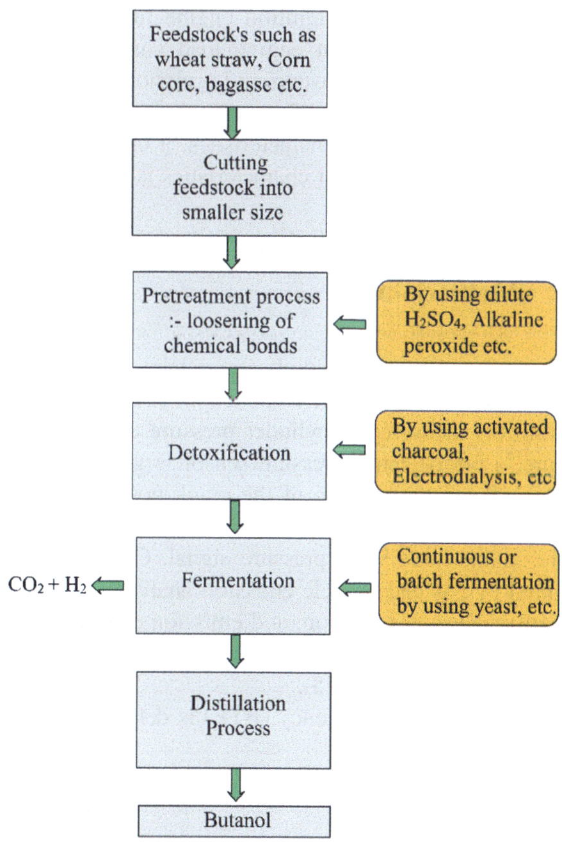

Fig. 1 Typical production process of butanol. Adapted from [31]

Table 1 Typical properties of methanol, ethanol, butanol, and diesel [7]

Property	Methanol	Ethanol	n-Butanol	Diesel
Molecular formula	CH_3OH	C_2H_5OH	C_4H_9OH	C_{12}-C_{25}
Cetane number	3	8	25	40–55
Octane number	111	108	96	20–30
Oxygen content (% weight)	50	34.8	21.6	–
Density (g/ml) at 20 °C	0.796	0.790	0.808	0.82–0.86
Auto-ignition temperature (°C)	470	434	385	210
Flash point (°C) at closed cup	12	9	35	65–88
Lower heating value (MJ/kg)	19.9	26.8	33.1	42.5
Boiling Point (°C)	64.5	78.4	117.7	180–370
Latent heating (kJ/kg) at 25°C	1109	904	582	270
Viscosity (mm^2/s) at 40°C	0.59	1.08	2.63	1.9–4.1

compression ignition engine using response surface methodology [15]. Another study investigated the effect of cyclic variations from a compression ignition engine fueled with butanol/diesel blends were investigated [16]. Nanoparticle emissions from the compression ignition engine fueled with butanol/diesel blends are also investigated at different engine load conditions [17]. This chapter provides the summary of the combustion and emissions characteristics of diesel engine fueled with butanol blends at different operating conditions. Before discussing the combustion and emission characteristics, a brief experimental methodology for determination of combustion characteristics is provided in next section.

2 Methodology

To analyze the engine combustion characteristics, combustion parameters (such as heat release rate, start of combustion, combustion phasing) are typically calculated from the measured in-cylinder pressure data. For measuring the in-cylinder pressure, a piezoelectric pressure sensor is generally mounted in the engine cylinder head. For determination of the crank position, a crank angle encoder is typically installed on the crankshaft. A high-speed data acquisition system is used for logging the online in-cylinder pressure signal. Gaseous and particles emissions are measured by gas and particle emission analyzers. Typical experimental setup used for performance, combustion, and emission characteristics analysis is shown in Fig. 2.

Performance and combustion characteristics are calculated by using following equations from (1) to (5).

Brake thermal efficiency (BTE) is determined by using Eq. (1)

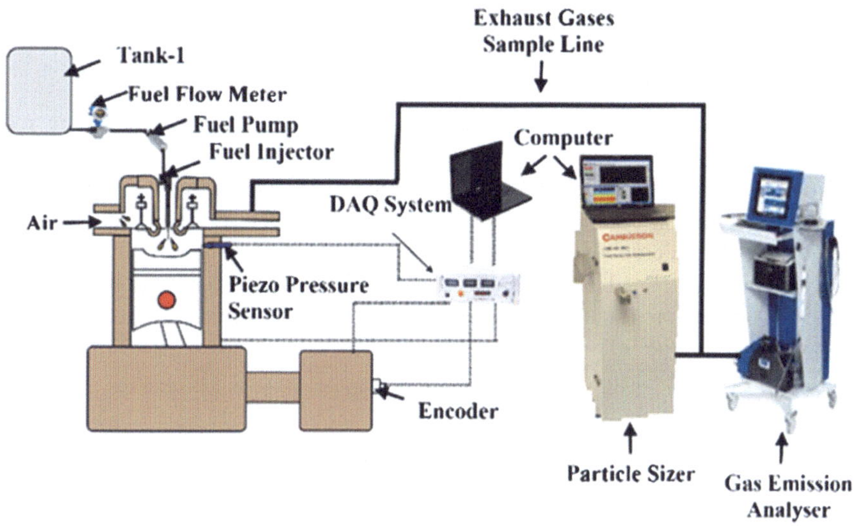

Fig. 2 Schematic of typical experimental setup for combustion and emission characteristics determination

$$\text{BTE}\ (\%) = \frac{\text{BP} \times 3600 \times 100}{\dot{m}_\text{f} \times \text{LHV}_\text{f}} \tag{1}$$

where '$\dot{m}_\text{f}$,' 'LHV_f,' and 'BP' are the mass flow rate of the fuel, lower heating value of fuel, and brake power, respectively. Brake power is determined by using Eq. (2)

$$\text{BP(kW)} = \frac{2 \times \pi \times N \times W \times R}{60 \times 1000} \tag{2}$$

where 'N', 'W', and 'R' are engine RPM, engine load, and dynamometer arm length respectively.

Brake specific fuel consumption is determined by Eq. (3)

$$\text{BSFC}\left(\frac{\text{kg}}{\text{kWh}}\right) = \frac{\dot{m}_\text{f} \times \text{LHV}_\text{f}}{\text{BP}} \tag{3}$$

Rate of heat release (ROHR) is determined by using Eq. (4)

$$\frac{\text{d}Q(\theta)}{\text{d}\theta} = \left(\frac{1}{\gamma - 1}\right) V(\theta) \frac{\text{d}P(\theta)}{\text{d}\theta} + \left(\frac{\gamma}{\gamma - 1}\right) P(\theta) \frac{\text{d}V(\theta)}{\text{d}\theta} \tag{4}$$

where '$\frac{\text{d}Q(\theta)}{\text{d}\theta}$,' '$\gamma$,' '$P(\theta)$,' and '$V(\theta)$' denote the ROHR, ratio of specific heat, in-cylinder pressure, and volume as a function of crank position, respectively.

$$V(\theta) = V_c + \frac{\pi}{4}B^2\left(L + R - R\cos(\theta) - \sqrt{L^2 - R^2\sin^2(\theta)}\right) \quad (5)$$

where 'B', 'L', 'R', and 'V_c' represent the bore of the cylinder, connecting rod length, radius of crank, and clearance volume, respectively.

Total heat release during cycle can be estimated by integrating the heat release rate curve. Different combustion phasing parameters (such as start of combustion (SOC), CA50, and combustion duration) are determined from heat release analysis. SOC is typically defined as position of crank angle where 10% of total heat release occurs. Combustion duration is typically defined as difference of crank angle position between the 90 and 10% of the total heat release.

3 Diesel Engine Characteristics Using Butanol Blends

This section presents the performance, combustion, and emission characteristics of the diesel engine fueled with neat diesel and butanol blends. This section is divided into two subsections. In the first subsection, performance, and combustion characteristics are discussed. In the second subsection, the emissions characteristics (gaseous as well as soot particles) of diesel engine fueled with butanol blends are discussed.

3.1 Performance and Combustion Characteristics

Performance and combustion characteristics of conventional diesel engine depends on several parameters such as injection events (like fuel injection timing, injection pressure, number of injections), physical, and chemical properties of the fuel, ambient condition (like air temperature and humidity). Addition of any biofuel in the diesel has significant effect on the combustion characteristics of the engine. Combustion characteristics are generally explained with the help of rate of heat release (ROHR) curve estimated from the measured in-cylinder pressure data. A study investigated the effect of butanol addition on the heat release characteristics of CI-engine fueled with n-butanol/diesel blend [26]. Figure 3 illustrates the effect of butanol addition (i.e., 16% butanol addition in neat diesel) on cylinder pressure and gross heat release rate at low (20%), medium (40%), and higher (60%) engine load conditions.

Figure 3 shows that with an increase in the engine load, the peak in-cylinder pressure, and ROHR increases for both the fuels (neat diesel and butanol/diesel blend). At higher engine load operation, relatively higher amount of fuel is injected and burns in the cylinder, which leads to increase the peak pressure and heat release. In addition to this, the peak in-cylinder pressure and rate of heat release are

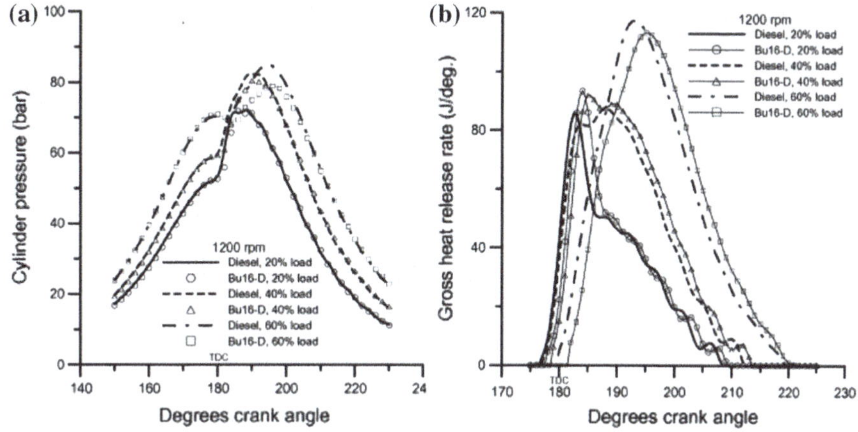

Fig. 3 Effect of butanol addition on **a** cylinder pressure; **b** gross heat release rate at different engine load condition [26]

slightly lower with butanol/diesel blend in comparison to neat diesel operation (Fig. 3). It is due to the delayed combustion because of longer ignition delay of butanol diesel blend. Longer ignition delay with butanol blends are observed due to relatively lower cetane number of butanol [26]. Effect of varying the butanol fraction in the diesel fuel on the combustion characteristics of conventional CI-engine is also investigated [17]. Results showed that with an increase in the butanol percentage in the fuel (from 10 to 20% butanol in the blend), the peak pressure, and ROHR increases. It is also attributed to longer ignition delay with an increase in the butanol percentage, which results in more amount of fuel–air mixture burn in the premixed combustion phase.

Figure 4 shows the effect of exhaust gas recirculation (EGR) rate on the in-cylinder pressure and ROHR for 40% butanol/diesel blend. Figure depicts that the peak of in-cylinder pressure decreases with an increase in the EGR rate, and

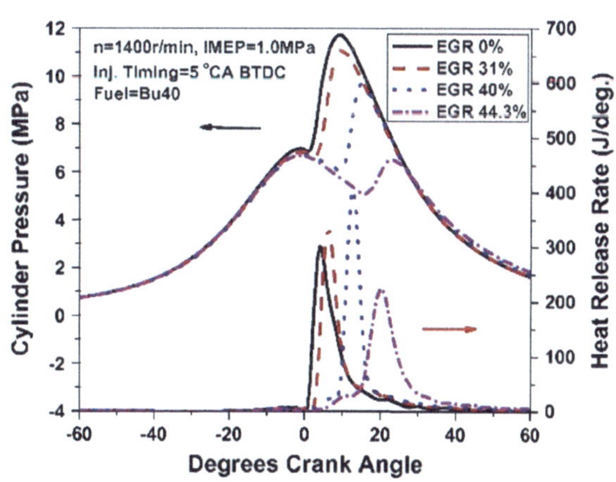

Fig. 4 Effect of EGR on cylinder pressure and heat release rate for 40% butanol-/diesel-blended fuel in compression ignition engine [27]

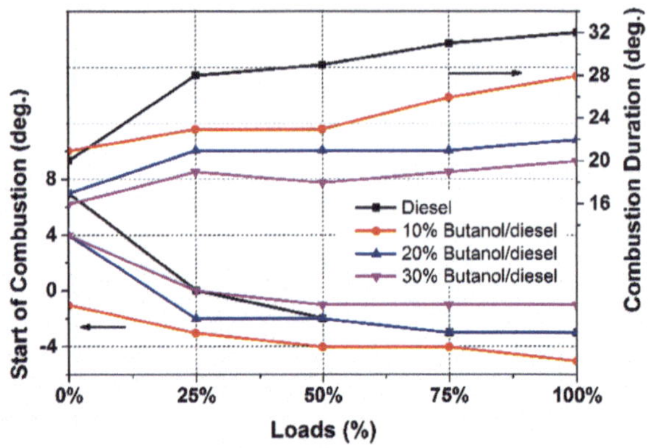

Fig. 5 Variation in SOC and combustion duration (CD) with engine operating load for neat diesel and butanol/diesel blends. Adapted from [17]

ROHR also has the similar trend. The peak of ROHR first increases with increase in EGR rate (for 31 and 40%), and further increase in EGR rapidly reduces the ROHR (for 44.3% EGR rate). With an increase in the EGR rate, the combustion phasing is retarded because of longer ignition delay so that more fraction of charge will burn in the premixed combustion phase and thus results in higher peak ROHR (in case of 31 and 40% EGR rate). With further increase in the EGR rate, ignition delay further increases, and charge may burn during the expansion stroke, which results in lower peak ROHR (mainly because of expansion cooling effect of cylinder volume expansion) [28].

Figure 5 demonstrate the effect of addition of butanol in the diesel fuel on start of combustion (SOC) and combustion duration (CD) at different engine load conditions. Figure 5 reveals that start of combustion advances with an increase in the engine load. With an increase in the engine load, the mean combustion temperature increases, which results in advance ignition timing. Figure reveals that with 10% addition of butanol in the blended fuel, the SOC advances. However, in case of 20 and 30% butanol fraction in the blended fuel, delayed SOC observed at all engine load condition. It might be due to the higher heat of vaporization of 20% and 30% butanol-/diesel-blended fuels in comparison to 10% butanol-/diesel-blended fuel. In addition, with 30% butanol-/diesel-blended fuel at engine load greater than 50% load condition, retarded SOC is observed in comparison to neat diesel fuel operation. It is possibly because of the higher heat of vaporization and lower cetane number of 30% butanol-/diesel-blended fuel in comparison to neat diesel.

Figure 5 also shows the effect of butanol blend on the combustion duration at different engine load conditions. The figure indicate that with an increase in the engine operating load, CD increases and it reduces with increase in the percentage of butanol for all the engine load conditions. Shorter combustion duration is observed with increase in the percentage of butanol in the blended fuel, which indicates the higher combustion rate in case of blended fuel in comparison to neat diesel operation. It is possibly due to the higher oxygen content in the blended fuel. Fuel–air mixing rate dominates the combustion during the diffusion combustion phase [18]. Thus, the combustion of higher oxygen content in the blended fuel leads to reduction in the combustion duration.

Impact of EGR on peak pressure rise rate (PPRR), ignition delay and burning duration for neat diesel, and 40% butanol/diesel blend are presented in Fig. 6. Ignition delay is defined as the duration (in terms of crank angle) between the fuel injection timing and the angle at which 5% mass burned fraction observed [27]. Combustion duration is defined as the duration between the 90% mass burned fraction and 5% mass burned fraction [27]. Figure 6 depicts that PPRR increases with increase in EGR rate (can be seen at 30% EGR rate) for both the fuels (i.e., neat diesel and butanol blend). In case of butanol blend operation, the PPRR increases with increase in EGR rate up to 30%. With further increase in EGR rate leads to steeply decline in PPRR. Figure reveals that the threshold value of EGR rate (after which PPRR start decreasing) for butanol blend is lower as compared to neat diesel. It attributes to lower cetane number and higher heat of evaporation of butanol blend, which leads to deterioration in the combustion and results in lower PPRR.

Figure 6 also indicates that with an increase in the EGR rate, the ignition delay and burning duration increases. Increase in the EGR rate leads to the reduction in the mean combustion temperature due to the combined effect of thermal, dilution, and chemical effects of EGR addition. Additionally, increase in the EGR rate leads to longer ignition delay and thus, more fraction of charge will burn during the premixed combustion phase. Therefore, it may also cause to increase the PPRR as shown in Fig. 6 (up to 30%). Further increase in EGR rate leads to longer delay period (Fig. 6), and smaller fraction of charge will burn during the expansion stroke leading to lower PPRR.

Combustion stability is an important issue, which need to be taken care, while employing new fuel. Combustion stability is typically characterized by cycle-to-cycle variations. In order to investigate the cyclic variations in the engine

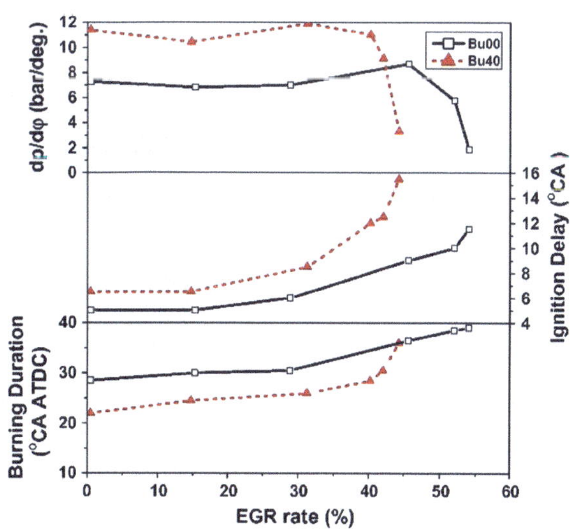

Fig. 6 Variations of peak pressure rise rate, ignition delay, and burning duration with EGR for 40% butanol-/diesel-blended fuel [27]

combustion cycles, statistical techniques (typically coefficient of variation) are used [29]. Cyclic variations are commonly characterized by evaluation of variations in the indicated mean effective pressure (IMEP). Figure 7 shows the effect of butanol addition in diesel fuel on the coefficient of variations IMEP of 2000 consecutive engine cycles. The figure depicts that at lower engine operating load, the cyclic variations are maximum, and cyclic variations reduce with an increase in the engine operating load. At lower load condition, the lower amount of fuel is burnt which results in lower mean combustion and residual gas temperature, which causes higher cyclic variations.

Figure 7 also shows that for engine load greater than 50% of full load, the cyclic variations in IMEP are lower for butanol addition up to 25% in the diesel fuel. Further increase in the butanol leads to drastic increase in cyclic variations of IMEP. It has been observed that addition of butanol in the diesel fuel is limited by cyclic variations. Up to a certain limit, addition of butanol reduces the variations in the combustion cycles and further increase in the butanol fraction increases the cyclic variability in combustion. As discussed, increase in the butanol percentage in the fuel blend retards the combustion phasing. Retarded combustion phasing results in higher cyclic variations due to higher sensitivity of reactions to combustion temperature. Temperature sensitivity increases at retarded combustion phasing.

Brake thermal efficiency is an important parameter used to evaluate the performance of engine at particular fuel. Impact of addition of butanol in the diesel fuel on the brake thermal efficiency (BTE) and brake specific fuel consumption (BSFC) at different engine load (i.e., 1.40, 2.57, and 5.37 bar brake mean effective pressure) conditions are investigated [25]. Results showed that BTE increases with increase in the engine load. The BTE slightly increases with increase in the butanol fraction in the blended fuel. Brake specific fuel consumption (BSFC) decreases with an increase in the engine load for all the test fuel blends. Relatively higher BSFC was observed for butanol/diesel blends due to lower calorific value of butanol as

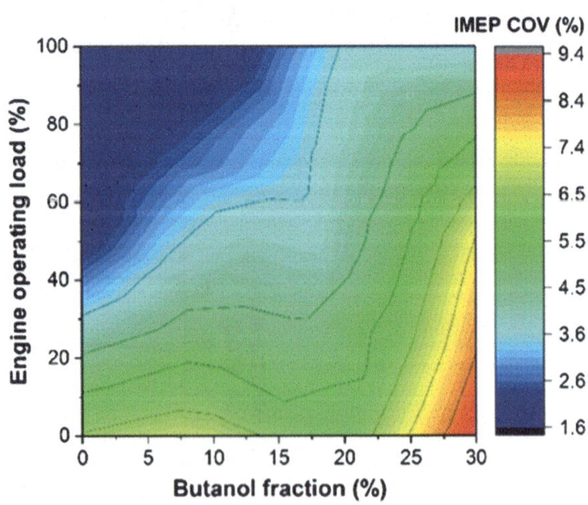

Fig. 7 Effect of butanol addition on the COV in IMEP at different engine load conditions at 1500 rpm

compared to diesel fuel. Lower calorific value of butanol in comparison to diesel leads to combustion of relatively higher amount of fuel for the same power output.

Another study investigated the effect of 10, 20, and 30% butanol addition in the diesel fuel on the performance of a stationary diesel engine at 25, 50, and 100% engine loads [30]. Results showed that BTE slightly decreases with 10 and 20% butanol/diesel blend as compared to neat diesel fuel operation for all the test engine loads (Fig. 8a). Butanol has a lower calorific value (heating value) as compared to diesel fuel (Table 1). Thus, higher fraction of butanol/diesel blend is required to inject for same engine load condition. Additionally, the test engine was coupled with a mechanical fuel injection system and the fuel injection pressure is constant and thus, fuel injection duration will be larger for injecting relatively larger quantity of fuel. The combustion phasing can be different for neat diesel fuel and butanol/diesel blend for a constant fuel injection timing. Another major factor is cetane number of butanol-/diesel-blended fuel, which results into different combustion phasing. Combustion phasing affects the fuel conversion efficiency of the engine.

Results also indicate that at 25 and 50% engine load, BTE is slightly increases with increase in butanol fraction. It is attributed to higher oxygen content in the butanol/diesel blends, which enhances the combustion [19]. Additionally, at 25 and 50% engine operating load, highest BTE is obtained with 30% butanol/diesel blend (Fig. 8a). It might be due to the longer ignition delay with 30% butanol/diesel blend operation, which results in more fraction of charge may be burning in the premixed combustion phase. More fraction of charge burn in the premixed combustion phase leads to higher BTE [20]. With an increase in the engine load, in-cylinder pressure, and temperature increases, which leads to shorter ignition delay, thus lower BTE observed with 30% butanol/diesel blend at 100% engine load condition. Figure 8b indicates that with an increase in the engine load, the BSFC decreases, and higher BSFC obtained for butanol/diesel blends. The brake thermal efficiency and brake specific fuel consumption have inverse relation, and inverse trend is also observed in Fig. 8.

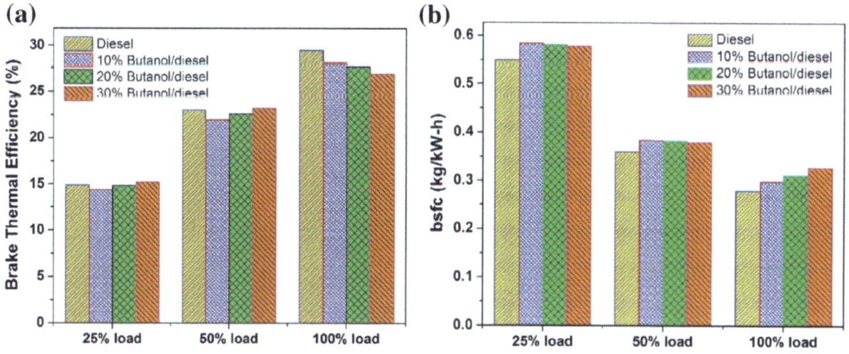

Fig. 8 Effect of butanol addition on the **a** brake thermal efficiency, and **b** brake specific fuel consumption at different engine load conditions. Adapted from [30]

3.2 Gaseous and Particle Emissions Characteristics

This section presents the gaseous and soot particle emission characteristics of diesel engine fueled with butanol blends. Figure 9 shows the gaseous emission characteristics of diesel engine fueled with neat diesel and butanol/diesel blends at different engine load conditions (i.e., 1.40, 2.57, and 5.37 bar break mean effective pressure). Effect of butanol addition in the diesel fuel on NO_x emissions at different engine operating loads are presented in Fig. 9a. The figure demonstrates that the NO_x emissions increases with an increase in the engine load. It is because of higher combustion temperature at higher engine load condition (due to more amount of fuel burnt). NO_x emissions decrease with an increase in the butanol fraction for all the engine load conditions. Butanol has higher heat of evaporation and thus

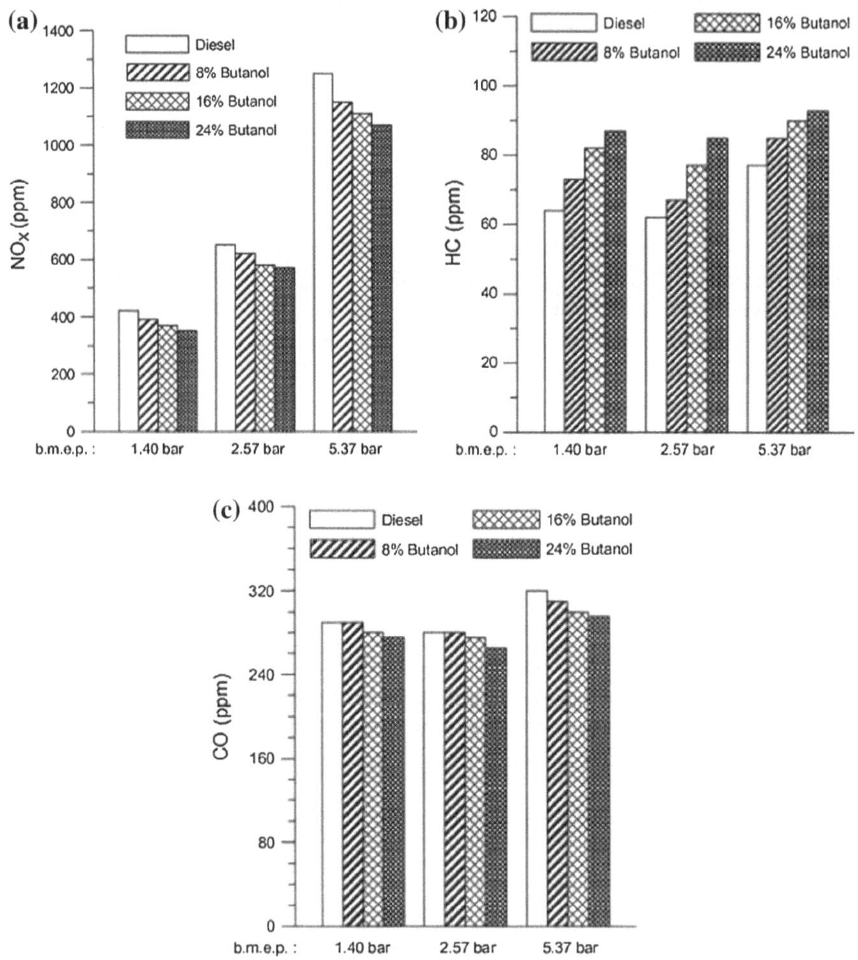

Fig. 9 Effect of butanol addition on **a** NO_x; **b** HC; **c** CO emissions of a diesel engine [25]

addition of butanol in the diesel fuel leads to reduction in the in-cylinder charge temperature, which results in lower NO_x emissions (Fig. 9a). Similar results were also obtained in another previous study [24]. Figure 9b indicates the effect of butanol addition in the diesel fuel on the HC emissions at different engine load conditions. The figure reveals that higher HC emissions observed with butanol/diesel operations in comparison to neat diesel operation. It might be due to the longer delay period (because of higher cetane number), slower droplet evaporation rate, and higher heat of evaporation [22, 23].

Figure 9c indicates that with an increase in engine load from 1.40 to 2.57 bar BMEP, CO emission decreases. At higher engine operating condition, the combustion temperature is higher that leads to more close to complete combustion [21]. Additionally, engine operation with butanol/diesel blend has a lower CO emission for all the test operating conditions in comparison to neat diesel. It is attributed to higher oxygen content of butanol/diesel blends, which leads to the oxidation of CO and converted it into CO_2. Formation of CO emission mainly depends upon the fuel–air ratio and combustion temperature. At 5.37 bar BMEP operation, the engine is running with locally rich mixture and thus, insufficient amount of oxygen is available for the oxidation of CO. Thus, higher CO emissions observed for all the test fuels in comparison to 1.40 bar and 2.57 bar BMEP operation.

Soot emissions emitted from conventional diesel engine is in the form of black smoke. Formation of soot depends upon the overall as well as local fuel–air ratio and combustion temperature. Conventional diesel engine operates at overall leaner fuel–air mixtures. However, direct injection of fuel near TDC leads to locally very rich fuel zone due to insufficient mixing timing. Locally rich combustion of diesel results into soot formation. Advanced premixed compression ignition engines emit very low soot due to combustion of premixed charge [6]. Typically, soot emission is qualitatively analyzed by smoke opacity. Smoke opacity implies the amount of light that is transmitted from the stream of smoke emitted from the engine. Figure 10 depicts the effect of engine load on the smoke opacity and smoke density in a diesel engine fueled with neat diesel and 10, 20, and 30% butanol/diesel blends. Figure 10 depicts that the smoke opacity and smoke density increases with engine

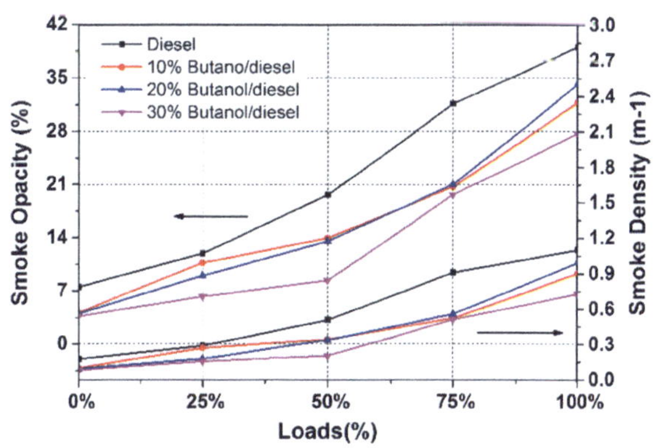

Fig. 10 Effect of butanol addition on **a** Smoke opacity and **b** Smoke density of CI engine exhaust. Adapted from [17]

load and lower smoke opacity observed with butanol-/diesel-blended fuel operation. It is mainly due to the lower carbon chain of butanol and higher oxygen content in the butanol-/diesel-blended fuel.

Recent emission legislation norms have included the limit on soot particle number emissions from automotive engines. Therefore, it is important to investigate the particle number emissions also. Figure 11 illustrates the particle size and its number distribution for neat diesel, 10, 20, and 30% butanol/diesel blends at 25, 50, and 100% engine load conditions. The figure reveals that the distribution trace follows the bimodal log normal distribution. In the distribution trace, first peak typically characterize the nucleation mode particles, while second peak characterizes for accumulation mode particles. Typically, particles having diameter less than 50 nm are known as nucleation size particles, while the particles having size more than 50 nm are known accumulation size particles. Figure 11 shows that the nucleation mode size particles are higher at 25% engine load condition but as the load increases from 25 to 100%, nucleation mode particles start decreasing, and higher accumulation mode particles obtained at 100% engine load condition. At 25% engine load, lower amount of fuel is burnt (as compared to 50 and 100% engine load), which leads to lower combustion temperature. Lower combustion temperature results in longer ignition delay and more fraction of charge will burn during the premixed combustion phase that result in higher nucleation mode particles.

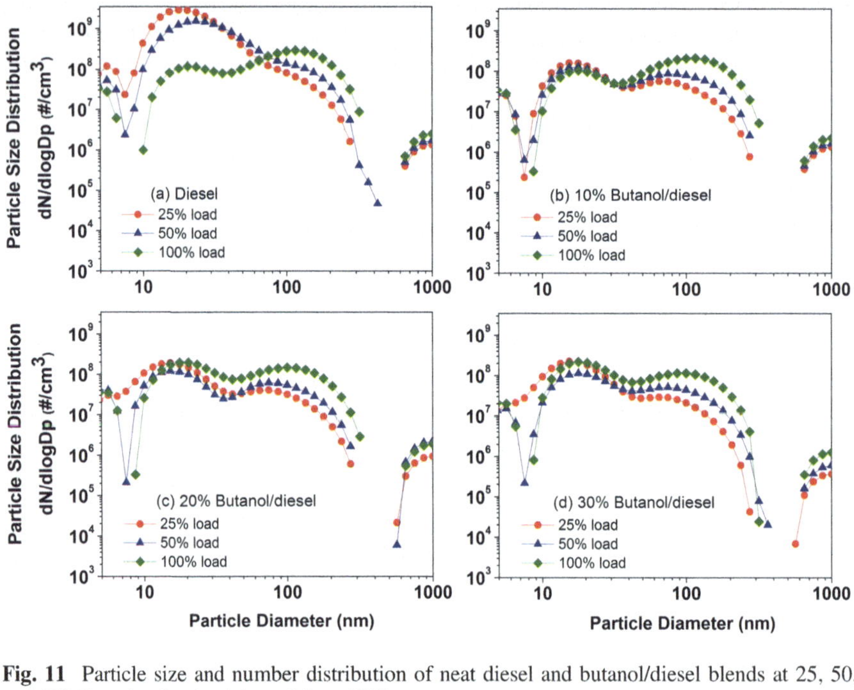

Fig. 11 Particle size and number distribution of neat diesel and butanol/diesel blends at 25, 50, and 100% engine loads. Adapted from [30]

At 50 and 100% engine load condition, the combustion temperature is higher (because of higher amount of fuel burn) in comparison to 25% engine load operation, which enhances the agglomeration rate, and thus results in higher accumulation mode particles. Figure 11 also reveals that nucleation mode particles are higher with neat diesel operation and reduces in case of butanol-/diesel-blended fuel operation. In case of butanol/diesel operation, the nucleation size particles slightly increase with an increase in the percentage of butanol in the blended fuel while accumulation size particles slightly reduces. As discussed in the previous Sect. 3.1, with the addition of butanol addition, ignition delay increases (because of lower cetane number); therefore, more fraction of charge will burn during the premixed combustion phase which leads to enhance the formation of nucleation mode particles. Particles of diameter more than 500 nm are also emitted in significant concentration from the test engine. These large size of particles are mainly because of the low pressure mechanical fuel injection system (~200 bar).

Figure 12 shows the effect of butanol addition on the total particle number concentration at different engine load concentration. At lower loads condition the total particle number concentration is higher and decreases with engine load (Fig. 12). An addition of butanol in the diesel fuel have a significant effect on the particle number emissions. Total particle number decrease with an increase in the butanol percentage in the diesel fuel. Figure 13 demonstrates the effect of butanol addition in the diesel fuel on the nucleation and accumulation mode particles at different engine load conditions. Figure indicates that the concentration of nucleation size particle decreases with engine load while accumulation mode particles are increasing with engine operating load.

Figure 13 also reveals that addition of butanol leads to increase the nucleation size particles while accumulation mode particles decrease. Thus, total particle concentrations are higher (as shown in Fig. 12) mainly due to the nucleation mode

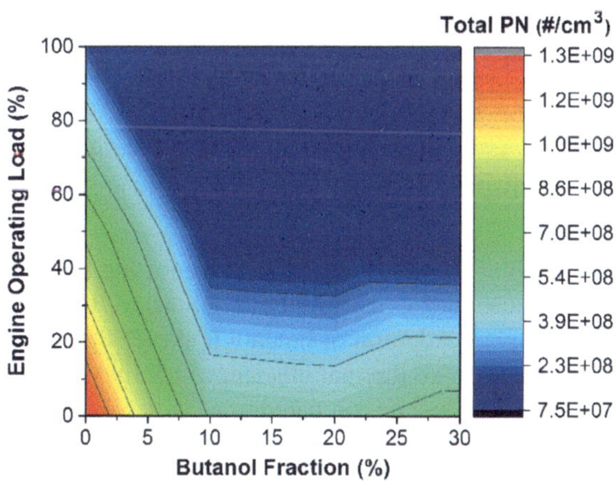

Fig. 12 Effect of butanol addition on the total particle number concentration at different engine load condition. Adapted from [17]

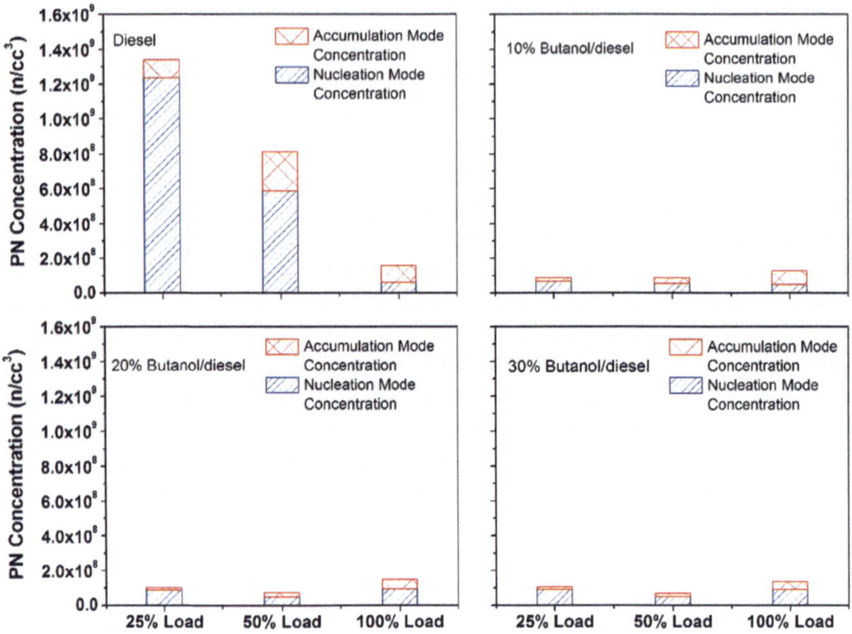

Fig. 13 Effect of butanol addition on nucleation mode and accumulation mode particle number concentration at different engine load condition. Adapted from [30]

particles. Nucleation size particles are possibly higher because of the condensation and coagulation of unburnt hydrocarbons, which are mainly generated because of the lower combustion temperature (in case of butanol-/diesel-blended fuel operation). HC emissions lead to generate the nucleation size particles and thus result in higher total particle concentration. Higher concentration of HC emissions can be observed from the Fig. 9b. It can be summarized that butanol addition in diesel has significant effect on particle number emission.

4 Summary

This chapter presents the effect of butanol addition in the diesel on the performance, combustion, and emissions (gaseous as well as nanoparticles) characteristics of diesel engine. Addition of butanol in diesel significantly affects the performance of the engine. Brake thermal efficiency slightly reduces with 10 and 20% addition of butanol in the fuel as compared to neat diesel operation at 25 and 50% engine load. However, higher brake thermal efficiency is achieved with 30% butanol/diesel blend as compared to neat diesel. Higher concentration of butanol leads to reduction in combustion stability. Addition of butanol in the diesel fuel has a potential to reduce the carbon monoxide (CO), nitrogen oxides (NO_x), smoke opacity, and nanoparticle emissions. However, HC emissions typically increase with the addition of butanol in the diesel fuel. It is found that total particle number reduces with an increase in engine load. Additionally, nucleation size particles are dominating at lower load condition and decreases with engine load. Engine operation with butanol blends lead to the reduction in the total particle number concentration. In summary, butanol can be a better alternative fuel for the partial replacement of diesel in combustion engines.

References

1. Kagawa J (2002) Health effects of diesel exhaust emissions—a mixture of air pollutants of worldwide concern. Toxicology 181:349–353
2. Ristovski ZD, Miljevic B, Surawski NC, Morawska L, Fong KM, Goh F, Yang IA (2012) Respiratory health effects of diesel particulate matter. Respirology 17(2):201–212
3. Jiaqiang E, Pham M, Zhao D, Deng Y, Le D, Zuo W, Zhang Z (2017) Effect of different technologies on combustion and emissions of the diesel engine fueled with biodiesel: a review. Renew Sustain Energ Rev 80:620–647
4. Shameer PM, Ramesh K, Sakthivel R, Purnachandran R (2017) Effects of fuel injection parameters on emission characteristics of diesel engines operating on various biodiesel: a review. Renew Sustain Energ Rev 67:1267–1281
5. Ghadikolaei MA (2016) Effect of alcohol blend and fumigation on regulated and unregulated emissions of IC engines—a review. Renew Sustain Energ Rev 57:1440–1495
6. Maurya RK (2018) Characteristics and control of low temperature combustion engines: employing gasoline, ethanol and methanol. Springer. ISBN 978-3-319-68507-6
7. Jin C, Yao M, Liu H, Chia fon FL, Ji J (2011) Progress in the production and application of n-butanol as a biofuel. Renew Sustain Energ Rev 15(8):4080–4106
8. Li Y, Gong J, Deng Y, Yuan W, Fu J, Zhang B (2017) Experimental comparative study on combustion, performance and emissions characteristics of methanol, ethanol and butanol in a spark ignition engine. Appl Therm Eng 115:53–63
9. Elfasakhany A (2016) Investigations on performance and pollutant emissions of spark-ignition engines fueled with n-butanol–, isobutanol–, ethanol–, methanol–, and acetone–gasoline blends: a comparative study. Renew Sustain Energ Rev
10. Yao D, Ling X, Wu F (2016) Experimental investigation on the emissions of a port fuel injection spark ignition engine fueled with methanol-gasoline blends. Energ Fuels 30 (9):7428–7434
11. Doğan B, Erol D, Yaman H, Kodanli E (2017) The effect of ethanol-gasoline blends on performance and exhaust emissions of a spark ignition engine through energy analysis. Appl Therm Eng 120:433–443
12. Delgado RC, Araujo AS, Fernandes VJ (2007) Properties of Brazilian gasoline mixed with hydrated ethanol for flex-fuel technology. Fuel Process Technol 88(4):365–368
13. López AF, Cadrazco M, Agudelo AF, Corredor LA, Vélez JA, Agudelo JR (2015) Impact of n-butanol and hydrous ethanol fumigation on the performance and pollutant emissions of an automotive diesel engine. Fuel 153:483–491
14. Şahin Z, Aksu ON (2015) Experimental investigation of the effects of using low ratio n-butanol/diesel fuel blends on engine performance and exhaust emissions in a turbocharged DI diesel engine. Renew Energ 77:279–290

15. Saravanan S, Kumar BR, Varadharajan A, Rana D, Sethuramasamyraja B (2017) Optimization of DI diesel engine parameters fueled with iso-butanol/diesel blends–response surface methodology approach. Fuel 203:658–670
16. Maurya RK, Saxena M R (2016) Investigation of effect of butanol addition on cyclic variability in a diesel engine using wavelets. In: The international symposium on intelligent systems technologies and applications Springer International Publishing, Berlin, pp 965–976
17. Saxena MR, Maurya RK (2016) Effect of butanol blends on nano particle emissions from a stationary conventional diesel engine. Aerosol Air Qual. Res 16:2255–2266
18. Chen Z, Liu J, Han Z, Du B, Liu Y, Lee C (2013) Study on performance and emissions of a passenger-car diesel engine fueled with butanol–diesel blends. Energy 55:638–646
19. Doğan O (2011) The influence of n-butanol/diesel fuel blends utilization on a small diesel engine performance and emissions. Fuel 90(7):2467–2472
20. Sayin C (2010) Engine performance and exhaust gas emissions of methanol and ethanol–diesel blends. Fuel 89(11):3410–3415
21. Choi CY, Reitz RD (1999) An experimental study on the effects of oxygenated fuel blends and multiple injection strategies on DI diesel engine emissions. Fuel 78(11):1303–1317
22. Ecklund EE, Bechtold R L, Timbario TJ, McCallum PW (1984) State-of-the-art report on the use of alcohols in diesel engines (No. 840118). SAE Technical Paper
23. Corkwell KC, Jackson, M. M., & Daly, D. T. (2003). Review of exhaust emissions of compression ignition engines operating on E diesel fuel blends (No. 2003-01-3283). SAE Technical Paper
24. Rakopoulos DC, Rakopoulos CD, Hountalas DT, Kakaras EC, Giakoumis EG, Papagiannakis RG (2010) Investigation of the performance and emissions of bus engine operating on butanol/diesel fuel blends. Fuel 89(10):2781–2790
25. Rakopoulos DC, Rakopoulos CD, Giakoumis EG, Dimaratos AM, Kyritsis DC (2010) Effects of butanol–diesel fuel blends on the performance and emissions of a high-speed DI diesel engine. Energ Convers Manag 51(10):1989–1997
26. Rakopoulos DC, Rakopoulos CD, Papagiannakis RG, Kyritsis DC (2011) Combustion heat release analysis of ethanol or n-butanol diesel fuel blends in heavy-duty DI diesel engine. Fuel 90(5):1855–1867
27. Chen Z, Wu Z, Liu J, Lee C (2014) Combustion and emissions characteristics of high n-butanol/diesel ratio blend in a heavy-duty diesel engine and EGR impact. Energy Convers Manag 78:787–795
28. Maiboom A, Tauzia X, Hétet JF (2008) Experimental study of various effects of exhaust gas recirculation (EGR) on combustion and emissions of an automotive direct injection diesel engine. Energy 33(1):22–34
29. Saxena MR, Maurya RK (2017) Investigation of cyclic variability in a non road diesel engine fueled with Diesel/butanol blends. Paper NO-SEEC-2017-60. In: International conference on sustainable energy and environmental challenges (SEEC-2017) 26–28 Feb 2017, Mohali, India
30. Saxena MR, Maurya RK (2017) Optimization of engine operating conditions and investigation of nano-particle emissions from a non-road engine fuelled with butanol/diesel blends. Accepted, Biofuels, 21st Aug 2017
31. Liu H, Wang G, Zhang J (2013) The promising fuel-biobutanol. In: Liquid, gaseous and solid biofuels-conversion techniques. InTech

Hydrogen-Enriched Compressed Natural Gas: An Alternate Fuel for IC Engines

Sadaraboina Moses Vidya Sagar and Avinash Kumar Agarwal

Abstract Depleting fossil fuel resources is forcing the transport sector to look for renewable fuels. CNG, being produced from fossil as well as natural resources, is a good alternative to liquid fossil fuels. It is relatively abundant and easily available compared to hydrogen. However, it has lower flame speed, shorter flammability range and other limitations, which make it a sub-optimum fuel for IC engines. Hydrogen, which can also be produced from renewable resources, is a possible solution to some of these issues. However, hydrogen has its own limitations in terms of low storage density. It occupies very large volume as a gas, and storing it in liquid form is extremely energy-intensive. There is a sharp contrast in vital properties of both these fuels; therefore, this study explores using mixtures of hydrogen and CNG as alternative fuel. This fuel exhibits merits of hydrogen as well as CNG. Hence, hydrogen-enriched CNG, also known as ***hythane*** *or* ***HCNG***, is being investigated worldwide. This fuel is storable, energy-efficient and emits fewer emissions compared to both constituent fuels individually. One way to produce HCNG blends is to mix the gases using Dalton's law of partial pressures and store them as premixed blend. This method is time-consuming and cumbersome. With this method, it becomes difficult to investigate all the HCNG blends. It does not have the flexibility to change the mixture ratio, while the engine is operating. Hence, in the current research, a dynamic gaseous fuel mixing system was developed by which one can change the proportions of hydrogen and CNG of the HCNG blends dynamically without necessarily stopping the engine. Validation of the system developed was done by theoretical methods and experimental investigations. We used this mixing system to investigate the technical feasibility of various HCNG blends ranging from 0% H_2 to 100% H_2 in HCNG. Combustion, performance end emission characteristics were compared. HCNG blend with 30% hydrogen showed better performance and superior anti-knocking characteristics.

S. M. V. Sagar
Indian Institute of Technology Kanpur, Kanpur, India

A. K. Agarwal (✉)
Engine Research Laboratory, Department of Mechanical Engineering,
Indian Institute of Technology Kanpur, Kanpur 208016, India
e-mail: akag@iitk.ac.in

D. K. Srivastava et al. (eds.), *Advances in Internal Combustion Engine Research*,
Energy, Environment, and Sustainability,
https://doi.org/10.1007/978-981-10-7575-9_6

Keywords HCNG · Hydrogen · CNG · Alternate fuels

1 Introduction

With rapid increase in global population in twentieth century, the number of vehicles and global fuel consumption have increased exponentially. In addition, reserves of non-renewable fossil fuel resources are limited and depleting at an alarming rate [1]. World is currently facing multiple challenges of environmental pollution, rapid climatic change and global warming, which are becoming increasingly severe. Emission of hazardous gases and particulate matter (PM) from internal combustion (IC) engines and vehicles is one of the prime man-made causes for these issues. For combating the environmental degradation, most countries across the world have adopted stricter emission standards. The number of engines/ vehicles releasing hazardous pollutants is growing rapidly [2, 3].

Hence, researchers are exploring alternate means to meet galloping energy demand, with fewer engine-out emissions. Various alternate fuels such as biodiesel, natural gas (NG), alcohols, hydrogen, liquefied petroleum gas (LPG) are being experimentally investigated for variety of engine/vehicle applications. In this context, gaseous fuels such as compressed natural gas (CNG) and hydrogen are of particular interest since they exhibit considerably cleaner engine combustion characteristics. Xu et al. [4] reported that CNG offers greater emission reduction than biodiesel (B20) in public transit buses. Natural gas has huge reserves available in the form of shale gas and gas hydrates worldwide which makes it an attractive alternative fuel. However, CNG suffers from combustion-related challenges due to slower flame speed, narrow flammability range and low energy density. On the other hand, hydrogen suffers from combustion-related challenges due to an order of magnitude of faster flame speed, which makes the heat release uncontrollable at times.

One of the fast emerging and acceptable ways to tackle these challenges is to use hydrogen-mixed compressed natural gas (HCNG) as a combustion fuel. Hydrogen enrichment enhances the fuel quality of HCNG mixtures. The potential of HCNG mixtures and their ability to replace conventional fuels without major hardware changes in the gas engines is the main motivation leading to evolution of HCNG mixtures as a potential fuel for large-scale applications in past decade.

2 History

HCNG mixtures are gaseous fuels with lower C/H ratio, emitting fewer emissions compared to fossil fuels. HCNG mixtures have the potential to replace conventional fossil fuels and are considered one of the best alternate gaseous fuels. This potential is the major contributing factor in developing HCNG-fuelled IC engines.

Conventional gasoline is replaced by hydrogen-enriched CNG in several experimental investigations using spark ignition engines. Application of hydrogen-enriched compressed natural gas represented as H_2CNG/ HCNG/ HANG (hydrogen-added natural gas) as an internal combustion (IC) engine fuel was done for the first time in 1983 [5]. Nagalingam et al. [5] studied the performance of CNG, H_2 and H_2/CNG blends in an AVL engine in 1983. Hydrogen Components Incorporation (HCI) was the first to utilize HCNG as an automotive fuel in 1989. hythane, a mixture of CNG and hydrogen, was invented and patented by Frank Lynch and Roger Marmaro according to US Patent #5,139,002 in 1992. Here, hythane was defined as 15% (v/v) blend of hydrogen in CNG. Since then, several researchers have investigated HCNG as an IC engine fuel.

Engine Research Lab (ERL) at IIT Kanpur University has conducted research on HCNG and development of HCNG engines which are cited in this chapter to help understand the characteristics of HCNG as an IC engine fuel.

3 Hydrogen-Mixed Compressed Natural Gas

Natural gas, because of its fewer emissions and low C/H ratio, is recognized as a superior alternative to conventional fossil fuels. Existing engine hardware of conventional IC engines requires minor modifications for using natural gas as a fuel. Natural gas is abundantly available in India. Natural gas is a clean fuel with low C/H ratio, resulting in fewer carbonaceous emissions compared to coal and other hydrocarbon-based liquid fuels. Table 1 shows the specifications of automotive CNG in India.

The main drawback with CNG is its low energy density. It has a narrow flammability range. Hence, in order to achieve efficient engine combustion, higher

Table 1 Automotive CNG specifications in India [6]

Constituents	Value	Tolerance
Free water (mg/m^3)	8	Max.
$N_2 + CO_2$ (vol.%)	3.5	Max.
Total sulphur including H_2S (mg/m^3)	20	Max.
Oil mist content (ppm)	Insignificant	
Oxygen (vol.%)	0.5	Max.
Methane (%)	87	Min.
Ethane (%)	6	Max.
C_3 and higher order HC (%)	3	Max.
C_6 and higher order HC (%)	1	Max.
Total unsaturated HC (%)	1	Max.
Hydrogen (mole %)	0.1	Max.
Carbon monoxide (mole %)	0.1	Max.

performance and fewer emissions, researchers explored the possibility of blending CNG with a gas that has higher flame speed, wider flammability limits and higher energy density. In this requirement, hydrogen turns out to be the best option. Hydrogen has certain special properties, which enhance the fuel quality, when mixed with CNG.

Some of the special properties of hydrogen are:

- Hydrogen has wide ignition limits: hydrogen enrichment increases the engine's lean operating limit compared to baseline CNG.
- Hydrogen has higher flame velocity than natural gas: flame propagation of hydrogen is nearly 8 times faster compared to that of methane (stoichiometric flame propagation speeds for hydrogen and methane are 3.4 and 0.4 m/s, respectively, at 298 K temperature at 1 bar). Hence, addition of hydrogen helps burn leaner mixtures, which provide better efficiency, when completely burnt.
- Hydrogen has more ideal characteristics than natural gas: the slower flame propagation speed of methane also increases cycle-to-cycle variations. It also adds to ignition lag. Addition of hydrogen to the mixture decreases both, the cycle-to-cycle variations and ignition lag. Hence, the combustion cycle approaches ideal air standard cycle.
- Hydrogen gives almost zero emissions of HC, CO and CO_2: it helps in reducing HC, CO and CO_2 levels, which are reasonably high during the combustion of natural gas.

Apart from these advantages, hydrogen also has certain disadvantages because of which hydrogen alone is not an effective fuel for IC engines. They are:

- Hydrogen engine provides less power for the same displacement than natural gas engines: for H_2, the stoichiometric fuel/air ratio is 0.418 (volumetric), and for CH_4, it is only 0.105. On mass basis, corresponding values are 0.0292 for H_2 and 0.0581 for CH_4. Even on a mass basis, the energy content of a stoichiometric mixture of hydrogen and air is $\sim$14% less than that for methane. Hence, by using a mixture of methane and hydrogen, one can get higher energy output than hydrogen without any major change in the properties of hydrogen.
- NO_x emissions are higher with hydrogen than with natural gas: because the peak temperature in hydrogen combustion is higher about 150 K than methane. Hence, blending it with methane gives us lesser emissions.
- The problem of flashback (backfire) is more pronounced in hydrogen: because of its lower energy of ignition and wider ignition limits. Since CNG is the major constituent of fuel mixture, this tendency is seldom observed.

From Table 2, it can be seen that the blended mixture of hydrogen and CNG has the better qualities compared to both hydrogen and CNG. The chemical reactions involved in combustion are vital in identifying the combustion products, energy and emissions from the test fuel. Listed below are the basic equations pertaining to HCNG combustion. Since methane is the major constituent of CNG, following combustion equations are considered for the HCNG:

Table 2 Properties of CNG, hydrogen and HCNG compared to gasoline [7]

Properties	H_2	HCNG	CNG	Gasoline
Flammability limit in air (vol.%)	4–75	5–35	5–15	1.0–7.6
Stoichiometric composition of fuel–air mixture (% v/v)	29.53	22.8	9.48	1.76
Minimum ignition energy in air (mJ)	0.02	0.21	0.29	0.24
Auto-ignition temperature (K)	858	825	813	501–744
Temperature of stoichiometric flames in air (K)	2318	2210	2148	2470
Velocity of burning in NTP air (cm/s)	325	110	45	37–43
Quenching gap in NTP air (cm)	0.064	0.152	0.203	0.2
Thermal energy radiated (%)	17–25	20.26	23–33	30–42
Diffusivity in air (cm^2/s)	0.63	0.31	0.2	0.08
Flame emissivity (normalized)	1.00	1.5	1.7	1.7
Equivalence ratio	0.1–7.1	0.5–5.4	0.7–4	0.7–3.8

$$CH_4 + 2O_2 \rightarrow (1 - x)CO_2 + 2H_2O$$

$$H_2 + \tfrac{1}{2}O_2 \rightarrow 2H_2O$$

Let "x" be the mole fraction of H_2 in the mixture. Now, using the molar masses of CH_4 and H_2 as 16 and 2 kg/mol, respectively, we get fraction of H_2 on mass basis "q" as

$$q = \frac{xM_{H_2}}{xM_{H_2} + (1 - x)M_{CH_4}} = \frac{2x}{2x + 16(1 - x)} = \frac{x}{8 - 7x}$$

Assuming that air contains 23.2% oxygen, the air–fuel ratio of HCNG mixture at stoichiometric conditions is "l_O"

$$l_O - 34.48\frac{4 - 3x}{8 - 7x} \quad [8]$$

4 HCNG Engine Development

HCNG mixtures can readily be used in existing CNG engines with minimal changes in engine hardware and distribution infrastructure and can be relatively easily handled compared to hydrogen. HCNG mixtures emit fewer hydrocarbons and NO_x in lean-burn combustion engines compared to CNG, and easily meet EURO-VI norms [9] without any exhaust gas after treatment. HCNG-fuelled engines can be operated up to extended lean limits without knocking [7, 10], with higher efficiency and lower coefficient of variation (COV_{imep}), i.e. cyclic variability [11].

Thipse et al. [12] and Helmut et al. [13] suggested necessary modifications in a SI engine hardware in order to operate it using CNG and hydrogen-enriched CNG. Kavathekar et al. [14] described development of a six-cylinder CNG engine using computational methods and experimental approaches to meet EURO-IV emission norms. His research groups also studied suitability of using gaseous fuel conversion kits for HCNG, similar to LPG and CNG kits, which were available commercially. Belchoir et al. [15] analysed four different "gasoline-to-CNG conversion kits" for assessing compatibility of HCNG at full load under stoichiometric conditions. They concluded that since HCNG can be used in conventional CNG engines, these conversion kits can be used without any significant technical issues. Most of these studies discussed above explore 15HCNG as test fuel. However, Chugh et al. [16] suggested that 18HCNG with 5–6% CO_2 is suitable for light-duty CNG engines in India. Other research groups also investigated combustion characteristics along with performance and emissions of HCNG on single-cylinder engines [17], multi-cylinder engines [12], heavy-duty engines [18, 19] and passenger vehicles [20, 21]. Ma et al. [11] studied cyclic fluctuations in SI engines caused by hydrogen enrichment of CNG. They observed that higher fractions of hydrogen in HCNG above a certain limit caused instability and knocking issues in conventional SI engines. Hydrogen also led to embrittlement and hardness of the materials exposed. In order to avoid embrittlement, materials such as austenitic stainless steel were experimentally investigated and were found to be suitable. Kavathekar et al. [14] investigated embrittlement by different HCNG mixture compositions and concluded that HCNG mixtures up to 30% (v/v) did not cause any significant embrittlement and the materials used for CNG engines could also be used for HCNG engines. Subramanian [22] conducted experiments to investigate the effect of catalytic converter and EGR on engine-out emissions. He reported that engine efficiency is improved more for catalytic converter and EGR together than catalytic converter and EGR individually.

At present, HCNG engine development is in its infancy and the technology is not so widespread worldwide. Hence, commercial availability of HCNG mixtures is rather limited worldwide. Very few nations have performed feasibility testing of HCNG-dispensing terminals, and even fewer have developed working prototypes. Various HCNG mixture preparation methods were developed to provide desired HCNG mixtures for experimental studies. Premixed HCNG bottles are the most used HCNG mixture preparation method [14]. Another method involves the use of a HCNG gas dispenser [20]. Such a dispensing terminal can be used to obtain HCNG mixtures of desired ratios, ranging from 5 to 50% v/v.

The most modern method for HCNG mixture preparation is the use of a dynamic HCNG mixing system. As the name suggests, this mixing system can produce various HCNG mixtures dynamically as desired by the user. Very few researchers described the development of such a versatile system. As per researcher's demand for experimental studies, this mixing system changes the mixture ratios of HCNG blends on demand dynamically, while the engine is being operated. This means that

HCNG mixture ratios can be changed without stopping the engine, thus enabling quick, reliable and realistic experimental studies. For a particular HCNG blending ratio, CNG's flow rate is measured and the corresponding hydrogen flow rate is provided by varying voltage signals to the flow controller. These signals are varied by a computer as per CNG flow rate to maintain a desired HCNG blending ratio throughout the experiment. Ma et al. [23] described an online HCNG mixture-preparing system used for their experimental studies. Spectroscopic analysis of mixtures was performed to validate the obtained HCNG compositions. However, a dynamic mixing system is quite complicated, costly and requires basic understanding of the fluid dynamics.

Due to the unavailability of such dynamic HCNG mixing systems commercially, majority of investigators confined their experimental studies on HCNG to 3 or 4 HCNG mixtures obtained from premixed bottles only. In order to avoid cumbersome process of premixed HCNG bottles for conducting experiments on different HCNG blends, a dynamic mixing system was needed. Using a dynamic mixing system, experimental studies can be conducted for different HCNG blends without stopping the engine. These studies help in more accurate comparison of the effect of different HCNG ratios on the engine performance and combustion to identify the optimized mixture ratio. For validating and determining the accuracy of such a dynamic mixing system, engine investigations can be performed and compared with those of premixed HCNG blends. Therefore, a dynamic mixing system (DMS) was discussed [24] along with its performance validation and accuracy in delivering required HCNG mixture ratios for experimental studies.

DMS consists of two gas cylinders and two gas flow controllers (GFC), one each for hydrogen and CNG. It also consists of a mixing chamber, which allows mixing of gases, and control valves to control gas flows, pressure regulators to regulate cylinder out pressures of each gas and safety devices to prevent backfire from reaching back to the gas cylinders. PU pipes are used as gas flow paths which can hold up to 5 bar pressure. In ratio mixing mode, the flow controllers automatically adjust the hydrogen flow rate, depending on natural gas flow rate while maintaining a predefined HCNG mixing ratio. Once the experiments are conducted, nitrogen is used to blow out the fuel gas traces. As per the engine load requirement, flow rate of the overall mixture is changed. Figure 1 shows the schematic of the DMS and its components.

A command module shown in Fig. 1 controls the flow rate of GFCs through a customized technique of "ratio mixing". In this mode, out of the two flow channels, one channel is set up as master, which is CNG here, and the other channel is set up as the slave, which is hydrogen here. A desired mixture ratio is given as the input to the command module through the computer. The command module adjusts the flow rates of the slave channel, depending on the master channel flow rate and the mixture ratio fed in by the user. In this entire process, the aim of the system is to maintain the mixture ratio of both the gas flow rates as close as possible to the input mixture ratio. For example, if CNG is the master and hydrogen is the slave and the

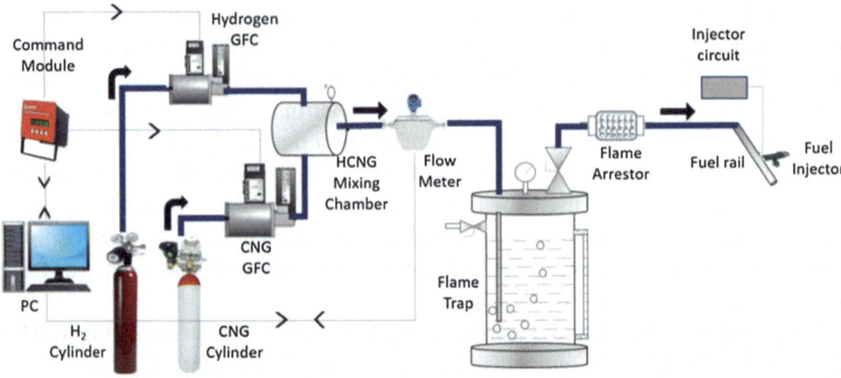

Fig. 1 Schematic of the fuel delivery system with DMS and safety devices [24]

desired mixture is 30HCNG, the ratio of slave channel flow rate to master channel flow rate is 0.429, which has to be given as input to the command module.

For the experimental studies, a two-stage pressure regulator was used for hydrogen to bring down the pressure from 140 bar (cylinder pressure) to 2 bar (fuel line pressure). However, CNG cools and freezes when its pressure decreases suddenly from 225 bar (cylinder pressure) to 2 bar (fuel line pressure), and this property is known as "Joule–Thompson effect". Because of this effect, a regulator-cum-intercooler was used to heat the CNG by circulating water across the regulator, thereby preventing freezing of the CNG. A pressure difference must be ensured across GFCs for gas flow to happen in the flow channels. A cylinder chamber is installed after GFCs, which allows CNG and hydrogen to mix well and form homogenous HCNG mixture. In order to measure the flow rate of the fuel mixture entering the engine, a Coriolis force-based flowmeter was set up across the fuel line. This flowmeter measures the flow rates of fuel mixture on volumetric basis as well as mass basis. Measured fuel flow rates were later used in engine's combustion and performance analyses.

One of the main challenges in dealing with HCNG mixture as fuel for engines is to design and develop a sophisticated and safe fuel handling and injection system. Conventional solenoid injectors are not capable of injecting gaseous fuels, which are required to be injected in significantly higher volumes. Hence, these injectors are replaced with customized high volume flow rate solenoid injectors.

Hydrogen's presence in the HCNG mixture is the main cause for increased backfiring tendency. Addition of hydrogen to CNG widens the knocking range; despite increasing operational limits of the engine, therefore appropriate safety devices needed to be in place in the fuel circuit in order to avoid any accidents. Two custom-built safety devices were installed in the experimental set-up for this purpose, namely "flame arrestor" and "flame trap". Since test fuel is gaseous, possibility of its leakage is also quite high. Detecting such leakages is rather difficult as both constituent gases are colourless, odourless and explosive. These characteristics also make fuel handling challenging. Hence, leakproof gas lines have to be used in the experimental set-up. Fuel storage is another major concern for these gaseous fuels. Conventional storage methods such as gas cylinders are used currently. However, more efficient methods of fuel storage such as metal hydrates, nano-spheres have to be developed for HCNG mixtures in future.

5 Engine Experimental Set-up

Unless mentioned otherwise, this chapter discusses the results of experimental studies conducted on a single-cylinder port-fuelled HCNG engine redesigned out of a baseline diesel engine (Kirloskar; DAF-10). The experimental set-up consisted of the prototype single-cylinder gas-fuelled SI engine, a transient dynamometer and dynamometer controller, an electronically controlled injector driver circuit, a solenoid gas injector and a customized fuel induction system. A specialized high volumetric flow rate solenoid-actuated injector was employed to inject HCNG mixtures into the intake manifold of the engine. The intake air system consists of a surge tank to scale down the oscillations and an air filter to screen dust particles from the airflow. In-cylinder pressure data, which is critical for combustion analyses, were acquired using a high-speed data acquisition system (Hi-Technique; MeDAQ). An engine exhaust gas analyser was employed to measure the regulated emissions from the engine. Two safety devices were installed along the HCNG fuel line, namely a flame trap and a flame arrestor, in order to restrict the backfire from reaching the gas cylinders. Schematic of the experimental set-up is shown in Fig. 2.

Modified engine was operated with fixed spark timing of 30° BTDC at constant speed of 1500 rpm. Specifications of the engine set-up are shown in Table 3.

Table 4 shows the instrumentation used in this experimental study and parameters measured or controlled. Lubrication is necessary for proper functioning of an engine and to avoid metal-to-metal contact between mating components. Lubricant

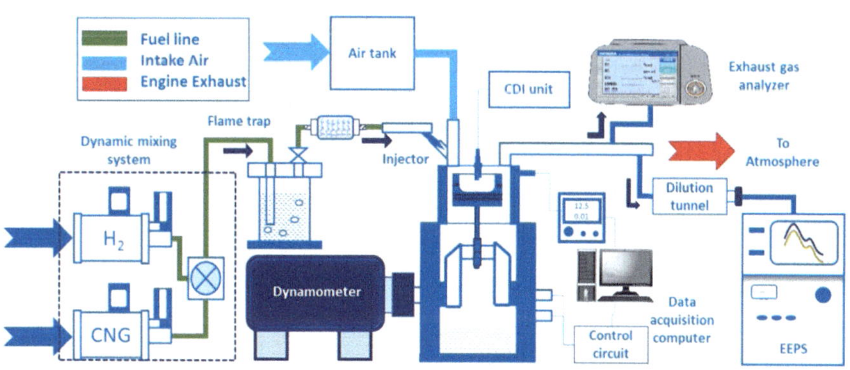

Fig. 2 Experimental set-up schematic [24]

Table 3 Specifications of engine

Make/model	Kirloskar/DAF10-modified
Relative air–fuel ratio (RAFR)	0.9
Compression ratio	11
Spark timing	30° BTDC
Fuel injection	Port fuel injection
Ignition	Spark (CDI)
Engine speed	1500 rpm

Table 4 Engine instrumentation

Parameter	Instrumentation
Crank angle measurement	Optical encoder (Encoders India)
Data acquisition system software	MeDAQ, Win600/REVelation (Hi-Techniques)
Dynamometer	Transient dynamometer (Dynomerk)
Fuel delivery system	Dynamic mixing system
Fuel injection control	Injector driver circuit
Fuel injector	Solenoid gas injector (Bosch-NGI2)
Ignition system	Capacitive discharge spark ignition (CDI) system
Pressure sensing	Piezoelectric transducer, amplifier (Kistler)
TDC sensing	Inductive proximity sensor
Temperature sensing	K-type thermocouple

is an important factor contributing to particulate emissions from CNG engines [25]. It enters the combustion chamber because of the piston motion in the cylinder. SAE40 grade engine oil was used in this study.

6 Validation of Dynamic Mixing System

GFCs were tested for their performance in "ratio mixing mode" under atmospheric conditions, and flow behaviours of CNG and hydrogen were measured for a desired mixture of 30HCNG. From Fig. 3, it can be seen that flow rates of CNG and hydrogen increased and stabilized after attaining the preset flow rates. Hydrogen being the lighter gas of the two stabilized its flow rate within 3 s after the flow initiation. However, CNG having relatively higher density took around 6 s to stabilize its flow rate. Hydrogen's flow rate stabilized earlier than that of CNG because of lower flow inertia due to lower density. It can also be observed that CNG flow fluctuations were lower compared to hydrogen. Because of these fluctuations, composition of HCNG mixture also had minor deviations from the desired 30HCNG mixture as shown in Fig. 4.

Following observations were made from this experimental study:

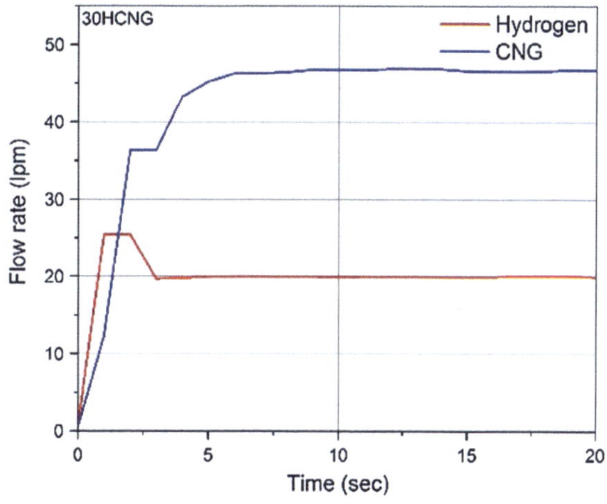

Fig. 3 Variations in gas flow rates with time for preset mixture ratio of 30HCNG [24]

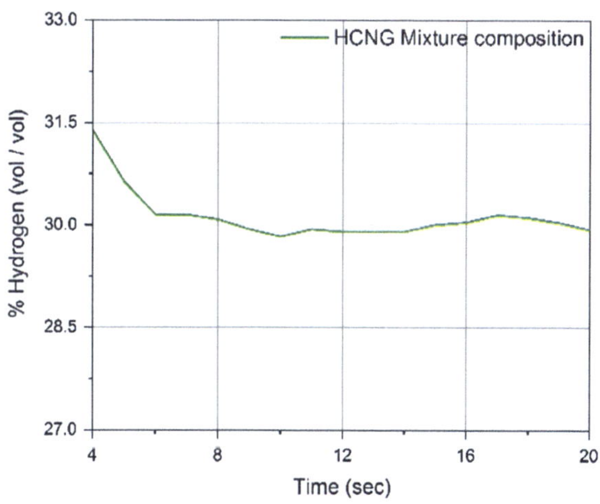

Fig. 4 Variations in produced HCNG composition with time for preset mixture ratio of 30HCNG [24]

1. Since CNG and Hydrogen carry contrasting thermophysical properties, ample time should be provided for their flow rates to stabilize.
2. At low engine loads, fuel–air mixture requirement is very less and gas flow rate values are in the order of accuracy limit of the GFCs. Hence, flow variations will be significant and may lead to erroneous measurements at these conditions.
3. For allowing smooth flow of gases, sufficient pressure difference must be maintained across the GFCs.
4. Volumetric flow rates given as inputs to GFCs must always be greater and nearer to the flow rates required for producing desired mixtures, in order to achieve best system accuracy.

For validating HCNG DMS, produced HCNG mixture compositions were compared with desired HCNG mixture compositions. This validation was carried out by two procedures discussed below. Here, mixture quality is nothing but the accuracy of DMS in producing desired HCNG mixtures. The higher the accuracy, the better the mixture quality because the mixture quality of produced HCNG blends demonstrates the functional accuracy and limitations of a DMS prototype.

6.1 Procedure-I for Function Accuracy Determination of HCNG DMS

Volumetric flow rates (VFRs) of constituent gases CNG and hydrogen were measured for different BMEPs from DMS. Mixture ratio of produced HCNG blend on volumetric basis is given by the ratio of hydrogen VFR to total VFR of the fuel mixture. The mixture ratios thus determined are compared against desired HCNG mixture ratio of 20HCNG, and deviations were calculated. These deviations were used to determine the accuracy of DMS in producing the desired HCNG mixtures.

$$\text{Accuracy}\ (\%) = 100 - \frac{\left|\text{Mixture ratio}_{\text{Actual}} - \text{Mixture ratio}_{\text{Desired}}\right| \times 100}{\text{Mixture ratio}_{\text{Desired}}}$$

From Table 5, a mean deviation of 0.75 was seen across four engine loads. The accuracy was greater than 95% for all engine loads, implying that DMS was actually delivering desired HCNG mixtures. It was observed that the deviations in mixture ratio increased to some extent at low operating loads. Deviation was maximum at 1.99 bar BMEP at 1.41. Fuel requirement was low at lower engine loads, and the fuel flow rates were in the order of accuracy limit of the GFCs. Diffusivity and density are different for CNG and hydrogen. Hence, flow fluctuations were higher at low BMEPs, which resulted in increased deviations.

Table 5 Determination of functional accuracy of DMS using individual VFRs

Engine load: BMEP (bar)	Produced CNG VFR (LPM)	Produced H_2 VFR (LPM)	Produced mixture ratio (v/v)	Deviation (mixture ratio)	Functional accuracy (%)
1.99	16.01	4.36	21.41	1.41	92.95
3.31	20.50	4.94	19.42	−0.58	97.1
4.64	26.73	6.33	19.14	−0.86	95.7
5.97	29.15	7.33	20.08	0.08	99.6
Mean deviation = 0.75					

6.2 Procedure-II for Function Accuracy Determination of HCNG DMS

In this procedure, total mass flow rate (MFR) of produced HCNG mixture was measured from a Coriolis mass flowmeter. Total VFR of the HCNG mixture was determined using GFCs. Using total VFR and total MFR, density of produced HCNG mixture was determined. The density thus determined was used to calculate the composition of the produced HCNG mixture.

$$\left.\begin{aligned} d_{\mathrm{HCNG}} = d_{\mathrm{CNG}} \times X_{\mathrm{CNG}} + d_{\mathrm{H_2}} \times X_{\mathrm{H_2}} \\ X_{\mathrm{CNG}} + X_{\mathrm{H_2}} = 1 \end{aligned}\right\} \tag{1}$$

Equation 1 yields the density of a HCNG mixture (d_{HCNG}), for given densities of the constituent gases (d_{CNG} and $d_{\mathrm{H_2}}$) and their mole fractions (X_{CNG} and $X_{\mathrm{H_2}}$). In the present experimental study, the measured densities of CNG and hydrogen were 0.73 and 0.082 kg/m^3, respectively. To find the mole fraction of hydrogen, substitute the densities of individual gases and produced HCNG density in Eq. 2. Correspondingly, HCNG mixture composition was determined from hydrogen mole fraction.

$$\left.\begin{aligned} X_{\mathrm{H_2}} = (d_{\mathrm{CNG}} - d_{\mathrm{HCNG}})/(d_{\mathrm{CNG}} - d_{\mathrm{H_2}}) \\ \%\mathrm{HCNG}_{\mathrm{Mixture}} = X_{\mathrm{H_2}} \times 100 \end{aligned}\right\} \tag{2}$$

Mixture ratio of produced blend was then compared with the desired mixture ratio of 20HCNG, in order to find the deviations in the mixture ratios. From Table 6, a mean deviation of 0.70 was observed across four engine loads with this procedure. The mean deviations from both the procedures were very similar and small. However, deviations were different at different engine loads for both procedures. In this procedure, maximum deviation was observed at 5.97 bar BMEP at 1.06. The differences among deviations from both procedures were because of flow device limitations and measurement errors. Accuracy of hydrogen GFC is 1% of its full-scale flow, which is 100 LPM, which means that flow fluctuations within 1 LPM cannot be controlled. Similarly, for CNG, GFC accuracy is 1.5% of the

Table 6 Determination of functional accuracy of DMS using total VFR and MFR

Engine load: BMEP (bar)	Total VFR (LPM)	Total MFR (g/min)	Produced HCNG density (kg/m^3)	Produced mixture ratio (v/v)	Deviation (mixture ratio)
1.99	20.37	12.30	0.60	19.47	−0.53
3.31	25.44	15.30	0.60	19.84	−0.16
4.64	33.06	19.63	0.59	21.02	1.02
5.97	36.48	21.65	0.59	21.06	1.06
Mean deviation = 0.70					

full-scale flow, which is 85 LPM, which implies that flow fluctuations within 1.28 LPM cannot be controlled. Deviations due to flow fluctuations within the above-specified limits were acceptable for this DMS.

Upon evaluating the functional accuracy of the DMS by these two procedures, it was deduced that compositions of produced HCNG mixtures from DMS were consistent with desired HCNG mixture compositions. Minor deviations in mixture ratios were observed at lower engine loads because of flow fluctuations. However, these flow fluctuations were within the accuracy limits of GFCs and were acceptable.

To further evaluate the performance and functional accuracy of DMS, experiments were conducted on a modified SI gas engine. Here, combustion, performance and emission characteristics of dynamic blends were compared with those of premixed blends of 20HCNG.

6.3 Combustion Analysis of Premixed and Dynamic 20HCNG Mixtures

For conducting the experiments, a single-cylinder diesel engine was modified to a port fuel-injected gas engine with spark ignition. Spark timing was kept constant at 30° BTDC, and relative air–fuel ratio (λ) was maintained at 0.9 throughout the experimental investigations. Engine was operated at four different load conditions, i.e. 1.99, 3.31, 4.64 and 5.97 bar BMEP. Combustion data were averaged over 250 combustion cycles to reduce cyclic variations. The combustion data thus averaged were further analysed for combustion characteristics.

In-cylinder pressure (*P*) and RoPR variations for premixed and dynamic 20HCNG mixtures were presented in Fig. 5. Peak values of in-cylinder pressure (P_{max}) and RoPR ($RoPR_{max}$) increased as BMEP increased. This increment in *P* and RoPR can be ascribed to advancement of "start of combustion" (SoC) at higher engine loads. As combustion started earlier, combustion efficiency enhanced and heat release rate (HRR) increased. Also, at higher engine loads, amount of fuel injected increased. Because of these, P_{max} and $RoPR_{max}$ peaks advanced towards the BTDC at higher BMEPs and resulted in higher P_{max} and $RoPR_{max}$. A comparison of P_{max} and $RoPR_{max}$ of premixed and dynamic 20HCNG mixtures is presented in Table 7.

As engine load increased from 1.99 to 5.97 bar BMEP, P_{max} increased from 21.76 to 54.47 bar for premixed 20HCNG, and for dynamic 20HCNG, P_{max} increased from 20.91 to 52.54 bar. P_{max} was slightly higher for premixed 20HCNG because of superior mixture uniformity of premixed 20HCNG. The difference between P_{max} of premixed and dynamic mixtures was negligible at low engine loads. However, at higher BMEPs, difference in P_{max} for both fuel mixtures increased. Non-homogeneity of dynamic 20HCNG mixtures might be a possible reason for these differences. At higher BMEPs, fuel flow rates increased and

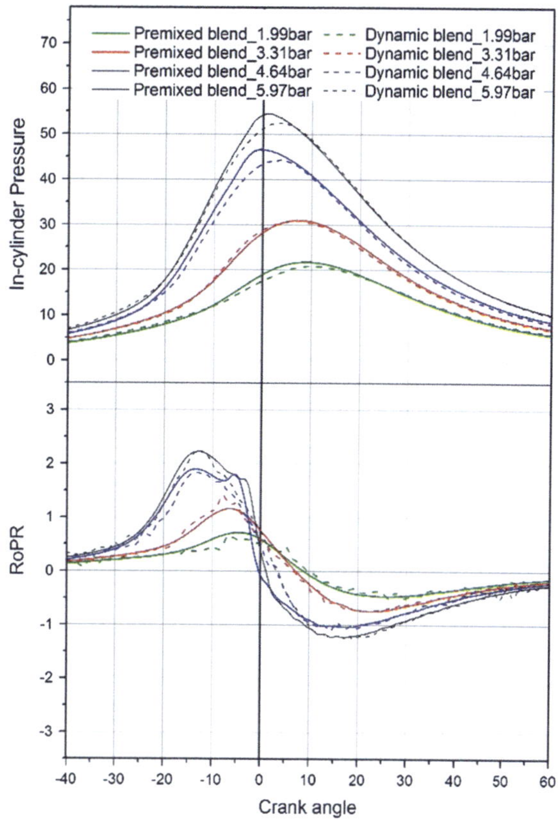

Fig. 5 In-cylinder pressure and RoPR variations with crank angle for premixed and dynamic 20HCNG mixtures [24]

Table 7 P_{max} and $RoPR_{max}$ for premixed and dynamic 20HCNG mixtures

BMEP (bar)	P_{max} (bar)		$RoPR_{max}$ (bar/deg)	
	Premixed	Dynamic	Premixed	Dynamic
1.99	21.76	20.91	0.72	0.72
3.31	31.06	30.92	1.17	1.59
4.64	46.62	44.36	1.90	1.85
5.97	54.47	52.54	2.23	2.24

adequate time was not available for homogeneous mixing of gases. Hence, variations in combustion parameters of both mixtures were observed. For premixed 20HCNG mixtures, $RoPR_{max}$ increased from 0.72 bar/deg at lowest engine load to 2.23 bar/deg at highest engine load. Similarly, for dynamic 20HCNG, $RoPR_{max}$ increased from 0.72 bar/deg at lowest engine load to 2.24 bar/deg at highest engine load. Both fuel types exhibited similar trends in RoPR, but for dynamic 20HCNG mixture, $RoPR_{max}$ was little higher compared to premixed 20HCNG mixture. Because of higher degree of non-homogeneity in dynamic mixture, hydrogen-rich and CNG-rich regions developed inside the engine cylinder. Due to the presence of

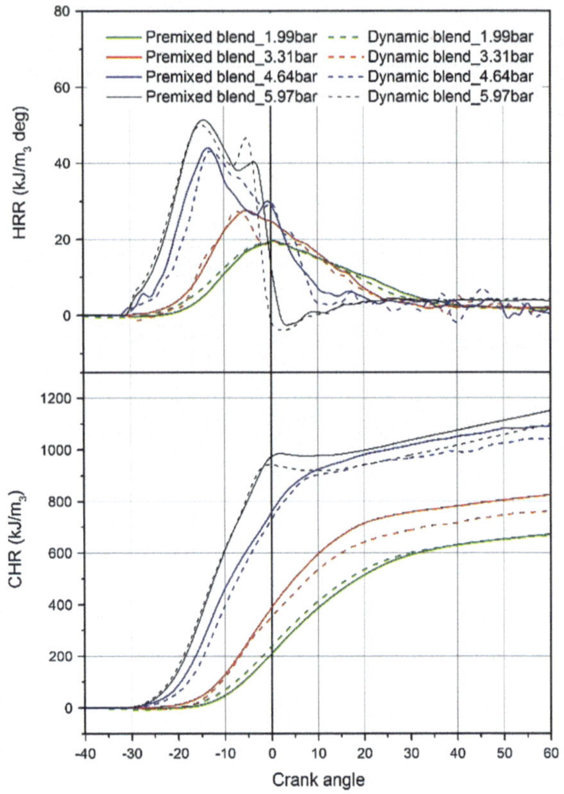

Fig. 6 HRR and CHR variations with crank angle for premixed and dynamic 20HCNG mixtures [24]

hydrogen, combustion was rapid in hydrogen-rich regions. Because of this rapid combustion, RoPR increased and $RoPR_{max}$ was relatively higher for dynamic 20HCNG mixture, whereas in CNG-rich regions, combustion was slower and extended beyond the TDC. Hence, P_{max} was comparatively lower for dynamic 20HCNG mixture.

Variations of HRR and cumulative heat release (CHR) for both mixtures are presented in Fig. 6. As engine load increased, an increasing trend in HRR_{max} was observed. HRR_{max} shifted towards BTDC at higher BMEPs. Large amount of fuel mixture injected at high engine loads was the primary reason for increment in HRR_{max}. CHR also increased with BMEP, which was in accordance with the large fuel quantities at higher engine load. As BMEP increased from 1.99 to 5.97 bar, CHR_{max} varied from 697.99 to 1217.76 kJ/m^3 for premixed 20HCNG; for dynamic 20HCNG, it varied from 707.72 to 1208.1 kJ/m^3. These results are presented in Table 8.

HRR_{max} varied from 0.61 kJ/m^3 deg at 1.99 bar BMEP to 1.62 kJ/m^3 deg at 5.97 bar BMEP for premixed 20HCNG mixture. Similarly, for dynamic 20HCNG mixture, HRR_{max} varied from 0.61 kJ/m^3 deg at 1.99 bar BMEP to 1.61 kJ/m^3 deg at 5.97 bar BMEP. It was deduced from these results that HRR and CHR trends

Table 8 HRR_{max} and CHR_{max} for premixed and dynamic 20HCNG mixtures

BMEP (bar)	HRR_{max} (kJ/m^{3}deg)		CHR_{max} (kJ/m^3)	
	Premixed	Premixed	Dynamic	Dynamic
1.99	0.61	697.99	707.72	0.61
3.31	0.86	839.08	825.66	0.86
4.64	1.61	1080.10	1069.62	1.60
5.97	1.62	1217.76	1208.10	1.61

were very much alike for both mixtures. However, CHR_{max} was slightly lower for dynamic 20HCNG mixture. Decreased combustion efficiency due to non-homogeneous mixing in dynamic mixtures might be the reason for this CHR_{max} decrement.

Mass fraction-burnt (MFB) analysis is a vital characteristic, which gives helpful information about the combustion through parameters such as start of combustion (SoC), end of combustion (EoC) and combustion duration (CD). In this experimental study, SoC is represented by the crank angle degree (CAD) at which 10% fuel mass is combusted (MFB_{10}). Likewise, the CAD, at which 90% of fuel mass is combusted, is represented by EoC (MFB_{90}). Combustion phasing is represented by the CAD, at which 50% fuel mass is combusted (MFB_{50}). Combustion duration is given by the CAD duration between SoC and EoC.

Figure 7 displays the MFB characteristics of premixed and dynamic 20HCNG mixtures. It was found that combustion started earlier as BMEP increased. At higher engine loads, fuel–air mixture reactivity increased because of larger fuel quantities. Because of this rapid combustion, ignition delay shrunk, which resulted in advancement of SoC in general. But as BMEP increased from 4.64 to 5.97 bar, SoC delayed slightly. Likewise, combustion phasing shifted towards bTDC for BMEP increment till 4.64 bar. However, as BMEP increased further to 5.97 bar, MFB_{50} almost remained constant. EoC was relatively constant showing only minor variations as BMEP increased. CD showed significantly increasing trend with BMEP increase up to 4.64 bar. Though as BMEP increased to 5.97 bar, CD remained unaffected. Combustion in SI engine occured in three major phases defined by the ignition delay, flame propagation and after-burning. At lower BMEPs, most fuel burns in the flame propagation phase because of lower amount of fuel injected. As BMEP increased, large amount of fuel was injected, requiring longer combustion duration. Therefore, combustion extended into after-burning phase. As a result, CD increased with BMEP up to 4.64 bar. At highest BMEP of 5.97 bar, improved fuel–air mixture reactivity led to quicker flame propagation leading to rapid combustion. This rapid combustion was reflected in EoC and CD trends at highest BMEP. Premixed and dynamic 20HCNG mixtures showed similar MFB characteristics, though CD was relatively lower for dynamic mixtures.

Overall, premixed and dynamic 20HCNG mixtures displayed similar combustion characteristics, emphasizing rich mixture quality of dynamic 20HCNG. However, little variations were observed for some combustion parameters primarily due to the non-homogenous nature of dynamic mixtures.

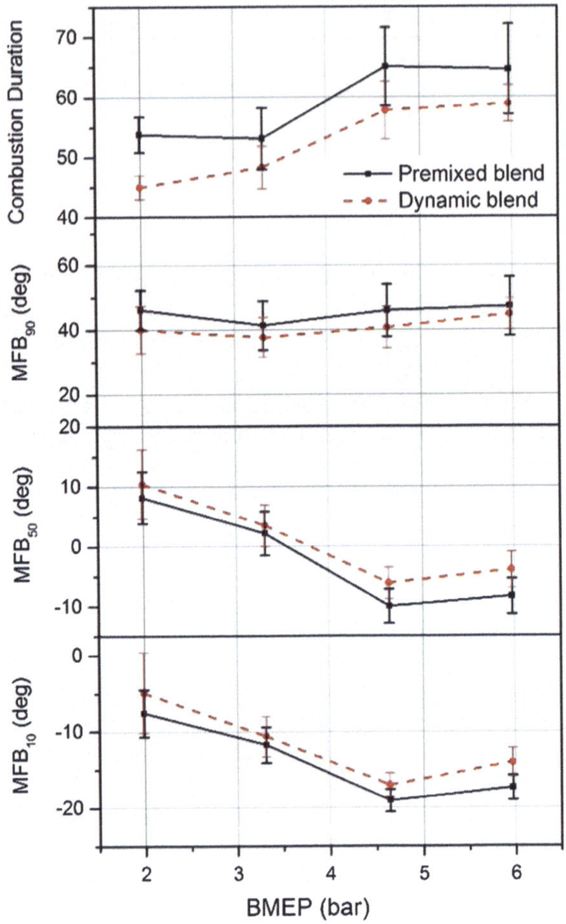

Fig. 7 SoC, MFB_{50}, EoC and CD variations for premixed and dynamic 20HCNG mixtures [24]

6.4 Performance Analysis of Premixed and Dynamic 20HCNG Mixtures

HCNG mixture quality influences the engine performance significantly. Hence, experiments were conducted to compare the performance of premixed and dynamic 20HCNG mixtures. Brake thermal efficiency (BTE), brake-specific energy consumption (BSEC) and exhaust gas temperature (EGT) were measured to estimate the performance of both mixtures. The results shown in Fig. 8 indicate that premixed and dynamic 20HCNG mixtures exhibited similar BSEC characteristics, which suggests similar calorific value of both mixtures. It also shows high functional accuracy of DMS in delivering desired HCNG mixtures.

BTE improved as engine load increased from 1.99 to 5.97 bar BMEP. This improvement can be ascribed to in-cylinder pressure and HRR increment at higher

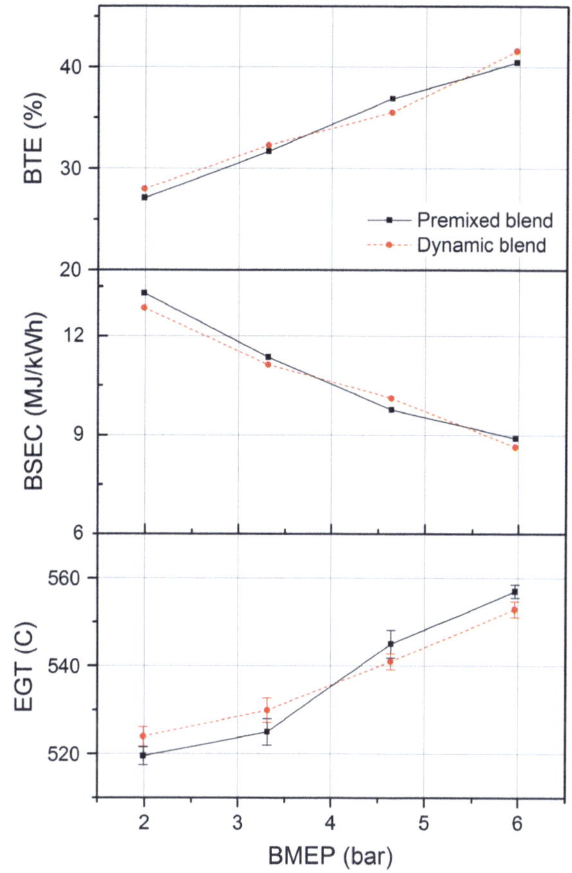

Fig. 8 BTE, BSEC and EGT variations for premixed and dynamic 20HCNG mixtures [24]

BMEPs. For premixed 20HCNG, BTE improved from 27.09% at 1.99 bar BMEP to 40.4% at 5.97 bar BMEP. For dynamic 20HCNG mixture also, BTE improved from 28.0% at 1.99 bar BMEP to 41.6% at 5.97 bar BMEP. BTE difference was higher at lower BMEPs of 1.99 and 3.67 bar. However, for higher BMEPs, BTE difference became negligible. BSEC showed decreasing trend with increasing engine load. By definition, as BSEC decreased, BTE increased. Hence, decreasing BSEC trend was in accordance with increasing BTE trend. For premixed 20HCNG mixture, BSEC decreased from 13.3 MJ/kWh at lowest BMEP to 8.9 MJ/kWh at highest the BMEP. Similarly, for dynamic 20HCNG mixture, BSEC decreased from 12.8 MJ/kWh at lowest BMEP to 8.68 MJ/kWh at the highest BMEP. As BMEP increased, in-cylinder pressure and temperature increased. These conditions coupled with improved air–fuel mixing increased the combustion efficiency at high engine loads. Hence, unburnt and residual fuel fractions decreased after the combustion, which resulted in lower BSEC and higher BTE. Increase in EGT was observed as BMEP increased. The reason was large amount of fuel injected at high engine loads, which lead to increased HRR, thereby resulting in higher EGT. For premixed 20HCNG, EGT increased from 519 °C at lowest BMEP to 557 °C at highest BMEP. Equivalently, for dynamic 20HCNG, EGT increased from 524 °C at lowest BMEP to 553 °C at highest BMEP. For both HCNG mixtures, engine performance characteristics were similar, emphasizing the high mixture quality produced by HCNG DMS.

6.5 *Emission Analysis of Premixed and Dynamic 20HCNG Mixtures*

Engine emission characteristics of premixed and dynamic 20HCNG mixtures were measured and compared in this experimental study. Brake-specific mass emissions of CO_2, NO_x, HC and CO were compared for both types of fuel mixtures. Results were shown in Fig. 9 for four different BMEPs. It was observed that there were no significant variations in CO_2 emissions across the four BMEPs. The presence of CO_2 in engine exhaust implied higher combustion efficiency of air–fuel mixtures. Concentration of $BSCO_2$ ranged between 0.30 and 0.35 kg/kWh. However, at low BMEPs, minor differences in CO_2 concentrations were observed for both fuels. The possible reason was the non-homogeneity of dynamic 20HCNG mixtures at low engine loads.

An increase in $BSNO_x$ emissions with engine loads was also observed from Fig. 9. This increment can be attributed to increased in-cylinder temperatures at higher engine loads. As BMEP increased from 1.99 bar, $BSNO_x$ concentrations increased up to 4.64 bar, but decreased again at 5.97 bar. Thermal NO_x is the main contributor for NO_x emissions in HCNG engines. Thermal NO_x is significantly dependent on in-cylinder temperature and available oxygen. As engine load

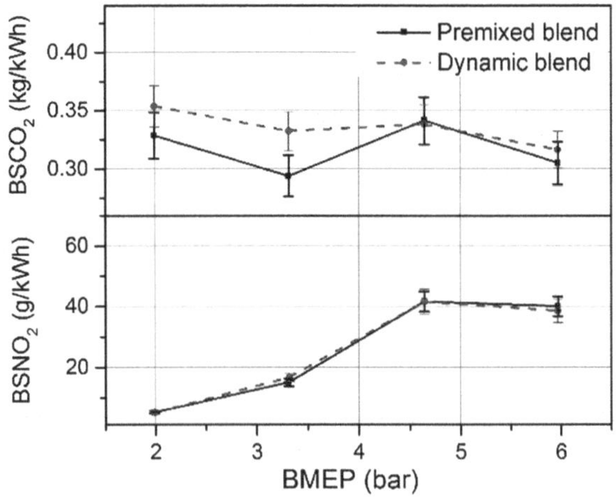

Fig. 9 Brake-specific CO_2 and NO_x emissions for premixed and dynamic 20HCNG mixtures [24]

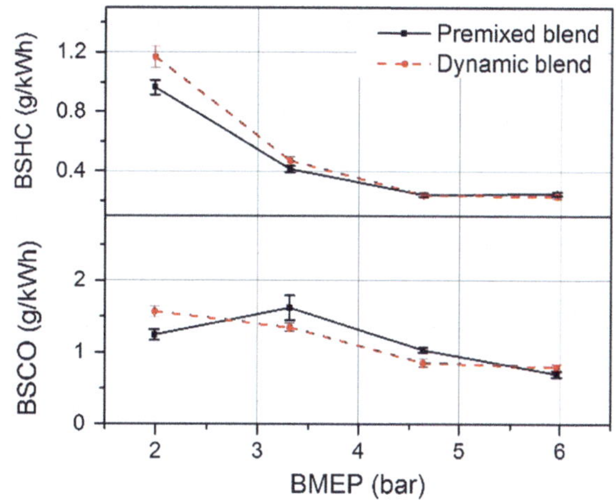

Fig. 10 Brake-specific HC and CO emissions for premixed and dynamic 20HCNG mixtures [24]

increased, in-cylinder temperature increased, and hence, NO_x emissions increased. NO_x emission trends and maximum NO_x were similar for both premixed and dynamic 20HCNG mixtures. These trends were in tune with HRR trends and EGR trends displayed by both mixture types.

Methane is a hydrocarbon, which is the major component of CNG. Hence, HCNG engines also emit HC. From Fig. 10, it was observed that BSHC concentrations decreased as engine load increased. BSHC was the maximum for lowest BMEP of 1.99 bar for both mixture types. For premixed 20HCNG mixture, maximum BSHC was 0.9 g/kWh, and for dynamic 20HCNG mixture, it was 1.1 g/kWh. Partial burning of hydrocarbons during combustion was the primary reason for HC emissions. Crevices and flame quenching inside the combustion chamber were the major contributors to such incomplete combustion. At lowest BMEP, temperatures were lower in the combustion chamber, which resulted in increased unburnt hydrocarbons and maximum BSHC emissions. However, at higher BMEPs, because of high in-cylinder temperatures, combustion efficiency improved. This improved combustion efficiency decreased BSHC concentrations considerably. CO was formed in the combustion chamber as a result of partial combustion of carbonaceous fuels. CO damages respiratory system, when inhaled even in small quantities and hence regulated by emission norms worldwide. From Fig. 10, it was observed that CO emission was the lowest for 5.97 bar BMEP. BSCO concentration decreased in general for both 20HCNG mixtures as BMEP increased. For premixed 20HCNG mixture, maximum BSCO was for 3.31 bar BMEP at 1.6 g/kWh, and for dynamic 20HCNG mixture, it was 1.57 g/kWh at 1.99 bar BMEP.

In general, both premixed and dynamic 20HCNG mixtures exhibited almost identical emission characteristics, especially at high BMEPs. Minor differences were observed in CO_2 and CO emissions, which were possibly because of non-homogeneous nature of dynamic 20HCNG mixture. Considering theoretical and experimental validations, it could be ascertained that HCNG DMS displayed high functional accuracy in delivering desired HCNG mixtures. This newly developed system therefore can be used for detailed engine investigations for wide range of HCNG mixtures.

7 Conclusions

In today's world, where population growth and environmental pollution are severe threats for a balanced ecosystem, fuels such as natural gas are a good alternate to depleting conventional fuel resources. Fewer carbonaceous emissions from natural gas help us reduce overall emissions from vehicles and transport sector. However, because of problems such as slow flame propagation and narrow flammability range, CNG alone cannot stand as the best alternate fuel. Hydrogen-enriched CNG is a prospective solution. HCNG mixture preparation is critical for an efficient HCNG engine development. A versatile dynamic gaseous mixing system which can change the mixing ratio of fuels as required is necessary. Dynamic mixing system discussed in this chapter provides optimum HCNG mixtures for engine experiments, as required. Experimental evaluations were performed to validate this mixing system's ability to provide desired mixtures and performance compared to premixed HCNG blends. Experiments were carried out to estimate the performance, combustion and emission characteristics of dynamic HCNG mixtures compared to premixed HCNG mixtures. It was deduced that dynamically prepared mixtures showed similar performance, combustion and emission trends as those of premixed HCNG blends, implying that the prototype HCNG DMS developed was successful in providing desired HCNG mixture compositions. Slight deviations were observed in the mixture composition provided by both mixing systems because of the limitations of flow control devices. A variation of 1 LPM in hydrogen flow rate and a variation of 1.28 LPM in CNG flow rate were allowed as acceptable limits during system validation.

From these experimental results, it can be concluded that the prototype HCNG DMS can readily be used to conduct full-scale experiments, enabling quicker and highly accurate investigations on wide range of HCNG mixture ratios.

References

1. Sorrell S, Speirs J, Bentley R, Brandt A, Miller R (2010) Global oil depletion. A review of the evidence. Energy Policy 38(9):5290–5295
2. Streets DG, Bond TC, Carmichael GR, Fernandes SD, Fu Q, He D, Kilmont Z, Nelson SM, Tsai NY, Wang MQ, Wo JH, Yarber KF (2003) An inventory of gaseous and primary aerosol emissions in Asia in the year 2000. J Geophys Res 108:8809

3. Uherek E, Halenka T, Borken-Kleefeld J, Balkanski Y, Berntsen T, Borrego C, Gauss M, Hoor P, Juda-Rezler K, Lelieveld J, Melas D, Rypdal K, Schmid S (2010) Transport impacts on atmosphere and climate. Land transport. Atmos Environ 44:4772–4816
4. Yanzhi Xu, Gbologah FE, Lee D, Liu H, Rodgers MO, Guensler RL (2015) Assessment of alternative fuel and powertrain transit bus options using real-world operations data. Life-cycle fuel and emissions modeling. Appl Energy 154:143–159
5. Nagalingam B, Duebel F, Schmillen K (1983) Performance study using natural gas; hydrogen-supplemented natural gas and hydrogen in AVL research engine. Int J Hydrogen Energy 8:715–720
6. Automotive CNG fuel specifications proposed by the committee constituted by EPCA. Environment pollution prevention control authority, India, 2007. EPCA report no. 29
7. Ma F, Naeve N, Wang M, Jiang L, Chen R, Zhao S (2010) Hydrogen-enriched compressed natural gas as a fuel for engines. In Nat Gas, InTech. https://doi.org/10.5772/9852
8. Bielaczyc P, Woodburn J, Szczotka A (2014) An assessment of regulated emissions and CO_2 emissions from a European light-duty CNG-fueled vehicle in the context of Euro 6 emissions regulations. Appl Energy 117:134–141
9. Patil KR, Khanwalkar PM, Thipse SS, Kavathekar KP, Rairikar SD (2009) Development of HCNG blended fuel engine with control of NO_x emissions. In: Second international conference of emerging trends in engineering and technology (ICETET). IEEE, pp 1068–1074
10. Verma G, Prasad RK, Agarwal RA, Jain S, Agarwal AK (2016) Experimental investigations of combustion; performance and emission characteristics of a hydrogen enriched natural gas fuelled prototype spark ignition engine. Fuel 178:209–217
11. Ma F, Ding S, Wang Y, Wang Y, Wang J, Zhao S (2008) Study on combustion behaviors and cycle-by-cycle variations in a turbocharged lean burn natural gas SI engine with hydrogen enrichment. Int J Hydrogen Energy 33:7245–7255
12. Thipse S, Rairikar S, Kavathekar K, Chitnis P (2009) Development of a six cylinder HCNG engine using an optimized lean burn concept. SAE Technical paper 2009-26-0031
13. Helmut E, Klaus S, Daniel L, Manfred K, Markus S (2009) Potential of synergies in a vehicle for variable mixtures of CNG and hydrogen. SAE Technical paper 2009-01-1420
14. Kavathekar K, Rairikar S, Thipse S (2007) Development of a CNG injection engine compliant to Euro-IV norms and development strategy for HCNG operation. SAE Technical paper 2007-26-029
15. Belchior CR, Barcellos WM, Pimentel VSDB, Pereira PP (2001) Analysis of vehicles converted from gasoline to CNG using conversion devises (kits). SAE Technical paper 2001-01-3883
16. Chugh S, Posina VA, Sonkar K, Srivatsava U, Sharma A, Acharya GK (2016) Modeling & simulation study to assess the effect of CO_2 on performance and emissions characteristics of 18% HCNG blend on a light duty SI engine. Int J Hydrogen Energy 41:6155–6161
17. Xu J, Zhang X, Liu J, Fan L (2010) Experimental study of a single-cylinder engine fueled with natural gas–hydrogen mixtures. Int J Hydrogen Energy 35:2909–2914
18. Bassi A (2011) Liquefied natural gas (LNG) as fuel for road heavy duty vehicles technologies and standardization. SAE Technical paper 2011-24-012
19. Collier K, Mulligan N, Shin D, Brandon S (2005) Emission results from the new development of a dedicated hydrogen-enriched natural gas heavy duty engine. SAE Technical paper 2005-01-0235
20. Khatri D, Singh V, Pal N, Maheshwari M, Singh S, Chug S, Singh R, Bhat A (2009) HCNG evaluation using a sequential gas injection system for a passenger car. SAE Technical paper 2009-26-0030
21. Flekiewicz B, Flekiewicz M, Kubica G (2012) Identification of optimal CNG-hydrogen enrichment ratio in the small SI engines. SAE Technical paper 2012-32-0015
22. Subramanian M (2011) Effect of hydrogen in CNG on small engine performance and emissions. SAE Technical paper 2009-26-0031

23. Ma F, Wang Y, Wang J, Zhao S, Yin Y, Cheng W, Zhou M (2008) Development and validation of an on-line hydrogen-natural gas mixing system for internal combustion engine testing. SAE Technical paper 2008-01-1580
24. Sagar SMV, Agarwal AK (2016) Experimental validation of accuracy of dynamic hydrogen-compressed natural gas mixing system using a single cylinder spark ignition engine. Int J Hydrogen Energy 41(32):14272–14282
25. Gangwar JN, Gupta T, Agarwal AK (2012) Composition and comparative toxicity of particulate matter emitted from a diesel and biodiesel fuelled CRDI engine. Atmos Environ 46:472–481

Characterization of Cycle-to-Cycle Variations in Conventional Diesel Engine Using Wavelets

Mohit Raj Saxena and Rakesh Kumar Maurya

Abstract Higher cycle-to-cycle variations in combustion engines lead to efficiency losses, engine roughness, lower power output, and higher exhaust emissions. Cycle-to-cycle variations in combustion engines are typically characterized by several techniques such as statistical method, symbol sequence statistics, chaotic methods, and wavelet analysis. Each strategy for cyclic variation characterization has its benefits and limitations depending on the application. Wavelet transform has a potential to analyze non-stationary signal in time domain as well as frequency domain simultaneously. This strategy has better temporal and spectral resolution; thus, wavelet analysis can be used to analyze the periodicities as well as magnitude of variations in the engine combustion cycles. This chapter presents the characterization of cycle-to-cycle variations in conventional diesel engine using statistical technique as well as wavelet technique. Cyclic variations in various combustion parameters (such as indicated mean effective pressure, total heat release rate, and peak pressure) are discussed in diesel engine operated at different operating conditions with diesel as well as butanol/diesel blends. Typically, cyclic variations in indicated mean effective pressure, peak pressure, and total heat release rate are found higher at lower engine loads and decrease with increase in engine load.

Keywords Diesel engine · Cyclic variations · Combustion · IMEP
THR · Wavelets

M. R. Saxena · R. K. Maurya (✉)
Advanced Engine and Fuel Research Laboratory, Department of Mechanical Engineering, Indian Institute of Technology Ropar, Rupnagar 140001, Punjab, India
e-mail: rakesh.maurya@iitrpr.ac.in

D. K. Srivastava et al. (eds.), *Advances in Internal Combustion Engine Research*, Energy, Environment, and Sustainability,
https://doi.org/10.1007/978-981-10-7575-9_7

1 Introduction

The World is facing the crises of depletion of fossil fuels and increase in the cost of petroleum products. To fulfill the energy demand, the development of clean and efficient combustion engines is required. Reciprocating internal combustion (IC) engines are the basic prime movers in transportation sector of the modern society. Two types of reciprocating IC engines are widely used for transportation in current society, i.e., spark ignition (SI) and compression ignition (CI) engines.

The CI engines (typically operated on diesel) are more preferred over the SI engines (typically operated on gasoline) due to its higher power output and thermal efficiency, particularly in heavy-duty vehicles and commercial applications. However, diesel engines emit nitrogen oxides (NO_x) and particulate matter (PM) in relatively higher concentration. In order to reduce the emissions and decrease the dependency on fossil fuels, various alternative fuels (such as methanol, ethanol, butanol, biodiesel) are proposed and demonstrated for the diesel engines [1–5]. However, some of these fuels are not prominently used for automotive engines due to some limitations (such as miscibility, higher viscosity).

Combustion process in a diesel engine depends on various factors such as mean in-cylinder temperature, fuel injection timing, injection pressure. Consecutive engine combustion cycles of IC engines are not exactly the same, like fingerprints of two human beings. The variations in the engine combustion cycles are known as cyclic variations. Variation in combustion characteristics leads to cyclic variations in the cylinder pressure of engine, which is the most widely measured and analyzed signal by researchers for the analysis of engine combustion. Figure 1 presents the variations in the cylinder pressure of 100 consecutive engine cycles of a four-stroke single cylinder conventional diesel engine.

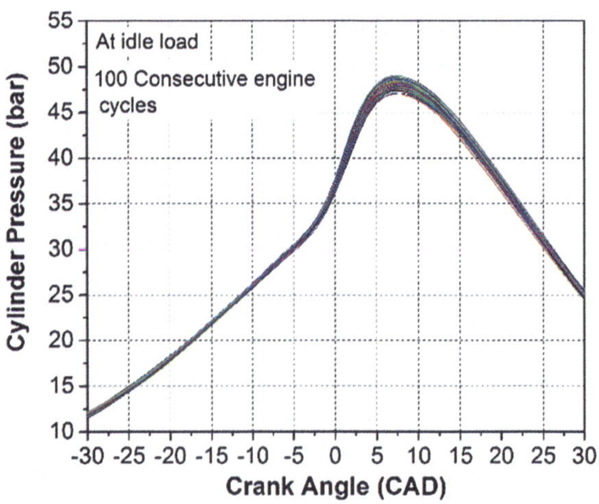

Fig. 1 In-cylinder pressure history of 100 consecutive engine cycles at idle condition

Since the two consecutive engine cycles are not similar (Fig. 1), it affects the combustion as well as emission characteristics of the engine. Higher cyclic variations in combustion lead to efficiency losses, lower fuel economy, variations in engine torque and speed, higher emissions, and limiting the actual operating range of the engine for practical applications. Excessive variations in engine cycles, particularly at higher engine loads, may cause damage to engine or reduce the engine life. To run an engine with stable combustion, these cyclic variations need to be minimized. The phenomenon of cyclic variability in combustion is more common in SI engines, which is mainly due to the variations in the charge burning rate of consecutive engine cycles. There are several factors responsible for cyclic variations in engines such as excessive dilution, variations in the amount of fuel and air inducted in each cycle, low ignition energy, variations in charge preparation (droplet size, cone angle, targeting, swirl, etc.), and variation in the in-cylinder flows. As compared to SI engine, cyclic variations are typically less significant in CI engines. In spark ignition, combustion is mainly governed by flame kernel development and flame propagation. These processes are highly sensitive to variations in engine operating conditions. In conventional diesel engines, the heterogeneous combustion (in which air–fuel mixing is governed by fuel injection in the cylinder and then combustion occurs with autoignition) occurs. Modern automotive diesel engine with electronic fuel injection systems has multiple fuel injection which reduces the ignition delay. Therefore, mostly in diesel engines, combustion occurs by diffusion combustion mode especially at higher engine load. However, there exist significantly high cyclic variations in diesel engines at lower engine loads, where ignition delay is longer. In advanced compression ignition engines known as low-temperature combustion engines, cyclic variations are very high at some of the operating conditions because of premixed charge engine operations [6].

A study demonstrated that cyclic variations in indicated mean effective pressure (IMEP) of diesel engine cycle occur due to the variations in the injected fuel quantity per cycle [7]. Another study investigated the fuel line pressure with respect to ignition delay, which indicates that fuel pump system has no significant effect on the cyclic pressure variations [8]. Rakopoulos et al. [9] experimentally investigated the effect of different biofuel blends with fossil diesel (i.e., 15% ethanol, 24% butanol, and 24% diethyl ether) on cycle-to-cycle variations in the CI engine using statistical techniques. In-cylinder pressure history data for 400 consecutive engine cycles was measured to diagnose the cyclic variations in peak pressure, peak pressure rise rate (P_{max}), and IMEP. Their results indicate that by using biofuels, the cyclic variations slightly increase as compared to fossil diesel. The coefficient of variation increases in the order of fossil diesel, n-butanol/diesel blend, ethanol/diesel blend, and diethyl ether/diesel blends. Increase in cyclic variations in biofuel blends is possibly due to corresponding increase in the ignition delay of blended fuel. In previous studies [9–14], different strategies were suggested to investigate the cyclic variations in the IC engine such as statistical technique, chaotic method, symbol sequence statistics analysis, wavelet analysis. Each of the techniques has its own benefits and limitations depending on the application. In past few years, several researchers prominently used wavelet transform (WT) to diagnose the cyclic variations in an IC engine cycle [14–21]. Wavelet transform has a potential to analyze non-stationary signal in frequency domain as well as time domain simultaneously. This strategy characterizes better temporal and spectral resolution; thus, wavelets can be used to analyze the periodicities as well as magnitude of variations in the engine combustion cycles. STFT (short-term Fourier transform) or WFT (windowed Fourier transform) can also be used for this purpose but these techniques have poor temporal resolution.

Sen et al. [14] used wavelet analysis to determine the cyclic variation in peak pressure at different operating loads of spark ignition engine. Their results indicate that for lower engine load conditions, the variation in the peak pressure is higher and it reduces with increase in engine load. In other studies, Sen et al. [16–18] investigated the different aspects of cyclic variations for different combustion parameters of SI engine. The cyclic variations in MIP (mean indicated pressure) by using wavelets at various engine speeds are investigated in compression ignition engine [15]. The results showed that with an increase in engine speed, the variations in IMEP reduce. It attributes to higher piston speed, which enhances the charge swirl motion and combustion process. Ali et al. [20] investigated the effect of diethyl ether addition in biodiesel/diesel blend on the cyclic variations for IMEP of diesel engine and found that with an increase in the diethyl ether fraction in the biodiesel/diesel blend, the cyclic variations increase. The cyclic variations possibly increase because of the higher volatility and lower flash point of the diethyl ether, which may lead to increase the variations. Maurya et al. [21] experimentally investigated the impact of engine load and the CR (compression ratio) on the cyclic variability in the combustion parameters of a stationary diesel engine. Their results demonstrated that cyclic variations in THR are higher at lower load and reduce with increase in operating load. Maximum cyclic variations were obtained at idle load with lower CR. Cyclic variations were also reduced with an increase in the CR.

As discussed above, since the cyclic variations have significant effect on the performance of the engine, it is essential to perceive the cyclic variations in the diesel engine to operate the engine efficiently at any particular fuel. This chapter focuses on the investigation of cyclic variations in diesel engine at different operating conditions using statistical technique as well as wavelet technique. The effect of engine load, compression ratio, and butanol addition in diesel fuel on the cyclic variations in combustion parameters such as total heat release (THR), IMEP, and maximum pressure (P_{max}) of compression ignition engine is discussed in Sect. 3. Before discussing the cyclic variation, the methodology of cyclic variation analysis is discussed in the following section.

2 Methodology of Cyclic Variation Analysis

To investigate and characterize the cyclic variations in combustion parameters, the estimation of combustion parameter is required on cycle-to-cycle basis. Typically, in-cylinder pressure is measured for analyzing combustion characteristics of

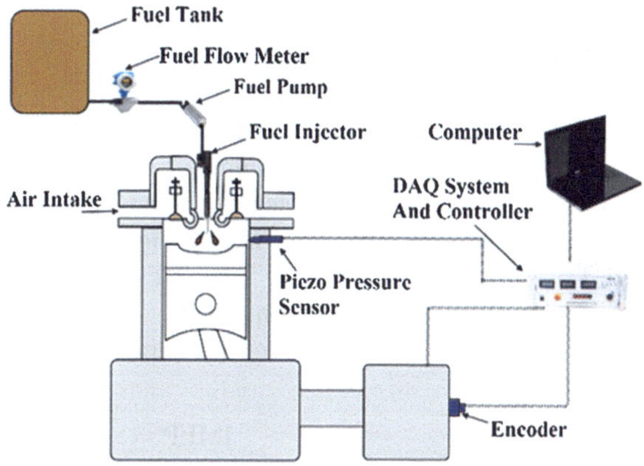

Fig. 2 Schematic diagram of typical cylinder pressure measurement setup

reciprocating engines. Current piezoelectric pressure sensors have fast response, which is required for engine combustion analysis. Heat release characteristics are calculated from experimentally measured data of cylinder pressure. Combustion parameters are calculated from heat release curve estimated from cylinder pressure data. For cycle-to-cycle variation analysis, in-cylinder pressure traces of several consecutive engine cycles for diesel engine are generally recorded. For the measurement of in-cylinder pressure, a piezoelectric pressure transducer is installed in the cylinder head of the engine. The crank angle position is measured by using a crank angle encoder which is installed on the axis of crankshaft. Typical experimental setup required for the measurement of cylinder pressure (which is used for the calculation of combustion parameters on cyclic basis) is illustrated in Fig. 2. A high-speed data acquisition system is used for online logging the experimental cylinder pressure data.

For the investigation of cycle-to-cycle variations, in-cylinder pressure of several (2000 in present study) consecutive engine cycles is recorded. For each cycle, combustion parameters are calculated by heat release analysis. Equations used for calculating the combustion parameters are given below:

Rate of heat release (ROHR) is determined by using Eq. (1)

$$\frac{\mathrm{d}Q(\theta)}{\mathrm{d}\theta} = \left(\frac{1}{\gamma - 1}\right) V(\theta) \frac{\mathrm{d}P(\theta)}{\mathrm{d}\theta} + \left(\frac{\gamma}{\gamma - 1}\right) P(\theta) \frac{\mathrm{d}V(\theta)}{d\theta} \tag{1}$$

where

'Q' represents the rate of heat release (ROHR),

'γ' denotes the specific heat ratio, and

'$P(\theta)$' and '$V(\theta)$' denote the in-cylinder pressure and volume as a function of crank position.

$$V(\theta) = V_c + \frac{\pi}{4} B^2 \left(L + R - R\cos(\theta) - \sqrt{L^2 - R^2 \sin^2(\theta)} \right) \tag{2}$$

where

'B' denotes the bore of the cylinder,
'L' is the connecting rod length,
'R' denotes the radius of crank, and
'V_c' denotes the clearance volume.

THR is determined by integrating the ROHR between start of combustion (SOC) and end of combustion (EOC). Figure 3c, b shows the time series of P_{max} and THR for 2000 consecutive engine cycles for diesel engine at idle load (0% engine operating load) condition.

IMEP is defined as the ratio of indicated work (W_{ind}) to the swept volume (V_s).

$$\text{IMEP} = \frac{W_{ind}}{V_s} \tag{3}$$

where

'W_{ind}' denotes the indicated work and
'V_s' is the swept or displacement volume.

$$W_{ind} = \frac{2\pi}{360} \int_{-180}^{180} \left(P(\theta) \frac{dV}{d\theta} \right) d\theta \tag{4}$$

Figure 3a illustrates the time series of IMEP for 2000 consecutive engine cycles for diesel engine at idle load (0% engine operating load) condition.

After calculating the time series of combustion parameters, cyclic variations are analyzed by statistical method and wavelet methods, which are described in the following subsections.

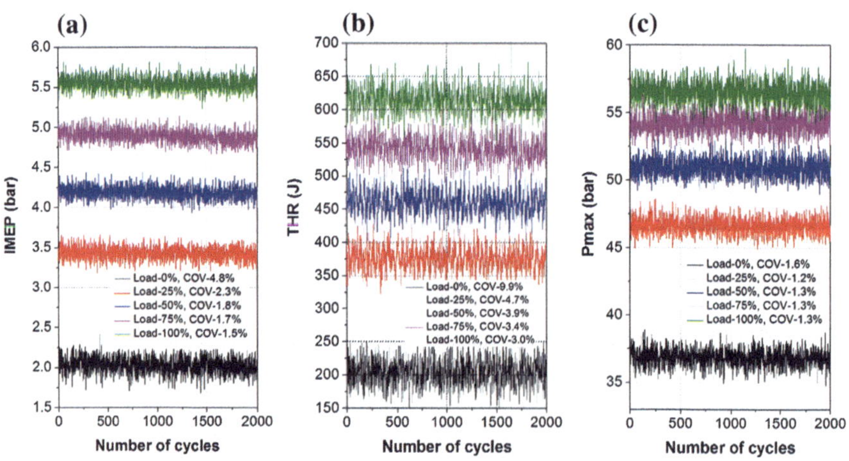

Fig. 3 Time series of **a** IMEP; **b** THR; **c** P_{max} for diesel fuel different engine operating load conditions at 17:1 compression ratio. Adapted from [24]

2.1 Statistical Methods

Statistical methods are used most commonly for the quantification of the cycle-to-cycle (cyclic) variations in the engine combustion parameters. Standard deviation and coefficient of variation are two commonly used parameters to quantify the variation in any combustion parameter. Standard deviation of the combustion parameter determines how far the values are spread from the mean values, and it is defined as follows:

$$\text{Std. deviation in 'X'} = \sqrt{n\sum_{i=1}^{n}\frac{(X_i - X)^2}{(n-1)}} \tag{5}$$

where

'X represents the combustion parameter,

'n is the number of samples, and

'i is the sample of interest.

Coefficient of variation (COV) for any combustion parameter represents the variations in that parameter and is calculated by using Eq. (6)

$$\text{COV}(\%) = \frac{\text{Std. deviation of 'X'}}{\text{mean of 'X'}} \times 100 \tag{6}$$

From statistical technique, temporal variations in the time series can be easily quantified. However, spectral variations in combustion parameters cannot be quantified by using statistical technique, which is the main limitation of this technique.

2.2 Wavelet Transform

Wavelet analysis is a common technique used for quantifying the local variations in a time series of non-stationary signal. In wavelet analysis, the time series is decomposed into frequency domain as well as time domain, which helps to quantify the leading mode of combustion variability. Additionally, with the help of this technique, one can also quantify how that mode varies with time [23]. A wavelet is considered as an instantly decaying oscillation or wave. That means amplitude of the wavelet starts increasing from zero, and within a short period of time, it comes back to initial condition. A wavelet is defined as a function, which has a finite energy and has zero mean. A continuous wavelet transform (CWT) for a wavelet function $\Psi(t)$ is represented as follows:

$$\text{CWT}(a,b) = \frac{1}{\sqrt{|a|}}\int_{-\infty}^{\infty} x(t)\Psi^*\left(\frac{t-b}{a}\right), \quad a,b \in R, \ a \neq 0 \tag{7}$$

where

'$\Psi(t)$' and '$\Psi^*(t)$' imply the mother wavelet and its conjugate,

'$x(t)$' implies a continuous signal, and

'a and 'b' denote the scaling and translating parameters.

To analyze the cyclic variations in combustion parameters, WPS (wavelet power spectrum) and GWS (global wavelet spectrum) are used. WPS gives the information about the fluctuation of variance at different frequencies or scales. The square modulus of CWT represents the magnitude of signal energy at a certain scale 'a' and at a particular location 'n' which is called WPS or scalogram. WPS is normalized by dividing with 'σ^2' so that power relative to white noise is obtained [23]. WPS is determined as follows:

$$\text{WPS} = |\text{CWT}_n(a)|^2 \tag{8}$$

WPS can be normalized by using Eq. (9)

$$\text{WPS}_n = \frac{|\text{CWT}_n(a)|^2}{\sigma^2} \tag{9}$$

where 'σ^2' denotes the standard deviation.

CWT is complex function; thus, modulus of CWT actually represents the amplitude of CWT. WPS depends on the scale (frequency) as well as time represented by a surface. Timescale representation of WPS can be obtained by plotting a contour of surface on a plane. From WPS, higher variance in the time series and their frequency of occurrence can also be easily determined. In addition, time duration of prevailing periodicities can be easily determined. GWS is defined as time average of WPS and represented by 'W_s'. Peak locations in the GWS represent the dominant periodicities in the given time series. GWS is calculated from Eq. (10).

$$\text{GWS} = W_s = \frac{1}{N}\sum_{n=1}^{N}|\text{CWT}_n(a)|^2 \tag{10}$$

3 Cycle-to-Cycle Variation Analysis

In this section, the effect of engine load, compression ratio, and butanol addition on cycle-to-cycle variations in combustion parameters is discussed. The cyclic variations are discussed using wavelet analysis as well as statistical methods.

3.1 Effect of Engine Load on Cyclic Variations

The cyclic variations have significant effect on the performance of diesel engine, and it leads to higher emissions and lower fuel economy. In this section, the effect of engine load on cyclic variations in combustion parameters of conventional diesel engines is analyzed using statistical approach as well as wavelet approach. Figure 3 shows the time series of IMEP, THR, and P_{max} for 2000 consecutive engine cycles. Time series is generated by the analysis of measured cylinder pressure (as discussed in Sect. 2) for each engine cycle. Figure 3 illustrates that IMEP, THR, and P_{max} increase with an increase in the engine operating load condition as higher amount of fuel is burned in the cylinder. Figure also reveals that with an increase in the engine operating load, the coefficient of variation (COV) decreases for all the combustion parameters. The decrease of the COV in combustion parameters with engine load indicates the reduction in the cycle-to-cycle variations. At 0% engine operating load, the amount of fuel injected is lower which results in lower combustion temperature. Lower combustion temperature leads to higher cyclic variations. At lower engine load, combustion phasing is relatively retarded and significant amount of combustion occurs during expansion stroke. Oxidation reactions occurring away from TDC are more sensitive to cyclic variations in temperature. Therefore, cyclic variation in combustion parameters is higher at lower engine loads. With increase in the engine load, more amount of fuel is injected in the cylinder, which leads to increase the mean in-cylinder combustion temperature and results in lower cyclic variations in combustion parameters.

Wavelet transform is used to analyze the frequency of cyclic variation with time in 2000 consecutive engine combustion cycles. It provides the time and frequency information simultaneously [19, 23]. With the help of CWT, wavelet power spectrum (WPS) and global wavelet spectrum (GWS) are created from the time series of IMEP, THR, and P_{max} of 2000 consecutive cycles. WPS reveals the energy of the signal, and it is the square modulus of CWT. In this chapter, spectrogram is mainly used to investigate the frequency band contribution to the signal energy. Additionally, it is used to determine how periodicity changes with time. The time average of WPS is called GWS.

Figure 4 illustrates the GWS and WPS of IMEP for neat diesel (D100) fuel at 0, 50, and 100% engine operating loads. Horizontal axis in WPS represents the engine combustion cycles, and the vertical axis indicates the periodicities of particular data series. Horizontal axis in GWS represents the power, and the vertical axis indicates the period. Peaks in the GWS plot reveal the prevalent periodicities in the data series of particular parameters. Colors shown in the WPS graph represent the energy of the signal, which represents the WT magnitude [23]. Red and blue colors indicate the maximum and minimum energy of the signal, respectively, over that interval. U-shaped curve in the bottom of WPS is known as the cone of influence (COI). It indicates the 5% significant level. It is the area where edge effect is important [23]. The area above the COI is only considered. Figure 4a depicts the GWS and WPS for IMEP at 0% engine operating load condition. Figure 4a

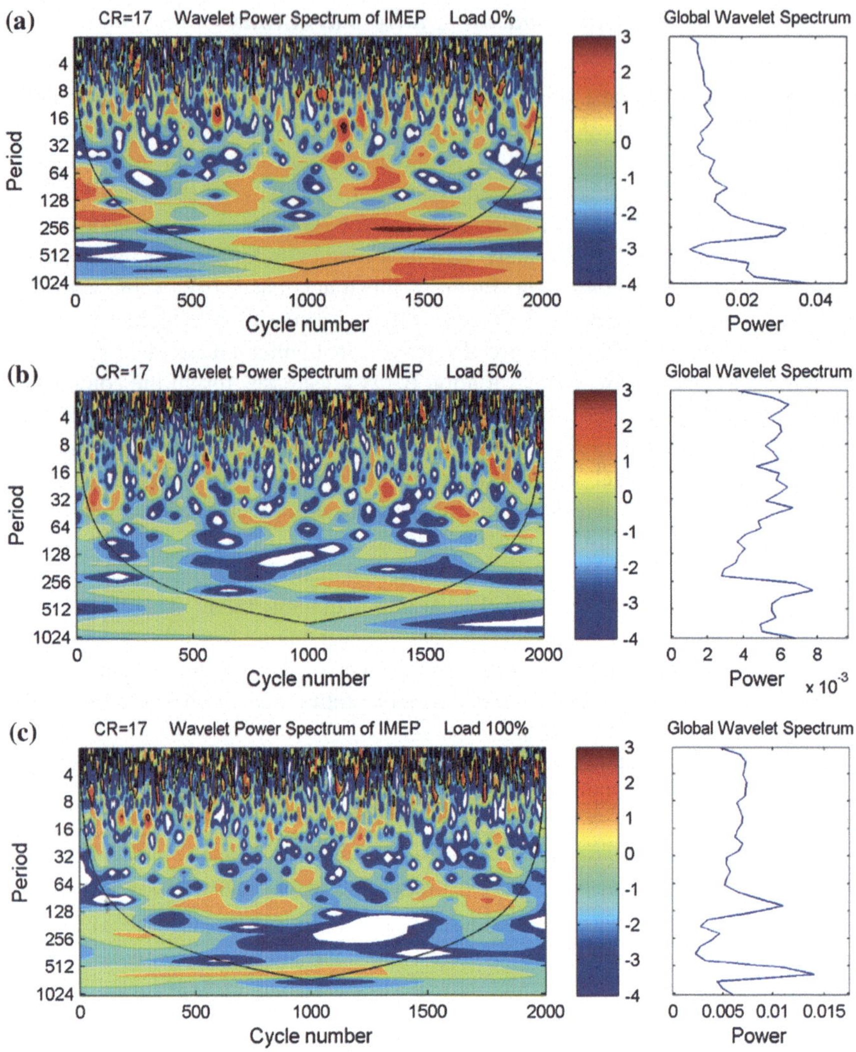

Fig. 4 WPS and GWS of IMEP for neat diesel **a** at 0% engine load; **b** at 50% engine load; **c** at 100% engine load for compression ratio of 17 in a CI engine. Adapted from [24]

indicates that band of periods 11–16, 16–28, 55–93, and 186–373 have higher cyclic variations in the cycles ranging between 596–639, 1128–1187, 1215–1405, and 982–1644, respectively. The strong intensity periodic band indicates higher cyclic variability at 0% engine operating load condition. Figure 4b shows the WPS and GWS for IMEP at 50% engine operating load condition. It has been observed from the Fig. 4b that fewer periodic bands of 20–32 and 32–39 with strong intensity of variance occur in the cycles ranging between 128–136 and 1576–1689, respectively. GWS at 50% engine load is also reduced in comparison with 0%

engine operating load which indicates reduction in cyclic variations. The decrease in cyclic variations is due to the higher mean in-cylinder combustion temperature at higher load condition (because more amount of fuel is injected) which leads to better combustion. Figure 4c shows the WPS and GWS for IMEP at 100% engine operating load condition. The GWS in Fig. 4c shows that the peak power of GWS is higher for 100% engine load condition which reveals higher cyclic variations. WPS at 100% engine load represents that a periodic band of 90–128 with higher intensity occurs between the cycles ranging from 1460 to 1870. Some weaker periodic bands are also observed in Fig. 4c. However, the power of GWS is higher at lower intensity periods. It is possibly due to the amplification of GWS power at higher periods [25].

Figure 5 shows the WPS and GWS of THR at 0, 50, and 100% engine operating load conditions for a fixed compression ratio of 17. Figure 5a shows the WPS and GWS for THR at 0% engine operating load condition. Figure 5a indicates a periodic band of 14–20 with strong intensity of variance observed during the cycles ranging between 51–410, 571–875, 927–950, 1015–1075, 1180–1555, and 1650–1870. The presence of strong periodicity in large number of engine cycles represents the higher cyclic variability. Additionally, some weaker bands of 46–54 and 128–157 periods with moderated or low intensity occur—in the cycle range of 145–160 and 625–770, respectively.

Figure 5b shows the WPS and GWS for THR at 50% engine operating load condition. Figure 5b shows that strong intensity of periodic band of 32–64 is observed in the cycles ranging from 72 to 1950. This periodic band of strong, moderated, and lower intensity (shown by decrease in the intensity of red color) of variance seems continuous throughout the cycles ranging from 72 to 1950. Lower intensity band represents the reduction in the cyclic variations. The decrease in the cycle-to-cycle variations attributes to higher mean in-cylinder combustion temperature at higher load condition which leads to better combustion. It can also be noticed from Fig. 5a, b that the peak power of the GWS at 50% engine load is higher in comparison to 0% (idle) engine load, which indicates higher cyclic variations at 50% engine load condition. A shift from lower periodicity to higher periodicity represents the lower frequency cyclic variations (as shown in WPS of Fig. 5a, b). However, GWS power decreases at 100% engine operating load condition, which indicates the decrease in the cyclic variations. Figure 5c shows the WPS and GWS for THR at 100% engine operating load condition. Figure indicates that a periodic band of 10–16 with strong intensity occurs intermittently in the cycles ranging between 140–255, 325–426,467–499, 552–580,642–714, 738–780, 1023–1194, 1260–1497, 1505–1580, and 1889–1960 cycles.

Figure 6 illustrates the WPS and GWS of P_{max} at different engine operating load conditions. Figure 6a shows that periodic bands of 157–313, 40–60, 55–93, and 55–78 with strong intensity of variance occur in the cycles between 1135–1620, 222–382, 821–953, and 1305–1440, respectively. Figure 6b depicts the WPS and GWS of P_{max} at 50% engine operating load condition. Figure indicates that the power of GWS increases in comparison with 0% engine operating load condition. It

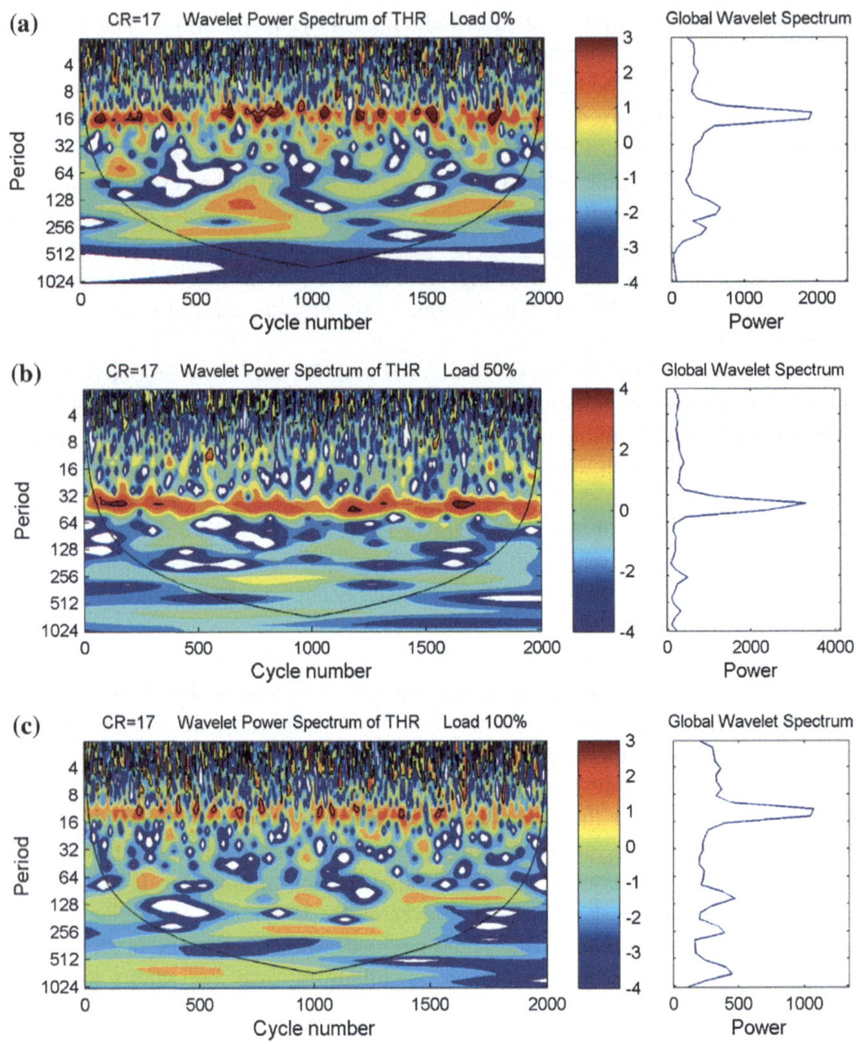

Fig. 5 WPS and GWS of THR for neat diesel **a** at 0% engine load; **b** at 50% engine load; **c** at 100% engine load for compression ratio of 17. Adapted from [24]

can be observed that peak GWS power is corresponding to intensity of period that is outside the cone of influence which is not significant. Fewer strong intensity periodic bands of 16–30, 21–28, and 30–38 occur in the cycle range of 818–845, 1333–1365, and 9420–1015 cycles, respectively. At 100% engine operating load condition, periodic bands of 112–225 and 39–64 with weak intensity of variance occur in the cycles ranging between 640–1055 and 960–1370 cycles, respectively.

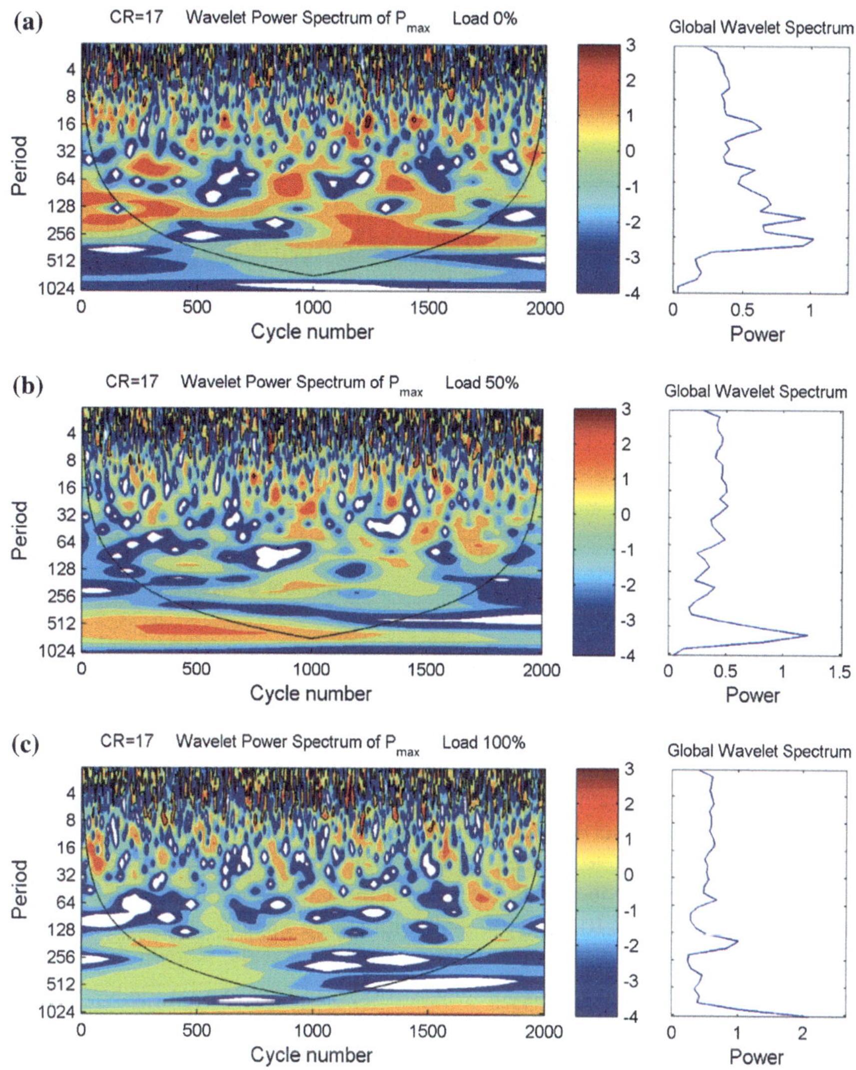

Fig. 6 WPS and GWS of P_{max} for neat diesel **a** at 0% engine load; **b** at 50% engine load; **c** at 100% engine load for compression ratio of 17. Adapted from [24]

3.2 *Effect of Compression Ratio on Cyclic Variations*

This section presents the effect of compression ratio on cyclic variation in combustion parameters of diesel engine. Figure 7 shows the time series of THR for different compression ratios (CR) at lower engine load (0.28 bar BMEP) condition.

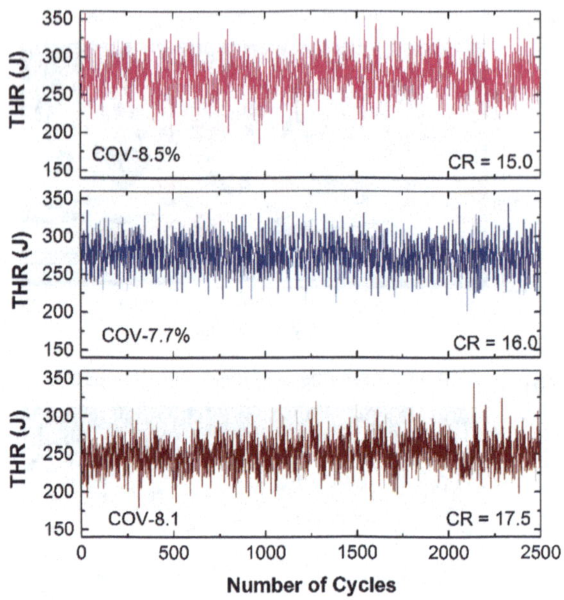

Fig. 7 Time series of THR for different compression ratios at 0.28 bar BMEP (lower engine operating load) [21]

Since the cyclic variations are higher at lower load condition, the effect of CR of compression ratio on the cyclic variation was investigated at lower load condition [21]. Cyclic variations in total heat release are highest at lowest compression ratio (Fig. 7).

Figure 8 illustrates the WPS and GWS of THR at compression ratios (CRs) of 15, 16, and 17.5 at lower load condition (0.28 bar BMEP). Figure 8 depicts that with an increase in the CR, the peak power of the GWS decreases which indicates that cyclic variations reduce with increase in the CR. At a CR of 17.5, periodic bands of 105–256, 64–110, and 100–540 with strong intensity occur between the cycles 380–890, 1090–1410, and 1290–2080, respectively (shown in Fig. 8a). Periodic bands with strong intensity and moderated intensity of variance indicate the higher cyclic variations. At CR of 16, a periodic band of 16–32 occurs intermittently throughout the cycles (shown in Fig. 8b). It can also be seen from Fig. 8b, a that with an increase in the CR from 16 to 17.5, a shift occurs from lower periodicity to higher periodicity which represents the variations occur with lower frequency. It can also be depicted from the figure that peak power in GWS slightly increases in case of CR of 16 in comparison with CR of 17. Figure 8c shows the WPS and GWS of THR at CR of 15. Figure 8c reveals that the periodic bands of 220–315 and 430–650 occur between the cycles 445–1054 and 981–1807, respectively. Figure 8c also reveals that power of GWS further increases with decrease in the CR which indicates the higher cyclic variations.

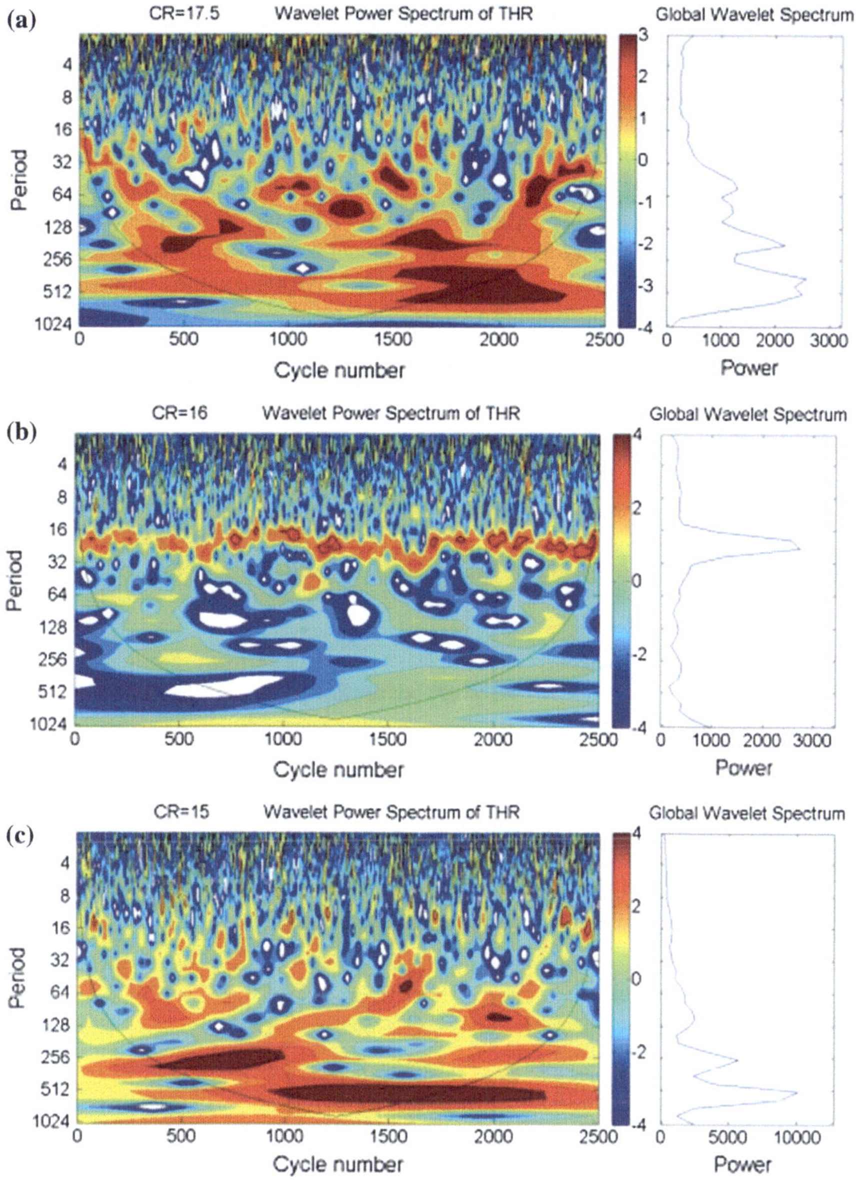

Fig. 8 WPS and GWS of THR for neat diesel at **a** CR of 17.5; **b** CR of 16; **c** CR of 15 for 0.28 bar BMEP [21]

3.3 *Effect of Butanol Blends on Cyclic Variations*

In recent years, butanol is evolved as an alternative fuel for compression ignition engine [1, 27–30]. The properties of butanol make it suitable candidate fuel for the replacement of conventional diesel fuel for diesel engines. Butanol can be easily mixed with diesel fuel, and no further additive is required for the preparation of stable blend (i.e., no fuel layer separation). Butanol also has a higher cetane number as compared to other primary alcohol fuels. Typical properties of butanol and diesel are provided in the study [22]. Previous study [26] investigated the cyclic variations in P_{max} for neat diesel and butanol blends at different engine operating load conditions. The results reveal that with an increase in the engine operating load, the peak power of GWS decreases which means that the cyclic variability reduces with increase in engine operating load. The increase in the butanol fraction in the blended fuel up to a certain limit leads to the reduction in the cyclic variations.

Figure 9 shows the time series of IMEP for neat diesel (D100), 10% butanol/diesel blend (B10), 20% butanol/diesel blend, and 30% butanol/diesel blend (B30) of 2000 consecutive engine cycles at 0%, 50%, and 100% engine load conditions. Figure 9 illustrates that the mean IMEP increases with an increase in the engine

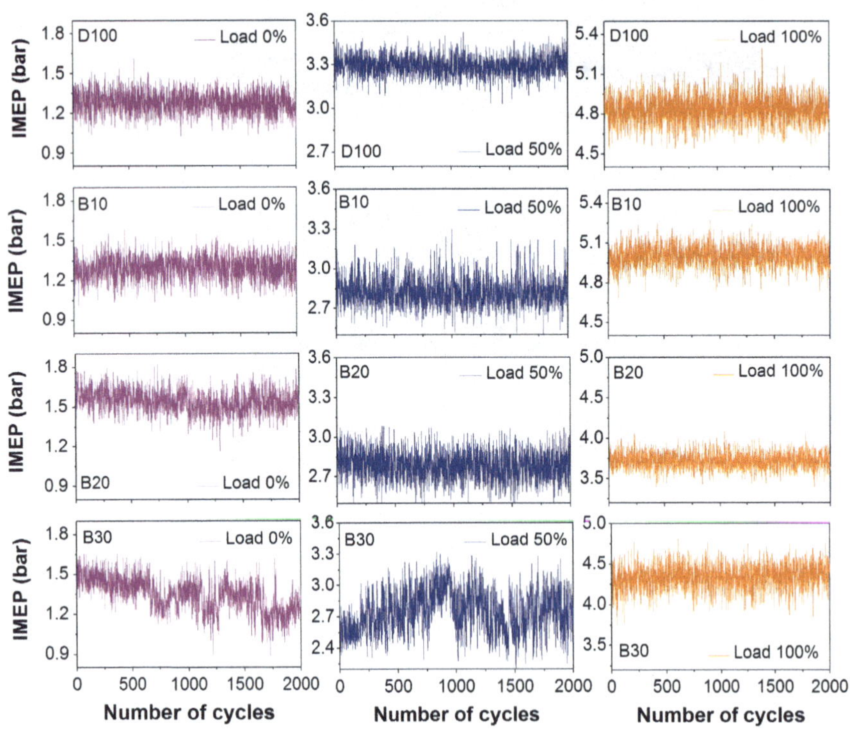

Fig. 9 Time series of IMEP for D100, B10, B20, and B30 at 0%, 50%, and 100% load conditions in CI engine

operating load condition. At higher engine load, more fraction of fuel is burned in each cycle which leads to increase the IMEP.

Figure 10 illustrates the COV in IMEP for 2000 consecutive engine combustion cycles at different load conditions for D100, B10, B20, and B30 blends. Figure 10 reveals that the higher cyclic variations were obtained at 0% load condition. At 0% engine operating load, the amount of fuel injected is lower which results in lower combustion temperature. Lower combustion temperature leads to increase the cyclic variations. Figure 10 also indicates that at lower load condition, up to 20 or 25% addition of butanol in the diesel fuel, COV in IMEP is close to COV in IMEP with neat diesel fuel. However, higher COV in IMEP was observed above 25% butanol fraction in the blended fuel. Figure reveals that the cyclic variations decrease with the addition of butanol fraction in the diesel fuel up to a certain limit. However, with further increase in butanol fraction, cyclic variation increases at 0% engine load condition. With increase in the butanol fraction in the blended fuel, the combustion phasing is retarded which results in higher cyclic variations. When combustion phasing is retarded, the velocity of the piston is comparatively higher, which leads to more sensitivity for temperature and pressure because of change in cylinder volume. Therefore, retarded combustion phasing results in higher cyclic variations. At higher load condition, higher fraction of butanol could be added in diesel fuel up to a certain limit which further needs to be investigated.

Since the cyclic variations were higher at 0% load, wavelet analysis is only presented for D100, B10, B20, and B30 at 0% load condition. Figure 11a indicates a periodic band of 16–32 with moderated intensity observed during the cycles 126–185. The presence of strong periodicity represents the higher cyclic variability. In addition to this, a periodic band of 4–16 with moderated or low intensity occurs intermittently till the 800 cycles. Figure 11b shows periodic bands of 16–32, 32–128, 64–128, and 256–512 with strong intensity observed during 458–524,

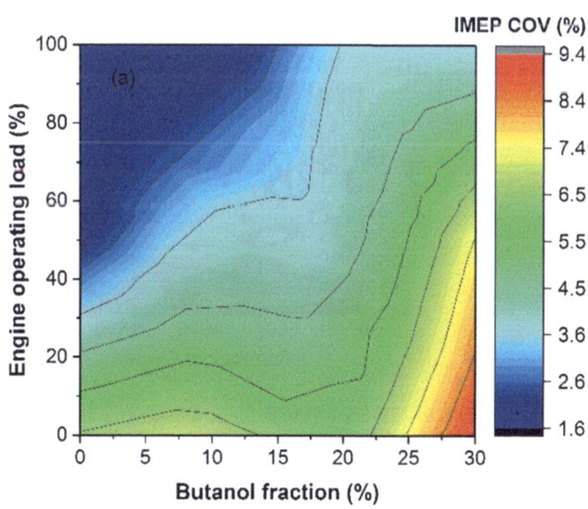

Fig. 10 Effect of engine load on the COV in **a** IMEP for different diesel/butanol blends in compression ignition engine

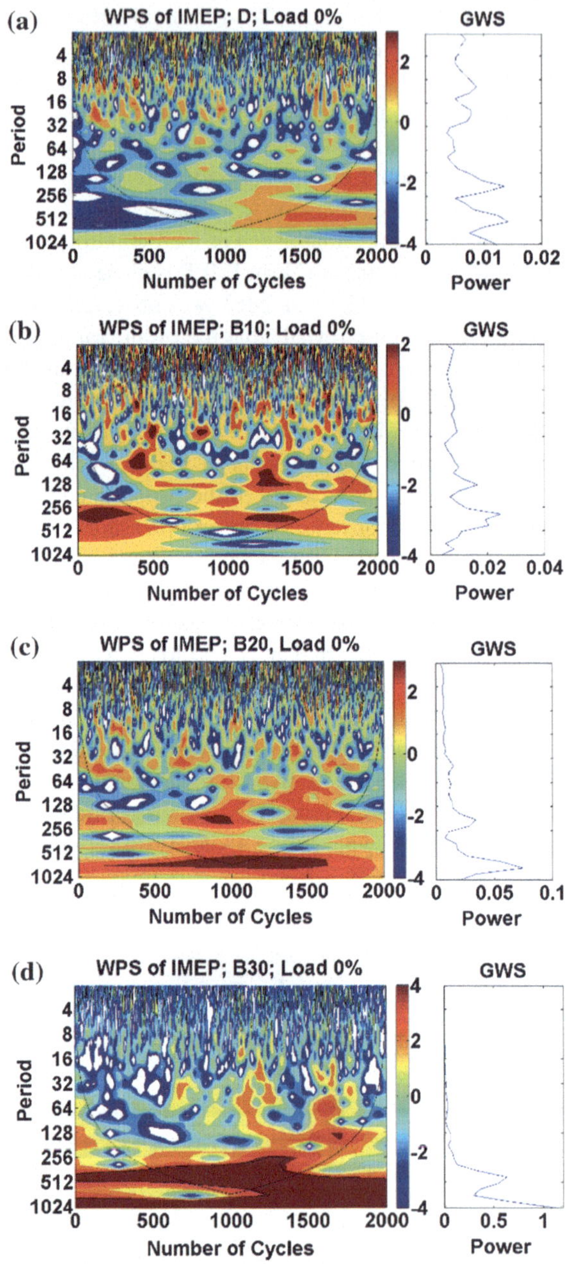

Fig. 11 WPS and GWS of IMEP for **a** D100; **b** B10; **c** B20; **d** B30 blended fuel at 0% load conditions in CI engine

304–479, 1168–1500, and 746–1622 cycles, respectively. Moderated intensity periodic bands also observed alternatingly throughout the cycles.

Similarly, in Fig. 11c, strong intensity periodic bands of 128–256, 64–256, and 512–1024 were observed during 672–1159, 1280–1752, and 845–1200 cycles, respectively. In case of 30% butanol/diesel blend, a strong intensity periodic band of 256–512 was observed during 494–1426 consecutive cycles (as shown in Fig. 11d).

In Fig. 11, GWS reveals that peak power in case of B10 and B20 blends is closer to the GWS peak power of neat diesel at 0% engine load condition. It represents that there was not too much difference in cyclic variation with 10 and 20% butanol blends as compared to neat diesel fuel. However, intense cyclic variations were obtained with B30 blended fuel. It might be because of the combined effect of lower cetane number and higher oxygen contents which retards the combustion phasing and results in unstable combustion.

4 Summary and Conclusion

This chapter discussed the characterization of cyclic variations in conventional diesel engine using statistical technique as well as wavelet technique. Combustion instability or cyclic variations in the engine cycle have significant deteriorating effect on the performance of the engine. Statistical and wavelet analysis results showed that cyclic variations are higher at lower load condition and decrease with increase in the engine load. Global wavelet spectrum (GWS) results indicate that with an increase in the engine operating load, peak power of GWS decreases. The decrease in GWS power suggests the lower cyclic variations. The GWS analysis also demonstrates the similar trend in cyclic variation with engine load as estimated by statistical methods. The WPS results also indicate that with an increase in the engine operating load, strong intensity variance shifts toward the higher frequencies from lower frequencies. At higher load conditions, fewer number of strong/moderated intensity periodic bands were found. It is also found that cyclic variability increases as compression ratio decreases. Butanol has a potential to reduce the cyclic variations in diesel engine. Statistical analysis of cyclic variations in IMEP indicates that at lower load condition, the cyclic variability is higher and it reduces with increase in the engine load. The results showed that up to 20% addition of butanol in the diesel fuel, cyclic variations in IMEP are close to the variation with neat diesel fuel. On further increasing the butanol fraction in the blended fuel, cyclic variations increase drastically. Finally, it can be summarized that wavelet analysis has a potential to analyze the cyclic variations in the combustion parameters and can provide additional information, which can be beneficial to control the cyclic variations. However, a more detailed exploration is required to utilize the full potential of wavelet analysis.

References

1. Ibrahim A (2016) Performance and combustion characteristics of a diesel engine fuelled by butanol–biodiesel–diesel blends. Appl Therm Eng 103:651–659
2. Imdadul HK, Masjuki HH, Kalam MA, Zulkifli NWM, Alabdulkarem A, Rashed MM, How HG (2016) Higher alcohol–biodiesel–diesel blends: an approach for improving the performance, emission, and combustion of a light-duty diesel engine. Energy Convers Manage 111:174–185
3. Agarwal AK, Singh AP, Agarwal A, Jeon J, Lee CS, Park S (2016) Spatial combustion analysis of biodiesel fueled engine using combustion chamber endoscopy and modeling. Renew Energy
4. Giakoumis EG, Rakopoulos DC, Rakopoulos CD (2016) Combustion noise radiation during dynamic diesel engine operation including effects of various biofuel blends: a review. Renew Sustain Energy Rev 54:1099–1113
5. Li N, Tang D, Xu Y, Zhao W (2015) Combustion and emission characteristics of an off-road diesel engine fuelled with biodiesel–diesel blends. Int J Sustain Energ 34(7):417–430
6. Maurya RK (2018) Characteristics and control of low temperature combustion engines: employing gasoline, ethanol and methanol. Springer, ISBN 978-3-319-68507-6
7. Koizum I, Gyakushi N, Takamoto Y (1977) Study on the cycle-by-cycle variation in diesel engines. Bulletin of JSME 20(145):869–876
8. Rakopoulos CD, Rakopoulos DC, Giakoumis EG, Kyritsis DC (2011) The combustion of n-butanol/diesel fuel blends and its cyclic variability in a direct injection diesel engine. Proc Inst Mech Eng [A] J Power Energy 225(3):289–308
9. Rakopoulos DC, Rakopoulos, CD, Giakoumis EG, Komninos NP, Kosmadakis GM, Papagiannakis RG (2016) Comparative evaluation of ethanol, n-butanol, and diethyl ether effects as biofuel supplements on combustion characteristics, cyclic variations, and emissions balance in light-duty diesel engine. J Energy Eng 04016044
10. Rakopoulos DC, Rakopoulos CD, Giakoumis EG, Dimaratos AM (2013) Studying combustion and cyclic irregularity of diethyl ether as supplement fuel in diesel engine. Fuel 109:325–335
11. Foakes AP, Pollard DG (1993) Investigation of a chaotic mechanism for cycle-to-cycle variations. Combust Sci Technol 90(1–4):281–287
12. Maurya RK, Agarwal AK (2012) Statistical analysis of the cyclic variations of heat release parameters in HCCI combustion of methanol and gasoline. Appl Energy 89(1):228–236
13. Maurya RK (2016) Investigation of deterministic and random cyclic patterns in a conventional diesel engine using symbol sequence analysis. In: Advanced computing and communication technologies, Springer, Singapore, pp 549–556
14. Sen AK, Litak G, Taccani R, Radu R (2008) Wavelet analysis of cycle-to-cycle pressure variations in an internal combustion engine. Chaos, Solitons Fractals 38(3):886–893
15. Sen AK, Longwic R, Litak G, Górski K (2008) Analysis of cycle-to-cycle pressure oscillations in a diesel engine. Mech Syst Signal Process 22(2):362–373
16. Sen AK, Litak G, Finney CE, Daw CS, Wagner RM (2010) Analysis of heat release dynamics in an internal combustion engine using multifractals and wavelets. Appl Energy 87(5):1736–1743
17. Sen AK, Litak G, Edwards KD, Finney CE, Daw CS, Wagner RM (2011) Characteristics of cyclic heat release variability in the transition from spark ignition to HCCI in a gasoline engine. Appl Energy 88(5):1649–1655
18. Sen AK, Zheng J, Huang Z (2011) Dynamics of cycle-to-cycle variations in a natural gas direct-injection spark-ignition engine. Appl Energy 88(7):2324–2334
19. Wu JD, Chen JC (2006) Continuous wavelet transform technique for fault signal diagnosis of internal combustion engines. NDT and E Int 39(4):304–311

20. Ali OM, Mamat R, Masjuki HH, Abdullah AA (2016) Analysis of blended fuel properties and cycle-to-cycle variation in a diesel engine with a diethyl ether additive. Energy Convers Manag 108:511–519
21. Maurya RK, Saxena MR, Akhil N (2016) Experimental investigation of cyclic variation in a diesel engine using wavelets. In: Intelligent systems technologies and applications, Springer International Publishing, pp 247–257
22. Kumar S, Cho JH, Park J, Moon I (2013) Advances in diesel–alcohol blends and their effects on the performance and emissions of diesel engines. Renew Sustain Energy Rev 22:46–72
23. Torrence C, Compo GP (1998) A practical guide to wavelet analysis. Bull Am Meteor Soc 79 (1):61–78
24. Maurya RK, Nekkanti A (2016) Combustion instability analysis using wavelets in conventional diesel engine. In: Mathematical concepts and applications in mechanical engineering and mechatronics, pp 390–413
25. Wu, S., & Liu, Q. (2005). Some problems on the global wavelet spectrum. J Ocean Univ Chin JOUC 4(4):398
26. Maurya RK, Saxena MR (2016, September) Investigation of effect of butanol addition on cyclic variability in a diesel engine using wavelets. In: The international symposium on intelligent systems technologies and applications, Springer International Publishing, pp 965–976
27. Saxena MR, Maurya RK (2016) Effect of butanol blends on nano particle emissions from a stationary conventional diesel engine. Aerosol Air Qual Res 16(9):2255–2266
28. Rakopoulos CD, Dimaratos AM, Giakoumis EG, Rakopoulos DC (2011) Study of turbocharged diesel engine operation, pollutant emissions and combustion noise radiation during starting with bio-diesel or n-butanol diesel fuel blends. Appl Energy 88(11):3905–3916
29. Mofijur M, Rasul MG, Hassan NMS (2017, May) Effect of butanol additive on the performance and emission of Australian macadamia biodiesel fuel in a diesel engine. In: Sustainable and renewable energy engineering (ICSREE), 2017 2nd international conference, IEEE, pp 33–37
30. Nayyar A, Sharma D, Soni SL, Mathur A (2017) Characterization of n-butanol diesel blends on a small size variable compression ratio diesel engine: modeling and experimental investigation. Energy Convers Manag 150:242–258

Part III
Exhaust After-Treatment and its Heat Recovery

Recent Advancements in After-Treatment Technology for Internal Combustion Engines—An Overview

Gaurav Tripathi, Atul Dhar and Amsini Sadiki

Abstract The increasing health problems due to engine exhaust and tightening of emission norms for engine exhaust force us to use exhaust after-treatment techniques. Carbon monoxide, carbon dioxide, unburnt hydrocarbon, particulate matter, and oxides of nitrogen are main automobile engine exhaust emissions. Most commonly, diesel oxidation catalysis effectively reduces unburnt hydrocarbon emission, diesel particulate filter reduces particulate matter emission, and selective catalytic reduction and NO_x trap technology reduce NO_x emissions. Recent advances include reduction with and without filter, reduction with catalyst and without catalyst, and some other after-treatment techniques such as plasma-assisted techniques, NO_x and soot combined reduction. This chapter provides overview of recent advancement in various after-treatment techniques and challenges of these technologies.

Keywords Exhaust after-treatment technology · Selective catalytic reduction Emissions · DPF · DOC

1 Introduction

Adversely changing environment is alarming us to work in field of engine exhaust emissions reduction. Engine exhaust can be reduced by modifying engine design, enhancing combustion, improving injection strategies, and using after-treatment technologies. The main engine exhaust emissions are nitrogen oxides, carbon monoxide, unburned hydrocarbons, and particulate matter along with carbon

G. Tripathi · A. Dhar (✉)
School of Engineering, IIT Mandi, Mandi, India
e-mail: add@iitmandi.ac.in

A. Sadiki
Technische Universität Darmstadt, Institute of Energy and Powerplant Technology, Darmstadt, Germany

D. K. Srivastava et al. (eds.), *Advances in Internal Combustion Engine Research*, Energy, Environment, and Sustainability,
https://doi.org/10.1007/978-981-10-7575-9_8

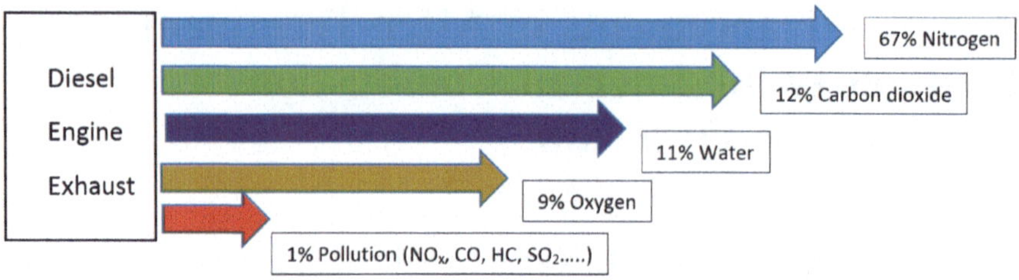

Fig. 1 Engine exhaust contents [1]

dioxide. Different after-treatment technologies and combinations are used for reducing the particular types of emission.

In Fig. 1, it can be seen that regulated pollutants from diesel engine are less than 1%. Nitrogen oxides and particulate matters are main pollutants in engine exhaust for CI engines. To deal with emissions of CO and unburned HC, an after-treatment device diesel oxidation catalyst is used to chemically convert carbon monoxide into carbon dioxide and hydrocarbon into water by oxidation process. Reduction of unburned hydrocarbons and carbon monoxides through diesel oxidation catalyst is comparatively simple process than control of NO_x and particulate matters. So, more focus is put on after-treatment of nitrogen oxides and particulate matter emission. In this chapter, attempt has been made to summarize the recent developments in various after-treatment technologies.

2 Control of Nitrogen Oxides (NO_x)

Nitrogen present in air reacts with oxygen in combustion chamber at temperature greater than 1600 °C [2] to form NO_x. Zeldovich mechanism is widely accepted to explain the formation of NO_x from atmospheric air [3]. Zeldovich mechanism is represented by the following set of equation

$$O + N_2 \leftrightarrow NO + N \tag{1}$$

$$N + O_2 \leftrightarrow NO + O \tag{2}$$

$$N + OH \leftrightarrow NO + H \tag{3}$$

Rate constants for reactions (1)–(3) are $k1$, $k2$, and $k3$ are described in Table 1. Most of them are either constant or function of temperature. The symbols f and r are the rate constants for forward and reverse reaction and their unit is cm^3/mol-s. 1600–1800 °C is important temperature range for Zeldovich mechanism for NO formation. While in reaction (3), NO formation is independent of temp range [4].

Rate of formation of NO is given by the following reaction (4),

Table 1 Reaction rate constants

S. No.	Rate constant	Description
1	Forward rate constant for reaction (1)	$k1,f = 7.6 \times 10^{13} \exp(-38000/T)$
2	Reverse rate constant for reaction (1)	$k1,r = 1.6 \times 10^{13}$
3	Forward rate constant for reaction (2)	$k2,f = 6.4 \times 10^{9} T \exp(-3150/T)$
4	Reverse rate constant for reaction (2)	$k2,r = 1.5 \times 10^{9} T \exp(-19500/T)$
5	Forward rate constant for reaction (3)	$k3,f = 4.1 \times 10^{13}$
6	Reverse rate constant for reaction (3)	$k3,r = 2.0 \times 10^{14} \exp(-23650/T)$

$$\frac{d[NO]}{dt} = k1,f[O][N_2] - k1,r[NO][N] + k2,f[N][O_2] - k2,r[NO][O] + k3,f[N][OH] - k3,r[NO][H] \tag{4}$$

$$NO + HO_2 \leftrightarrow NO_2 + OH \tag{5}$$

$$NO_2 + O \leftrightarrow NO + O_2 \tag{6}$$

Here, reaction (5) shows conversion of NO into NO_2. This reaction occurs in the cooler region of flame where temperature is below 1000 K. At temperatures higher than 1200 K, this NO_2 further breaks into NO according to reaction (6). Here, air nitrogen is main source of NO formation than fuel Nitrogen.

Formation of NO_x depends on combustion temperature, residence time, and oxygen content. On 100 °C increase in combustion temperature, NO_x formation becomes three times [5]. Major part of NO_x is formed in diffusion flame, and small part of NO_x is formed in after combustion region [6].

2.1 De-NO_x After-Treatment Approaches

In this approach, NO_x is decomposed into nitrogen in the presence of some catalyst with the help of some reductant. Selective catalytic reduction and non-thermal plasma technology are important examples of de-NO_x after-treatment technologies.

2.1.1 Non-thermal Plasma (NTP)

In NTP after-treatment approach, with the help of electricity, exhaust gas is ionized. That ionized exhaust gas is known as plasma. Highly energetic electron from plasma converts the oxygen into oxygen radical. Those radicals react with NO and convert into NO_2. Then NO_2 converted into N_2 with the help of catalyst as shown in reactions (7) and (8) [7, 8].

$$\text{Plasma} + \text{NO} + \text{HC} + \text{O}_2 \rightarrow \text{NO}_2 + \text{HC} - \text{product} \tag{7}$$

$$\text{Catalyst} + \text{NO}_2 + \text{HC} \rightarrow \text{N}_2 + \text{CO}_2 + \text{H}_2\text{O} \tag{8}$$

In this way, NO reduction of 70% is achieved in large temperature range [9]. NTP after-treatment approach is preferred over de-NO_x after-treatment approach because it does not require any external catalyst or resource such as ammonia and it does not produce secondary waste product.

Plasma in NTP process can be generated with the help of electron beam or by electric discharge. In Fig. 2, plasma generated with the help of electron beam is shown. When a high-voltage source is applied to cathode, the electrons are emitted. These electrons are accelerated by anode and vacuum. These electron beams fall on engine exhaust gas and ionize the gas, and this ionized gas produces number of ionized electrons. This huge ionized electron cloud is known as plasma. This plasma used for conversion of NO into NO_2 and that NO_2 converted into nitrogen gas with the help of catalyst. NO_x removal achieved is 80%. By-products are sent to electrostatic precipitator (ESP) that is helpful in making fertilizer.

Figure 3 shows the NTP treatment by electric discharge method. In electric discharge method, one electrode is connected with positive terminal of power source and another electrode is connected with negative terminal of power source. When both electrodes connect with power source, then electron starts flowing from positive terminal to negative terminal and ionizes the exhaust gas in between. These electrons produce more and more ions, and in this way, plasma is created, which is used for NO_x reduction present in exhaust gas. There are different electric discharge methods for plasma generation such as corona electric discharge, dielectric barrier discharge, microwave electric discharge, and plasma jet electric discharge. In corona electric discharge method, one electrode shape is sharp or pinpointed and

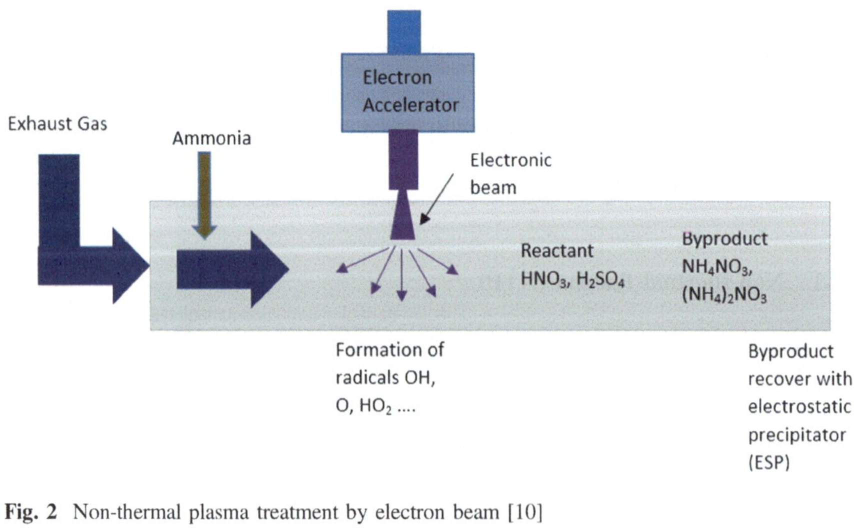

Fig. 2 Non-thermal plasma treatment by electron beam [10]

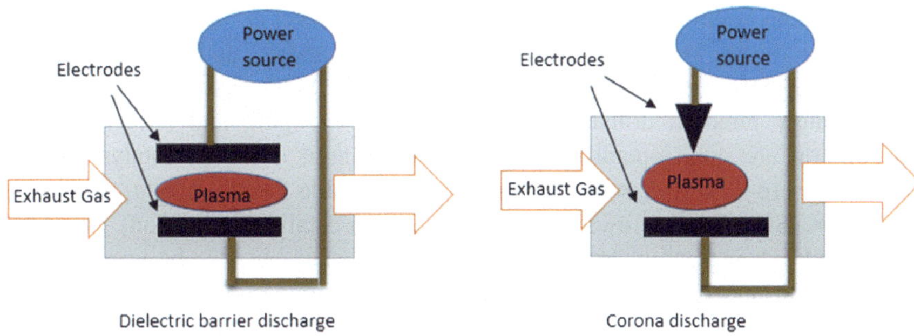

Fig. 3 Plasma generated by electric discharge

other in form of plate. But in dielectric barrier discharge method, both electrodes are in plate form. In plasma jet electric discharge method, plasma is generated in the form of jet so it is known as plasma jet electric discharge method. In microwave electric discharge method, plasma is generated with the help of microwaves [11]. The chemical reaction which takes place is given as (9).

$$2NO_x + 2N \rightarrow 2N_2 + xO_2 \tag{9}$$

This reaction occurs at room temperature. In which, NO_x decomposed into nitrogen and oxygen with the help of nitrogen radical.

NTP is in less use in road transportation because it has high energy requirement and not economical for commercial purpose. NTP is preferred when NO_x amount is less (up to 500 ppm). In 14 kW engine, 70% NO_x removal was achieved when engine exhaust gas flow rate was 1200 l per min [12].

2.1.2 Selective Catalytic Reduction (SCR) de-NO_x After-Treatment Approach

In this process, some catalyst is used to reduce the activation energy required for NO_x reduction. SCR system is preferred over other de-NO_x after-treatment approach because it has high de-NO_x efficiency

In SCR system, required reductant is injected into engine exhaust line to reduce NO_x, and it is selectively oxidized by oxides of nitrogen over the surface of catalyst without reacting with oxygen and reduces NO_x into nitrogen molecule. For lower temperature range (<300 °C), ammonia as a reductant is preferred with noble metal catalyst. For medium temperature range (<425 °C), ammonia is preferred with vanadia/titania catalyst. For higher temperature range, N-containing reductant is preferred with zeolite catalyst [13]. SCR system is generally used after diesel oxidation catalyst in order to protect injected reductant from oxidation before catalytic bed [14]. Performance of SCR system depends on fuel quality such as sulfur content, nature of use catalyst, temperature of exhaust gas [15].

SCR system commissioning cost depends on NO_x level at input, NO_x reduction required at outlet, type of fuel, reactor arrangement, and other controls. Cichanowicz [16] reported the operating cost of SCR system for boiler is approximately hundred dollars per kW. Catalyst changing and reducing agent consumption are main operating cost of SCR system, in which catalyst cost is dominating [16]. SCR system is generally applicable for high NO_x emission. Sometimes, SCR is also used with selective non-catalytic reduction, low NO_x burner, and exhaust gas recirculation [17]. SCR system can also be used for low NO_x emission and operate on both small and large temperature differences [18]. Based on used reducing agent, SCR systems can be classified into following categories.

Ammonia SCR After-Treatment

Here, ammonia is used as reductant for decomposition of NO_x. The following are the main chemical reactions in ammonia SCR system.

$$4NH_3 + 4NO + O_2 \rightarrow 4N_2 + 6H_2O \tag{10}$$

$$NO + NO_2 + 2NH_3 \rightarrow 2N_2 + 3H_2O \tag{11}$$

$$6NO_2 + 8NH_3 \rightarrow 7N_2 + 12H_2O \tag{12}$$

Reactions (10)–(12) are reactions of ammonia and nitrogen oxides in the presence and absence of oxygen that result in decomposition of nitrogen oxide [19].

Figure 4 shows the arrangement of ammonia SCR after-treatment system on engine exhaust side. 70% reduction in NO_x emissions is achieved in diesel–ethanol blend-fueled DI engine when ammonia SCR after-treatment device is used with 0.75 l/h aqueous urea supply at 5 bar and uses titanium oxide-coated catalyst [20].

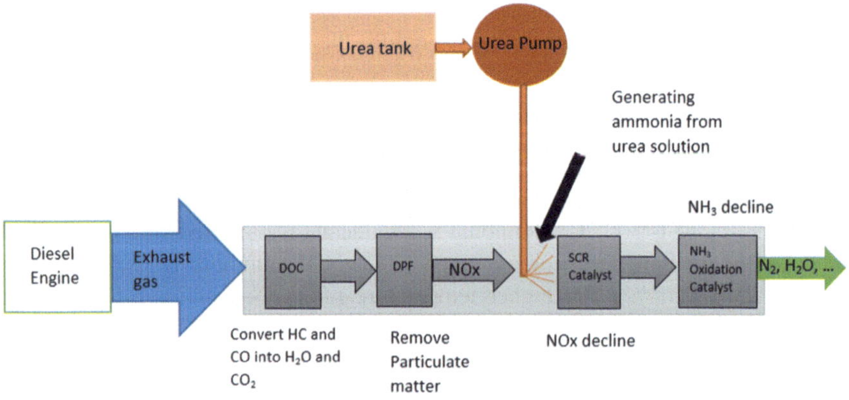

Fig. 4 Arrangement of NH_3-SCR system on engine exhaust

There is 82–84% conversion in NO_x after applying 20 l, 300 cpsi SCR system on heavy-duty diesel engine [21]. Ammonia storage, ammonia transport, ammonia slip, and poisonous by-products are some problems related to ammonia SCR after-treatment devices.

SCR system can be made more effective by combining it with low NO_x trap (LNT), ammonia slip catalyst (ASC) or ammonia oxidation catalyst (AMOC), or by coating SCR catalyst on DPF. Formation of NH_3 and N_2O as by-products is a major problem of NH_3-SCR system. NH_3 by-product can be reduced by using ammonia slip catalyst (ASC) and ammonia oxidation catalyst (AMOC). In which, NH_3 oxidizes to N_2 and H_2O. Three main reasons for N_2O formation are reduction of NO_x by hydrocarbon emission in DOC, oxidation of ammonia in ammonia slip catalyst (ASC), and excess NO_2 and ammonium nitrate in SCR. At low temperature (<200 °C), ammonium nitrate breaks into N_2O and water as described in reaction (13). At high temperature (>450 °C), oxidation of ammonia results in N_2O formation as described in reaction (14). In the presence of excess NO_2 in SCR, ammonium nitrate is formed that further decomposes into N_2O as described in reaction (15).

$$NH_4NO_3 \rightarrow N_2O + 2H_2O \tag{13}$$

$$2NH_3 + 2O_2 \rightarrow N_2O + 3H_2O \tag{14}$$

$$2NH_3 + 2NO_2 \rightarrow NH_4NO_3 + N_2 + H_2O \tag{15}$$

N_2O is a stable species at lower temperatures, and for decomposing N_2O, temperature higher than 450 °C is needed. N_2O formation can decrease in DOC by effective designing. More conversion of NH_3 and high nitrogen selectivity can reduce N_2O formation in ASC. By using N_2O reducing catalyst like FeZ, N_2O formation can be reduced in SCR system [22].

Hydrocarbon SCR After-Treatment

In this SCR method, hydrocarbon is used as reductant for decomposition of NO_x. The main chemical reaction in hydrocarbon SCR system is (16).

$$4NO + C_xH_y + (4x + y - 4)/4O_2 \rightarrow 2N_2 + xCO_2 + y/2H_2O \tag{16}$$

Reaction (16) shows decomposition of nitrogen oxide into nitrogen [23].

There is 95% conversion of NO at 200 °C and 35% conversion at 400 °C when heptane and propene are used as reductants and Pt/Al_2O_3 as catalyst [24]. Use of hydrocarbon as a reductant again has problem of fuel penalty.

Carbon Monoxide SCR After-Treatment

Here, CO is used as reductant in SCR after-treatment device. The chemical reaction in CO-SCR after-treatment device is as follows:

$$2NO + 2CO \rightarrow N_2 + 2CO_2 \tag{17}$$

Reaction (17) is decomposition of NO into nitrogen with the help of CO [23].

There is 98% conversion of NO when CO reductant is combined with NH_3 reductant in SCR after-treatment. The arrangement of NH_3 and CO combined reductant SCR system can be easily understood with the help of following Fig. 5.

Figure 5 shows that NH_3-SCR system is put above with Cu/Beta catalytic bed. Here, operating temperature range is large, i.e., 160–600 °C. NO reduction is 88%. Below 350 °C, there is no oxidation of CO over Cu/Beta bed. CO-SCR system is

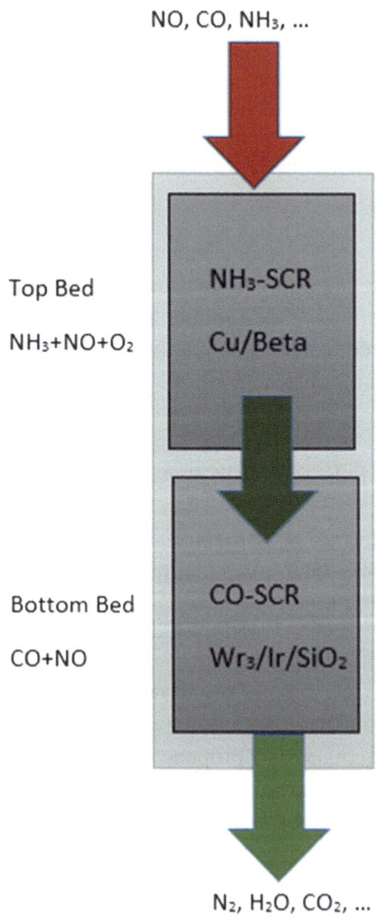

Fig. 5 Arrangement of combined NH_3 and CO-SCR system

put below NH_3-SCR system. Here, $Wr_3/Ir/SIO_2$ is catalytic bed and operating temperature range is low (260–400 °C). After using NH_3-SCR system with CO-SCR system, there is more than 90% conversion of NO [25].

After combining both SCR systems, there is increase in NO conversion efficiency but that is small while the system becomes bulky. Also, the CO-SCR system has low temperature operating range.

Hydrogen SCR After-Treatment System

Hydrogen is an effective reductant in SCR after-treatment device. The following kinetic model is used for representing NO_x reduction by H_2 on noble metal-supported alumina.

$$NO + S \rightarrow NO_{(s)} \tag{18}$$

$$H_2 + 2S \rightarrow 2H_{(s)} \tag{19}$$

$$NO_{(s)} + H_{(s)} \rightarrow N_{(s)} + OH_{(s)} \tag{20}$$

$$NO_{(s)} + S \rightarrow N_{(s)} + O_{(s)} \tag{21}$$

$$N_{(s)} + N_{(s)} \rightarrow N_2 + 2S \tag{22}$$

$$NO_{(s)} + N_{(s)} \rightarrow N_2 + O_{(s)} + S \tag{23}$$

$$NO_{(s)} + N_{(s)} \rightarrow N_2O + 2S \tag{24}$$

$$NO_{(s)} + NO_{(s)} \rightarrow N_2O + O_{(s)} + S \tag{25}$$

$$O_{(s)} + H_{(s)} \rightarrow OH_{(s)} + S \tag{26}$$

$$OH_{(s)} + H_{(s)} \rightarrow H_2O + 2S \tag{27}$$

In reaction (18), S represents vacant catalytic site. When NO molecule comes in contact with S, it forms adsorbed NO. In reaction (19), hydrogen is adsorbed on catalytic surface and forms adsorbed hydrogen. In reaction (20), hydrogen reduces adsorbed NO and forms N atom. N atom can also be formed by excess adsorption of NO, given by reaction (21). Reactions (22) and (23) describe the formation of stable nitrogen molecule with the help of two adsorbed N atom or with the help of adsorbed N atom and NO. If temperature reaches below 150 °C, then N_2O species will form either by combination of N atom and adsorbed NO (reaction 24) or by combination of two adsorbed NO (reaction 25). Reaction (26) shows that adsorbed oxygen and hydrogen atom combines and again vacate the catalytic site. Reaction

(27) shows the formation of stable species that is water. Calculation of rate of important species is given by the following equation.

$$R_{NO} = \frac{2k_{20}\lambda_{NO}p_{NO}\sqrt{\lambda_{H}p_{H_2}}}{\left(1 + \lambda_{NO}p_{NO} + \sqrt{\lambda_{H}p_{H_2}}\right)^2} \quad (28)$$

$$R_{H_2} = \frac{2k_{23} + k_{24}}{k_{23} + k_{24}} k_{20}\theta_{NO}\theta_{H} \quad (29)$$

$$R_{N_2} = \frac{k_{23}}{k_{23} + k_{24}} k_{20}\theta_{NO}\theta_{H} \quad (30)$$

$$R_{N_2O} = \frac{k_{24}}{k_{23} + k_{24}} k_{20}\theta_{NO}\theta_{H} \quad (31)$$

$$R_{H_2O} = \frac{2k_{23} + k_{24}}{k_{23} + k_{24}} k_{20}\theta_{NO}\theta_{H} \quad (32)$$

Rate of formation of NO, H_2, N_2, N_2O, and H_2O are given by Eqs. (28)–(32), where λ is exponential decay constant, p is partial pressure of corresponding gas, θ is concentration of corresponding species, and k is kinetic constant of corresponding reaction. These constants are calculated by using the following equations:

$$k = A\exp\left(\frac{-E}{RT}\right) \quad (33)$$

$$\lambda = f\exp\left(\frac{-\Delta_{abs}H}{RT}\right) \quad (34)$$

Equations (33) and (34) are used to calculate kinetic constant of reaction and exponential decay constant, where A is frequency factor, E is activation energy constant, $\Delta_{abs}H$ is enthalpy of reaction, f is initial constant, T is temperature in Kelvin, and R is universal gas constant (=8.314 J/K-mol). For Pd (1% by weight)/Al_2O_3 catalyst, values of these constants are A is 2.69×10^{13} mol/m^3/s, E is 82×10^3 J/mol, f_{NO} is 7.11×10^{-10} Pa^{-1}, f_{H} is 3.45×10^{-9}, $\Delta_{abs}H^{NO}$ is -42.6×10^3 J/mol, $\Delta_{abs}H^{H}$ is -16.6×10^3 J/mol [26, 27].

There is more than 95% conversion of NO when hydrogen is used as reductant in SCR after-treatment device over Ni/Al_2O_3 catalyst at 220 °C [28].

Hydrogen efficiently reduces NO_x in the presence of various catalysts. Table 2 shows conversion of NO_x by hydrogen with different operating conditions and parameters.

In Table 2, GHSV is gas hourly space velocity. On investigating different H-SCR system data, it is known that hydrogen supplied is four times the NO present for complete reduction of NO. Noble metal-supported alumina and

Table 2 Investigation of different H-SCR systems

NO amount	H_2 amount	Other species	Catalyst	Maximum NO conversion	Temperature (°C)	Refs.
500-ppm	2000-ppm	6% O_2	Pt/Al_2O_3	78%	60 (catalyst)	[29]
500-ppm	2000-ppm	6% O_2	Pt/SiO_2	52%	146 (catalyst)	[29]
0.08%	0.28%	10% O_2 and He	Pt/TiO_2–ZrO2 (reduced catalyst)	90%	90	[30]
0.08%	0.28%	10% O_2 and He	Pt/TiO_2–ZrO_2 (oxidized catalyst)	50%	175	[30]
1000-ppm	8000-ppm	2% O_2 and GHSV = 90,000 h^{-1}	Pd/MFI	82%	200	[31]
0.25%	1%	5% O_2 and GHSV 80,000 h^{-1}	Pt/MgO–CeO_2	N_2 formation rate is 0.43 mmole per gram of Pt per sec	150	[32]
1000-ppm	1%	2% O_2	Pt/Al_2O_3	98%	100	[33]
1000-ppm	1%	1–10% O_2 and flow rate 100 cm^3/min	Pt/Al_2O_3	100%	100	[34]
1000-ppm	1%	1–10% O_2 and flow rate 100 cm^3/min	Pt/ZrO_2	100%	125	[34]
1000-ppm	1%	1–10% O_2 and flow rate 100 cm^3/min	Pt/TiO_2	100%	100	[34]
150-ppm	0.8%	2% O_2 and GHSV = 33,000 h^{-1}	Pt/MgO–CeO_2	100%	180	[35]
150-ppm	0.6%	2% O_2 and GHSV = 33,000 h^{-1}	Pt/MgO–CeO_2	97%	165	[35]
H_2/NO = 10/1	H_2/NO = 10/1	–	$NiFe_{1.95}Pd_{0.05}O_4$	100%	230	[36]

zirconium oxide catalyst show better potential. In optimum case, maximum NO_x conversion occurs in 100–200 °C temperature range.

Alcohol-SCR After-Treatment System

Alcohols can be used as reductant in SCR after-treatment system but it has poor reducing capacity [37]. The reaction mechanism of NO_x reduction with the help of Ag/Al_2O_3 can be easily understood by Figs. 6 and 7. The formation of enolic species and acetate from alcohol and formation of nitrates by NO and O_2 start the reduction mechanism. There is also the formation of intermediate species NCO either by adsorbed species directly or by R–ONO and R–NO_2 that are organo-nitrogen compound. NCO species yield to N_2 by reacting with nitrates and NO and O_2. There is more than 80% conversion of NO_x by ethanol over Ag/Al_2O_3 in temperature range of 300–400 °C [38].

SCR system is effective in temperature range of 200–600 °C [2]. Use of SCR system below 200 °C is still a challenge, because below 200 °C, catalyst becomes inactive [39].

2.2 *Use of NO_x Trap After-Treatment Device Approaches*

NO_x trap after-treatment device uses a base metal oxide and a blend of noble metal group for NO_x reduction. Base metal oxide is used to trap NO_2 during lean operation and emits it during rich operation. Noble metal surface helps in oxidation and reduction reaction. On base metal surface, NO_x is reduced to N_2. The physical arrangement of NO_x trap can be seen in Fig. 8. In this case, DOC and NO_x trap have

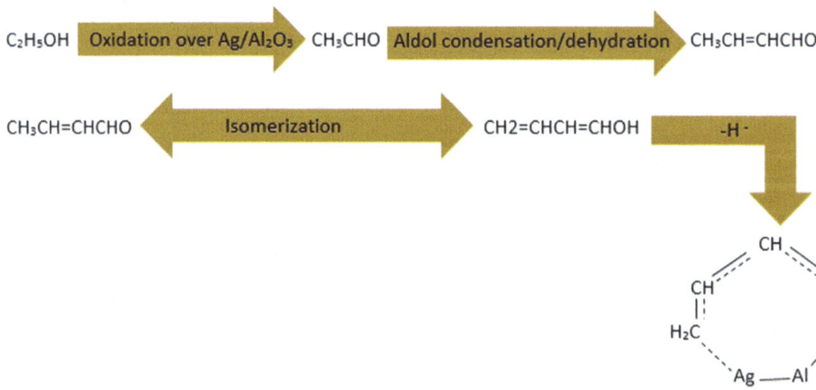

Fig. 6 Primary step in NO_x reduction, i.e., formation of enolic species from alcohol over Ag/Al_2O_3

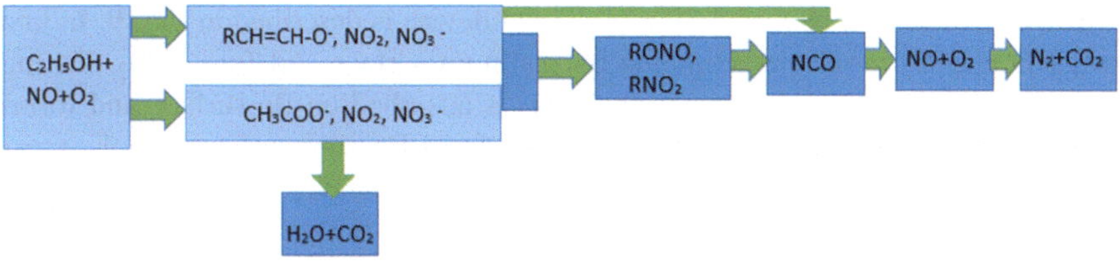

Fig. 7 Selective catalytic reduction of NO_x by ethanol

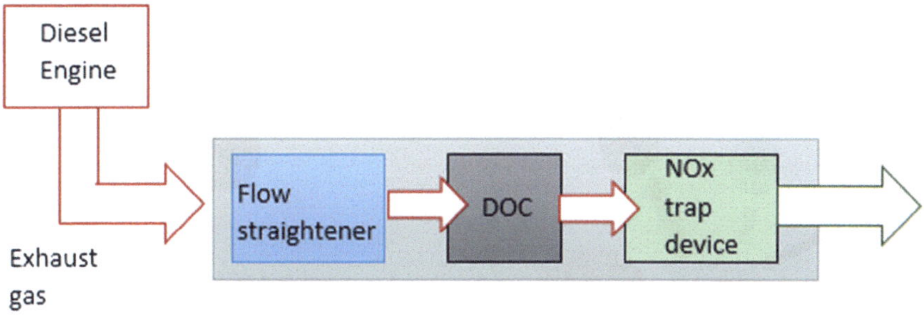

Fig. 8 Arrangement of NO_x trap after-treatment device

same diameter (143.76 mm) and same cell density (400 cpsi) but different length (152.4 mm in NO_x trap and 76.2 mm in DOC) [40].

The reaction mechanisms of NO_x trap are divided into two parts: storage mechanism (reactions 35 and 36) and purge mechanism (reactions 37–41).

$$2NO + O_2 \rightarrow 2NO_2 \quad (35)$$

$$2BaCO_3 + 4NO_2 + O_2 \rightarrow 2Ba(NO_3)_2 + 2CO_2 \quad (36)$$

$$2Ba(NO_3)_2 \rightarrow 2BaO + 4NO_2 + O_2 \quad (37)$$

$$4C_3H_6 + 18NO_2 \rightarrow 12CO_2 + 12H_2O + 9N_2 \quad (38)$$

$$4CO + 2NO_2 \rightarrow 4CO_2 + N_2 \quad (39)$$

$$4H_2 + 2NO_2 \rightarrow 8H_2O + N_2 \quad (40)$$

$$BaO + CO_2 \rightarrow BaCO_3 \quad (41)$$

Reaction (35) shows the oxidation of NO and formation of NO_2 over Pt/Rh surface. Reaction (36) shows the adsorption of NO_2 over BaO surface during lean burning operation. Reaction (37) shows the release of NO_2 during rich burning mixture. Reactions (38)–(40) show the reduction of NO_2 and formation of elementary nitrogen. $BaCO_3$ that is used during reaction (36) is restored in reaction (41) [41].

The mechanism of NO_x trap after-treatment device is described in Fig. 9. In first part, NO and O_2 molecules combine on platinum surface and form NO_2. During lean mode of engine (second part), this NO_2 is adsorbed on Ba surface and forms barium nitrate. In third part, NO_2 is released from Ba surface because of excess adsorption of NO_2 during rich mode of engine. This released NO_2 is reduced into N_2 on platinum surface with the help of engine exhaust (CO, H_2, HC, and so on) as shown in fourth part of figure [42].

NO_x trap has some problems such as (i) it can trap NO_2 up to a limit, after that, NO_x slippage occurs, (ii) NO_x trap is efficient in small temperature range, (iii) sulfur sensitivity of NO_x trap etc.

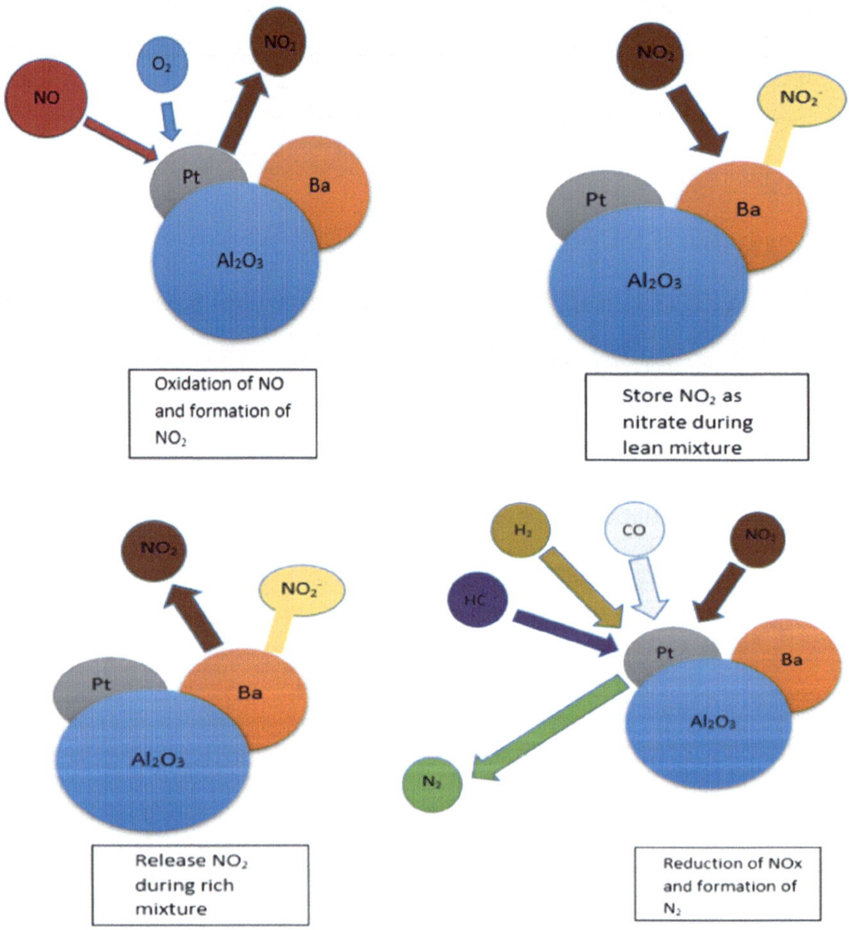

Fig. 9 Mechanism of NO_x trap after-treatment

3 Control of Particulate Matter (PM) Emissions

PM is combination of soluble material, insoluble material, and dry fraction in engine exhaust line. Soot is a part of PM which comes under insoluble material exhaust emissions category. Approximately 50% of PM is soot while remaining 50% PM is unburnt hydrocarbon, partially burnt hydrocarbon, unburnt or partially burnt lubricating oil, engine wear material particle, and sulfate derived from fuel [43, 44].

A heavy-duty diesel engine has 41% carbon as soot, 14% sulfates and water, 13% ash, and remaining are soluble organic compound in which 25% unburnt oil, and 7% unburnt fuel. Along with mass and chemical characterization of particulates, its size number distribution is also important, which plays crucial role in deciding its residence time and consequent toxicity. Based on size distribution, there are mainly three modes of particulate matter. First is nuclei mode in which PM diameter is 5–50 nanometer. This mode contains 1–20% PM mass and 90% PM number. Second is accumulation mode that contains most of the PM mass in which diameter of PM is in the range of 100–1000 nanometer size. Third is coarse mode in which PM diameter is in the range of 1–10-micrometer size and it contains 1–20% particulate matter mass [45]. Generally, soot is generated between 10 to 20° after top dead center (ATDC) [46].

There are six steps in soot formation as shown in Fig. 10. These six steps are pyrolysis, nucleation, coalescence, surface growth, agglomeration, and oxidation. Soot formation occurs in two phases (gas phase and solid phase). In gas phase, formation of small fuel molecules and radicals by the process of pyrolysis or oxidation occurs. The main step in pyrolysis is the formation of aromatic ring that is mostly benzene or phenyl. This aromatic ring results in the formation of polycyclic aromatic hydrocarbons (PAH) at the nucleus. Then, growth of aromatics by hydrogen abstraction and carbon addition (HACA) and by other species takes place. In second solid phase, after enough growth of PAH, PAH is treated as solid particle in place of treating it as molecule. This process is known as soot particle inception. Then soot particle growth occurs with the help of acetylene and PAH. During this, sticking of smaller soot particles is known as coagulation. Soot oxidation occurs parallel to growth. Main oxidants are O, OH, O_2, and CO_2. After certain growth, coagulation is not possible because no more surface growth is possible. Then, 10–100 particles accumulate together. This process is known as soot agglomeration [48].

Following sub-sections describe the after-treatment devices for reduction of particulate matter.

3.1 *Diesel Oxidation Catalyst (DOC) and PM Control*

For the sake of completeness and role of DOC in the regeneration of DPFs, the short description of DOC in the context of PM reduction is needed though their

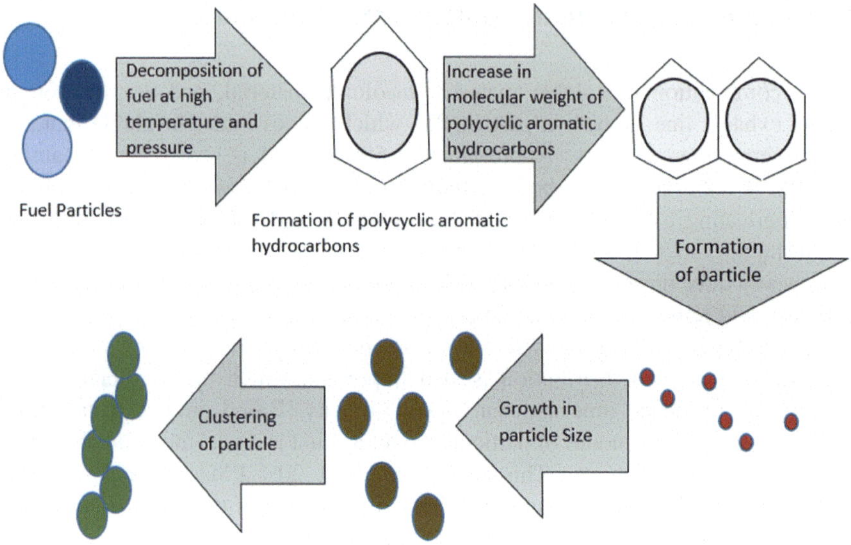

Fig. 10 Steps involved in formation of soot [47]

main role is oxidation of CO and unburned hydrocarbons. DOC is a stainless steel cylindrical-shaped material that contains honeycomb monolithic substrate. On substrate, wash coating is done with catalyst of platinum base metal group [49]. When raw exhaust passes through DOC, carbon monoxide, hydrocarbon, and nitrogen oxides are oxidized. DOC after-treatment also deals with approximately 32% mass of PM, that is, soluble organic fraction; in this way, DOC also helps DPF to filter PM. The following are the main reactions in DOC.

$$CO + \frac{1}{2}O_2 \rightarrow CO_2 \tag{42}$$

$$C_3H_6 + \frac{9}{6}O_2 \rightarrow 3CO_2 + 3H_2O \tag{43}$$

$$NO + \frac{1}{2}O_2 \rightarrow NO_2 \tag{44}$$

Reactions (42) and (43) show the oxidation of CO and HC results in formation of stable species CO_2 and H_2O. Reaction (44) shows the conversion of NO to NO_2. In reactions (42) and (43), with increase in temperature, conversion increases. Conversion of CO_2 and H_2O below 250 °C is less because less reactivity of palladium–rhodium catalyst [50] while reaction (44) occurs between temperature range of 270–430 °C [51]. There is 30% reduction in CO emission and 50% reduction in HC emission by the use of oxidation catalyst on engine exhaust [52]. Selection of catalyst is very important in case of DOC. Use of nanostructured perovskite-based

Nanoxite catalyst makes DOC effective at low temperature with high catalytic efficiency [53]. Use of PGM-free formulation shows 20% higher carbon monoxide reduction and 10% higher HC reduction than Pt catalyst [54]. Future DOC designs focus on making balance between CO and HC conversion along with higher NO to NO_2 conversion [55].

3.2 Diesel Particulate Filter (DPF) After-Treatment Device

DPF is a ceramic monolithic wall filter. When exhaust gas passes through DPF, soot particles stick inside wall of DPF. Inertial impaction, interception, diffusion, and sedimentation are the basic principles in DPF process. The inertial impaction is the process in which particles in engine exhaust are unable to remain in main gas stream; particles change its direction and strike with some stationary obstacle. In interception process, particles remain in main exhaust stream but strike some stationary part because of its dimension of body or part. In sedimentation process, particles in main exhaust stream strike with bottom part of body and settle down because of gravity effect. In diffusion process, particles strike with each other because of thermal motion and get away from main exhaust stream [56]. DPF after-treatment device should have high filtration efficiency; less pressure drop with high soot-storing capacity along with this DPF device should be mechanically, thermally, and chemically stable and compatible with regeneration methods [57]. There is approximately 90% filtration efficiency by the use of honeycomb ceramic wall DPF [58].

Due to continuous use of DPF, there is a chance of clogging of DPF. To avoid clogging, there is a need to oxidize the soot particle. But for oxidation, soot requires higher temperature (approximately 550 °C). The temperature of engine exhaust line is approximately 200 °C. So, there is a need for some external source for heating. Active regeneration and passive regeneration methods are used for cleaning the DPFs.

3.2.1 Regeneration of DPF

In active regeneration of DPFs, the temperature required in DPF for soot oxidation is generated through DOC (Fig. 11). In DOC, heat is generated through complete burning of hydrocarbon and sent into DPF. For high soot emissions as in passenger cars and commercial vehicles, active regeneration after-treatment device is used.

Passive regeneration after-treatment device is also known as continuous regeneration traps (CRT). As shown in Fig. 12, here, combination of DOC and DPF is used. For soot oxidation, NO_2 species is utilized, which is formed in DOC. Since there is no external heat requirement, so it is termed as passive regeneration technique. Since soot is continuously oxidized with the help of NO_2, so it is also known as CRT after-treatment device.

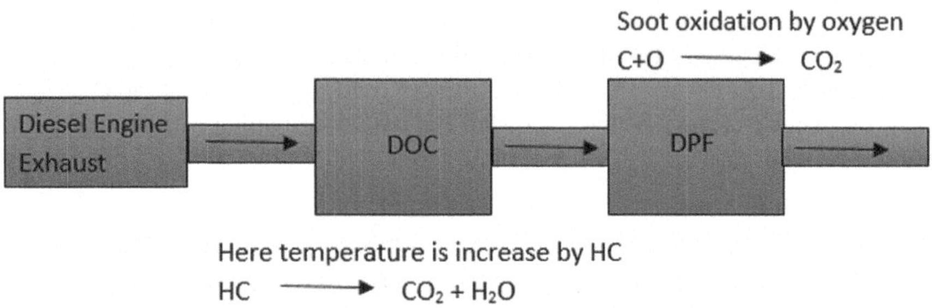

Fig. 11 Arrangement of active regeneration after-treatment devices [59]

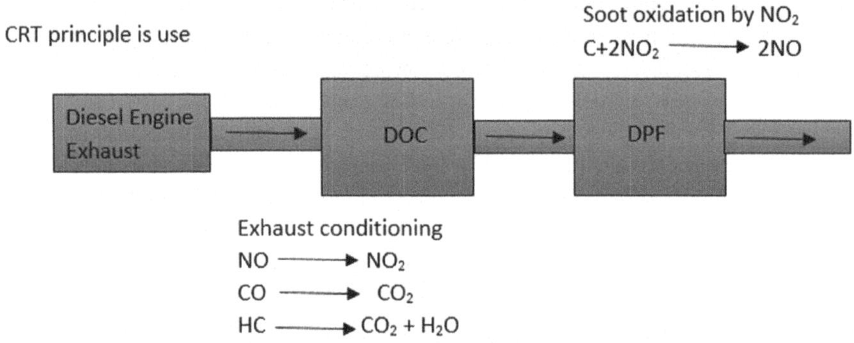

Fig. 12 Arrangement of passive regeneration after-treatment devices [59]

3.3 Summary

SCR, NTP, and NO_x trap are modern after-treatment devices for NO_x reduction. All after-treatment devices have some merit and demerit. NO_x trap is a cost-effective device. It has low light-off temperature and high NO_x conversion activity. On the other hand, it has penalty of fuel consumption, poisonous behavior because of no de-sulfureting process. It has limited temperature range and limited NO_x trap capacity. SCR system has better fuel efficiency, it is used in exhaust line without any modification in combustion unit, and it has higher NO_x reduction capacity. But it has problems such as poor efficiency at lower temperature operations, reductant refilling station, secondary pollutant, and some by-products are poisonous. While dealing with SCR and NTP after-treatment device, it is found that in reducing NO_x emission, NTP is better than SCR because it does not use any reducing agent such as urea, hydrocarbon. NTP after-treatment device is able to reach zero secondary emission. But NTP after-treatment device requires higher energy consumption for production of plasma. NTP after-treatment device requires costlier components and arrangement. SCR after-treatment device is considered as overall better after-treatment device. In future, use of SCR with other available techniques such as selective non-catalytic reduction (SNCR), low NO_x burner (LNB), exhaust gas recirculation (EGR), NO_x traps, and ammonia slip catalyst (ASP) can result in more effective output.

DOC, DPF, and DPF with regeneration are PM-reducing after-treatment devices. Use of DOC effectively reduces CO, HC, PM, and NO. But it has problem with sulfur content in fuel and variation of exhaust gas temperature. Development efforts for DOC are attempting to find some catalyst that increases the conversion of CO, HC, and PM and also ensures conversion of NO into NO_2. DPF is an effective after-treatment device for reducing the PM emission. But it has problem of pressure drop and clogging. DPF with regeneration is effective in reducing PM emission and also dealing with clogging problem. But there is a need to modify and develop the geometry of DPF to make it more effective. For combined reduction of PM and NO_x, there is a need to use combination of DOC, DPF, and SCR devices. Coating of SCR on DPF is a new concept, which can effectively reduce PM as well as NO_x.

References

1. Majewski WA, Khair MK (2006) Diesel emissions and their control, vol 303. SAE Technical Paper
2. Reşitoğlu İA, Altinişik K, Keskin A (2015) The pollutant emissions from diesel-engine vehicles and exhaust aftertreatment systems. Clean Technol Environ Policy 17(1):15–27
3. Heywood JB (1988) Internal combustion engine fundamentals, vol 930. McGraw-Hill, New York
4. Hill SC, Smoot LD (2000) Modeling of nitrogen oxides formation and destruction in combustion systems. Prog Energy Combust Sci 26(4):417–458
5. Lee T, Park J, Kwon S, Lee J, Kim J (2013) Variability in operation-based NO_x emission factors with different test routes, and its effects on the real-driving emissions of light diesel vehicles. Sci Total Environ 461:377–385
6. Flynn PF, Durrett RP, Hunter GL, zur Loye AO, Akinyemi OC, Dec JE, Westbrook CK (1999) Diesel combustion: an integrated view combining laser diagnostics, chemical kinetics, and empirical validation (No. 1999-01-0509). SAE Technical Paper
7. Puchkarev V, Kharlov A, Gundersen M, Roth G (1999). Application of pulsed corona discharge to diesel exhaust remediation. In: 12th IEEE international conference on pulsed power, 1999. Digest of technical papers, vol 1. IEEE, pp 511–514
8. Talebizadeh P, Babaie M, Brown R, Rahimzadeh H, Ristovski Z, Arai M (2014) The role of non-thermal plasma technique in NO_x treatment: a review. Renew Sustain Energy Rev 40:886–901
9. Hoard J (2001) Plasma-catalysis for diesel exhaust treatment: current state of the art (No. 2001-01-0185). SAE Technical Paper
10. Matsumoto T, Wang D, Namihira T, Akiyama H (2012) Non-thermal plasma technic for air pollution control. In: air pollution-a comprehensive perspective. InTech
11. Scholtz V, Pazlarova J, Souskova H, Khun J, Julak J (2015) Nonthermal plasma—a tool for decontamination and disinfection. Biotechnol Adv 33(6):1108–1119
12. Penetrante BM, Vogtlin GE, Bardsley JN, Vitello PA, Wallman PH (1993). Application of non-thermal plasmas to pollution control. In: Proceedings of 2nd international Plasma symposium. World progress in plasma applications, Palo Alto, California, pp 1–11
13. Shelef M (1995) Selective catalytic reduction of NO_x with N-free reductants. Chem Rev 95 (1):209–225

14. Twigg MV (2011) Catalytic control of emissions from cars. Catal Today 163(1):33–41
15. Azzara A, Rutherford D, Wang H (2014) Feasibility of IMO annex VI tier III implementation using selective catalytic reduction. International Council on Clean Transportation
16. Cichanowicz JE (2010) Current capital cost and cost-effectiveness of power plant emissions control technologies. Utility Air Regulatory Group
17. Sorrels JL, Randall DD, Schaffner KS, Fry CR (2015) Selective catalytic reduction
18. USA Env. Protection Agency. Air pollution control technology fact sheet, EPA—452/F-03-032
19. Skalska K, Miller JS, Ledakowicz S (2010) Trends in NO_x abatement: a review. Sci Total Environ 408(19):3976–3989
20. Johnson JH, Parker GG (2013) Experimental studies for DPF and SCR model, control system, and OBD development for engines using diesel and biodiesel fuels. Michigan Technological University
21. Gekas I, Gabrielsson P, Johansen K, Bjørn I, Kjær JH, Reczek W, Cartellieri W (2002) Performance of a urea SCR system combined with a PM and fuel optimized heavy-duty diesel engine able to achieve the Euro V emission limits (No. 2002-01-2885). SAE Technical Paper
22. Guan B, Zhan R, Lin H, Huang Z (2014) Review of state of the art technologies of selective catalytic reduction of NO_x from diesel engine exhaust. Appl Therm Eng 66(1):395–414
23. Cîmpean M (2015) Catalytic reduction of nitrogen oxides from the residual gases of the 15 N separation column. Babes-Bolyai University
24. Burch R, Ottery D (1996) Selective catalytic reduction of NO_x by hydrocarbons on Pt/Al_2O_3 catalysts at low temperatures without the formation of N_2O. Appl Catal B 9(1–4):L19–L24
25. Sultana A, Haneda M, Hamada H (2009) A new concept of combined NH_3–CO-SCR system for efficient NO reduction in excess oxygen. Appl Catal B 88(1):180–184
26. Dhainaut F, Pietrzyk S, Granger P (2007) Kinetic investigation of the NO reduction by H_2 over noble metal based catalysts. Catal Today 119(1):94–99
27. Dhainaut F, Pietrzyk S, Granger P (2007) Kinetics of the NO + H_2 reaction over supported noble metal based catalysts: support effect on their adsorption properties. Appl Catal B 70 (1):100–110
28. Mihet M, Lazar MD (2014) Effect of Pd and Rh promotion on Ni/Al_2O_3 for NO reduction by hydrogen for stationary applications. Chem Eng J 251:310–318
29. Burch R, Coleman MD (1999) An investigation of the $NO/H_2/O_2$ reaction on noble-metal catalysts at low temperatures under lean-burn conditions. Appl Catal B 23(2):115–121
30. Machida M, Ikeda S, Kurogi D, Kijima T (2001) Low temperature catalytic NO_x–H_2 reactions over Pt/TiO_2–ZrO_2 in an excess oxygen. Appl Catal B 35(2):107–116
31. Wen B (2002) NO reduction with H_2 in the presence of excess O_2 over Pd/MFI catalyst. Fuel 81(14):1841–1846
32. Costa CN, Efstathiou AM (2007) Low-temperature H_2-SCR of NO on a novel Pt/MgO–CeO_2 catalyst. Appl Catal B 72(3):240–252
33. Li L, Zhang F, Guan N, Schreier E, Richter M (2008) NO selective reduction by hydrogen on potassium titanate supported palladium catalyst. Catal Commun 9(9):1827–1832
34. Nanba T, Kohno C, Masukawa S, Uchisawa J, Nakayama N, Obuchi A (2003) Improvements in the N_2 selectivity of Pt catalysts in the NO–H_2–O_2 reaction at low temperatures. Appl Catal B 46(2):353–364
35. Olympiou GG, Efstathiou AM (2011) Industrial NO_x control via H_2-SCR on a novel supported-Pt nanocatalyst. Chem Eng J 170(2):424–432
36. Tsujimura M, Furusawa T, Kunii D (1983) Catalytic reduction of nitric oxide by hydrogen over calcined limestone. J Chem Eng Jpn 16(6):524–526
37. Thomas JF, Lewis SA, Bunting BG, Storey JM, Graves RL, Park PW (2005) Hydrocarbon selective catalytic reduction using a silver-alumina catalyst with light alcohols and other reductants (No. 2005-01-1082). SAE Technical Paper
38. He H, Yu Y (2005) Selective catalytic reduction of NO_x over Ag/Al_2O_3 catalyst: from reaction mechanism to diesel engine test. Catal Today 100(1):37–47

39. Fu M, Li C, Lu P, Qu L, Zhang M, Zhou Y, Yu M, Fang Y (2014) A review on selective catalytic reduction of NO_x by supported catalysts at 100–300 °C—catalysts, mechanism, kinetics. Catal Sci Technol 4(1):14–25
40. Alimin AJ, Benjamin SF, Roberts CA (2009) Lean NO_x trap study on a light-duty diesel engine using fast-response emission analysers. Int J Engine Res 10(3):149–164
41. Ye Z, Li L (2003) Control options for exhaust gas aftertreatment and fuel economy of GDI engine systems. In: Proceedings of 42nd IEEE conference on decision and control, 2003, vol 2. IEEE, pp 1783–1788
42. Matsumoto SI (1997) Recent advances in automobile exhaust catalyst. Catal Surv Jpn 1 (1):111–117
43. Ullman TL (1989) Investigation of the effects of fuel composition on heavy-duty diesel engine emissions (No. 892072). SAE Technical Paper
44. Peirce DM, Alozie NSI, Hatherill DW, Ganippa LC (2013) Premixed burn fraction: its relation to the variation in NO_x emissions between petro-and biodiesel. Energy Fuels 27 (7):3838–3852
45. Kittelson D (1997) Engines and nanaoparticles: a review. J Aerosol Sci 29(5):575–588
46. Johnson TV (2001) Diesel emission control in review (No. 2001-01-0184). SAE Technical Paper
47. Srivastava DK, Agarwal AK (2008) Particulate matter emissions from single cylinder diesel engine: effect of engine load on size and number distribution. SAE Int. https://doi.org/10.1017/CBO9781107415324.004
48. Faculties C, Sciences N (2007) Soot formation modeling during hydrocarbon pyrolysis and oxidation behind shock waves soot formation modeling during hydrocarbon pyrolysis and oxidation behind shock waves
49. Nett Technology INC. What is a diesel oxidation catalyst
50. Zheng M, Banerjee S (2009) Diesel oxidation catalyst and particulate filter modeling in active-Flow configurations. Appl Therm Eng 29(14):3021–3035
51. Banerjee S (2007) Thermal analysis of active catalytic diesel particulate filter regeneration, MASc thesis, University of Windsor, Canada, May 2007
52. Hou WJ (2011) A experimental study on SCR system of diesel engine to reduce NO_X emissions. Dalian University of Technology, China
53. Seegopaul P, Bassir M, Alamdari H, Neste AV (2006) Nanostructured perovskitebased oxidation catalysts for improved environmental emission control. NSTINanotech 1. ISBN 0-9767985-6-5. www.nsti.org
54. Ishizaki K, Mitsuda N, Ohya N, Ohno H, Naka T, Abe A, Takagi H, Sugimoto A (2012) A study of PGM-free oxidation catalyst $YMnO_3$ for diesel exhaust after treatment. SAE Technical Paper 2012-01-0365. https://doi.org/10.4271/2012-01-0365
55. Johnson T (2013) Vehicular emissions in review. SAE Int J Eng 6(2013-01-0538):699–715
56. Liang X (2013) Research on treatment of diesel engine combined with DOC and DPF. Scientific Technol Innov Appl 11:24
57. Guan B, Zhan R, Lin H, Huang Z (2015) Review of the state-of-the-art of exhaust particulate filter technology in internal combustion engines. J Environ Manage 154:225–258
58. Liu TT (2012) Analysis on diesel after-treatment system fault diagnosis for CRT aging and failure. Beijing Transportation University, China
59. Schäffner G (2014) Diesel particulate filter: exhaust aftertreatment for the reduction of soot emissions

Calcium Oxide Nanoparticles as An Effective Filtration Aid for Purification of Vehicle Gas Exhaust

B. Bharathiraja, M. Sutha, K. Sowndarya, M. Chandran, D. Yuvaraj and R. Praveen Kumar

Abstract Calcium oxide nanoparticles and its potential towards purification of vehicle gas exhaust were investigated in this work. Calcium oxide nanoparticles were synthesised and its efficiency in absorbing constituents of vehicle gas exhaust has been estimated. Calcium oxide nanoparticles were synthesised by chemical coprecipitation and thermal decomposition of chicken egg shells. The obtained powdered particles were characterised using Fourier transform infrared spectroscopy, SEM and X-ray diffraction techniques. The powdered products on the addition of polyvinyl alcohol (PVA) have been fabricated into a nanoporous membrane using electrospinning technique. The thin film was made with varying weight per cent of CaO particles. The efficiency of these membranes was estimated by passing exhaust gas through it using exhaust gas analyser/smoke metre. The FT-IR analysis confirms the presence of CaO particles. The SEM and XRD of the obtained samples showed the crystalline nature of CaO nanoparticles, and the size of the powdered CaO was found to be ~15–20 nm. The nanoporous electrospun membrane has low filtration capacity. The other two CaO nanoparticles with varying weight percentage showed great potential towards purifying vehicle gas exhaust. It reduces the quantity of HC, CO, CO_2 in high rates. It is also proved that increase in the concentration of CaO increased the efficiency of filtration.

Keywords Calcium oxide nanoparticles · Electrospinning · Nanofilm Nanomembrane · Vehicle gas exhaust

---

B. Bharathiraja (✉) · M. Sutha · K. Sowndarya · M. Chandran · D. Yuvaraj
Vel Tech High Tech Dr. Rangarajan Dr. Sakunthala Engineering College, Chennai 62, India
e-mail: btrbio@gmail.com

R. Praveen Kumar
Arunai Engineering College, Tiruvannamalai, India

D. K. Srivastava et al. (eds.), *Advances in Internal Combustion Engine Research*, Energy, Environment, and Sustainability,
https://doi.org/10.1007/978-981-10-7575-9_9

1 Introduction

Generally, metal oxide nanoparticles are found to possess good structural characteristics, and they are found to be very stable [1]. They are being employed in air and water purification due to their non-toxic nature and high oxidation potential [2]. Ca-based materials are found to be better sorbents [3], and hence, they can be used as Co_2 acceptor aiming towards air purification. Calcium oxide, when synthesized in nanoscale will act as an excellent sorbent because when particles are synthesized in nanoscale, the reactivity of CaO with CO_2, SO_2, NO_X may increase. CaO nanoparticles are reported to be synthesized by many methods which include sol–gel method [4], thermal decomposition method [5], microwave processing [6], sonochemical method [7], solution combustion method [8], coprecipitation method [9]. In our work, we synthesized CaO nanoparticles using coprecipitation method and thermal decomposition of chicken egg shells. In case of chemical coprecipitation method, polymers such as PVP and PVA are used as capping agents for CaO nanoparticles and those polymers will maintain small particle size and prevent agglomeration of CaO nanoparticles [9]. Both the coprecipitation and thermal decomposition methods are easier and cost-effective methods for producing CaO nanoparticles. And by this method, an efficient pollution control aid can be made using waste egg shells can be eliminated [5]. CaO is found to be a cheap antibacterial agent [10], and photocatalyst for dye degradation [11], and nanosize semiconductor [8], and catalyst in transesterification [12], as moisture adsorbents in OLEDS [13] and in wall paint conservation [4]. Moreover, CaO was reported to be used in low-cost desulfurization [14] and in detoxification of pollutant such as 2-CEPS [9]. Since CaO has such potential reactivity and adsorption characteristics, we chose to use CaO for purification of the exhaust. Electrospinning is generally used for fabricating nanoparticles into nanofibrous membranes. Poly vinyl alcohol, gelatine and polycaprolactone were reported to be suitable polymers for CaO for electrospinning. In the present work, we fabricated CaO both by electrospinning and conventional thin film formation technique for the purification of vehicle gas exhaust. The filtration efficiency of the membranes and the films were analysed using gas analyser or smoke metre.

2 Materials and Methods

2.1 Materials

Chicken egg shells, calcium nitrate, sodium hydroxide, polyvinylalcohol, moulds

2.2 Methodology

Calcium oxide nanoparticles were synthesized via both chemical coprecipitation and thermal decomposition of egg shells.

2.2.1 Chemical Coprecipitation

0.5 M of $Ca(NO_3)_2 \cdot 4H_2O$ solution was prepared. To the above solution, 0.7 M of NaOH was added in drops and stirring was done simultaneously until the resulting final solution turned white. The above solution was subjected to microwave heating for 5 min at 160 °C and. Then, the solution was filtered to collect the white precipitate. The collected white precipitate was dried in hot air oven at 80 °C for 1 hour. Calcination at 700 °C for 5 hours ensures the purity of CaO nanoparticles.

2.2.2 Thermal Decomposition of Egg Shells

Waste chicken egg shells are a cost-effective source of obtaining CaO nanoparticles. Waste egg shells were collected, washed and dried for about 48 h. Using pestle and mortar, the egg shells were crushed into fine powder. This fine powder material was calcined at 700 °C for 7 h yielding pure CaO nanoparticles.

2.2.3 Fabrication of CaO

Fabrication of CaO nanoparticles was carried out by two methods (1) Electrospinning technique (2) Conventional method of thin film formation.

Electrospinning

Before electrospinning, the solution for electrospinning must be prepared. Three steps are involved in the preparation of solution required for electrospinning.

(a) Preparation of solution of CaO nanoparticles
(b) Preparation of polymer solution of PVA
(c) Mixing of polymer with nanoparticle solution and fabrication of membrane by electrospinning.

(a) Preparation of solution of CaO nanoparticles

CaO nanoparticles at varying concentrations (say 0, 0.1, 0.3, 0.5 g) was added to 10 ml of distilled water and ultrasonicated for about 4 h to obtain a well-dispersed solution of nanoparticles.

(b) Preparation of polymer solution of PVA

15 wt% of PVA solution was prepared by adding 15 g of PVA in 100 ml distilled water. The above solution was continuously stirred with the help of magnetic stirrer set at a temperature of 60 °C initially. After we get a well-dissolved PVA solution, the temperature was set to zero when the solution foams. Stirring was maintained until a clear solution of PVA develops.

(c) Mixing of PVA solution With CaO nanoparticle solution and fabrication of membrane by electrospinning

The supernatant of the solution containing nanoparticles was drawn out and 2 ml of this supernatant was mixed with double the volume of PVA solution and overnight stirring was maintained to obtain a homogeneous solution. The above procedure was carried out with nanoparticles obtained by both methods. 5 ml of the above homogenously mixed solution was loaded into a 5 ml syringe which was then fixed in the electrospinning apparatus. The apparatus was run with parameters such as 0.5 g/ml flow rate 1500 rpm drum rotation speed 25 kv voltage, and a 15 cm distance between the syringe and collector was maintained. The membrane was obtained only for CaO from egg shell as shown in Fig. 1.

Thin Film Formation (Conventional Method)

8 wt% of PVA polymer solution was prepared as specified above. Different wt% of CaO nanoparticles say 0.5, 1.0, 1.5, 2.0 g were added to PVA solution and mixed with the help of magnetic stirrer. After the particles get well dispersed, the solution was poured into a circular mould, and it was allowed to dry in a hot air oven at 100 °C. The dried thin films of CaO nanoparticles were then peeled from the moulds.

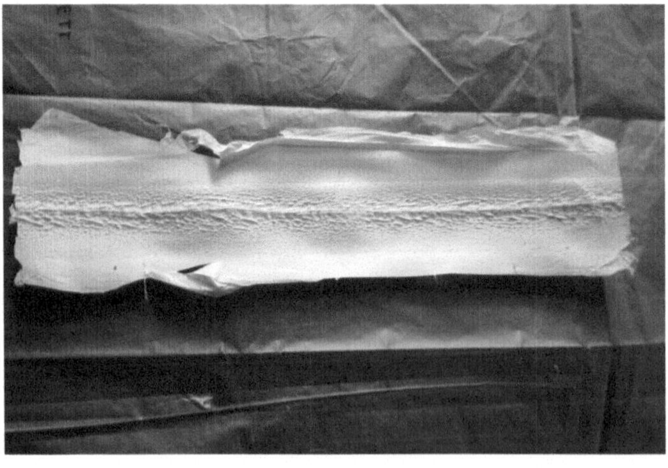

Fig. 1 Electrospun membrane of CaO from egg shells

2.2.4 Characterisation of CaO Nanoparticles

CaO nanoparticles synthesized were confirmed for functional groups by FT-IR analysis. The structure, shape and size of the CaO nanoparticles were analysed by SEM and XRD analysis.

Analysis of Thin Films and Membranes

For evaluating the potential of CaO thin films and membranes in the absorption of harmful gases of vehicle gas exhaust, a setup was made as in Fig. 1. This setup includes a long cylindrical metal tube with an opening at both ends, and it also contains an opening covered with a lid at the centre where the membrane/film of CaO was fixed and closed with the lid. One end of the tube was connected to the silencer of a vehicle and the other end to the heat resistance hose of gas analyser/ smoke metre. Once the setup was arranged, the engine of the vehicle was accelerated to release exhaust gas which as known passes through the membrane/film of CaO. The gas that has passed via the membrane and film of CaO was analysed for its composition by gas analyser, and its values are recorded.

3 Results and Discussion

Calcium oxide nanoparticles were synthesized by coprecipitation of calcium nitrate and thermal decomposition of chicken eggshell. The powdered products obtained from coprecipitation and thermal decomposition were white and greyish white in colour, respectively.

3.1 Confirmatory Studies for CaO

3.1.1 Fourier Transform Infrared Spectroscopy (FT-IR)

The presence of calcium oxide in the obtained powdered products was confirmed by Fourier Transform Infrared Spectroscopy. Figure 2 represents the FT-IR spectra of CaO nanoparticle obtained from coprecipitation method and thermal decomposition method. The powdered product obtained from coprecipitation of calcium nitrate showed only peaks at 875.524 cm^{-1} which represents the presence of Ca–O–Ca bond, and the powdered product from calcined egg shell showed a peak at both 874.524 and 711.604 cm^{-1}. This indicates the presence of both Ca–O–Ca bond and Ca–O. The results obtained is almost similar to the results for CaO from green synthesis using green tea leaves and papaya leaves and thermal decomposition of duck egg shells by Aswini Anatharamam et al. and Tangboriboon et al.,

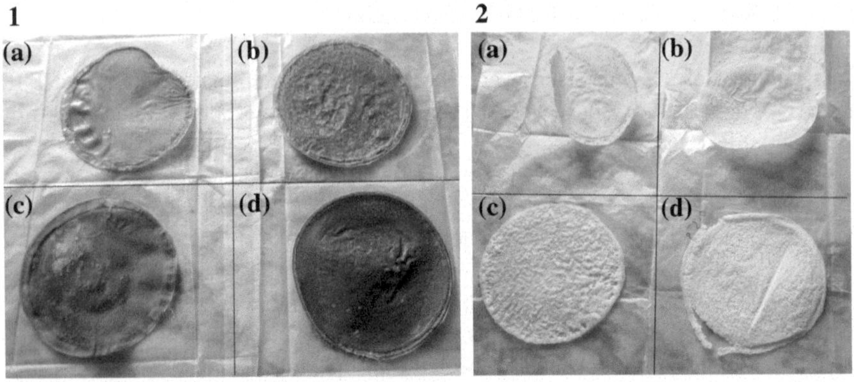

Fig. 2 1: Thin films of CaO from egg shells at varying wt% of **a** 0.5 g **b** 1.0 g **c** 1.5 g **d** 2.0 g. 2: Thin films of coprecipitated CaO at varying wt% of **a** 0.5 g **b** 1.0 g **c** 1.5 g **d** 2.0 g

respectively. This shows that CaO from egg shell will have more efficiency than coprecipitated CaO while filtration of vehicle exhaust. FT-IR analysis of different samples with different calcination temperature showed that calcination with increasing temperature increases the purity of CaO obtained by removing the unwanted OH bond. Thus, calcination ensures the purity of the products obtained.

3.2 Physical Characterisation of CaO Nanoparticles

3.2.1 Powder X-Ray Diffraction Studies

Using X-ray diffraction studies the size of the particles was found to be $\sim$21.33 and $\sim$15 nm using Debye Scherrer equation for the coprecipitated CaO and CaO from chicken egg shells, respectively. This showed that the obtained powdered product is in nanosize and nanocrystalline nature. The XRD images of CaO from coprecipitation and thermal decomposition of eggshells were shown in Fig. 3.

3.3 Morphological Studies Using Scanning Electron Microscope

SEM images of CaO from calcium nitrate and egg shells at different scales of 500 nm, 1, 3, 5 μm were shown in Figs. 4 and 5, respectively. The morphology of the powdered CaO products obtained by coprecipitation and decomposition of eggshells were analysed by scanning electron microscope. These results show that

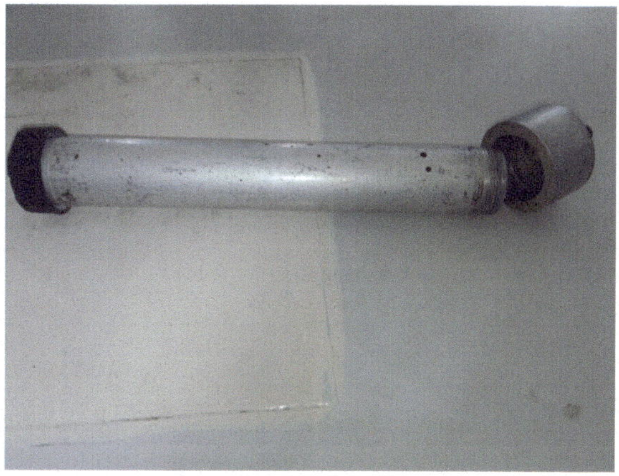

Fig. 3 Experimental setup for exhaust filtration

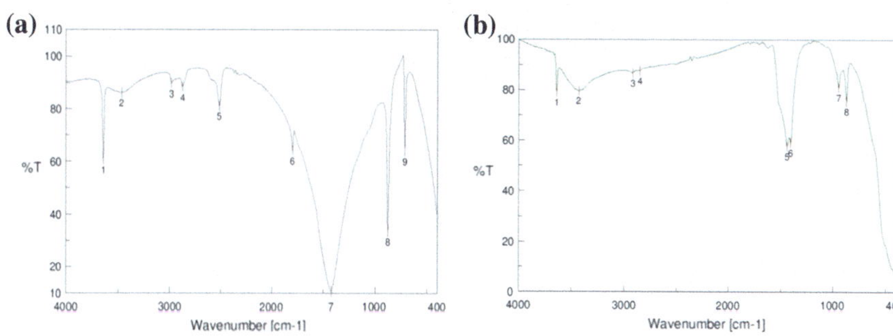

Fig. 4 FT-IR spectra of CaO from **a** coprecipitation of calcium nitrate and **b** thermal decomposition of eggshells

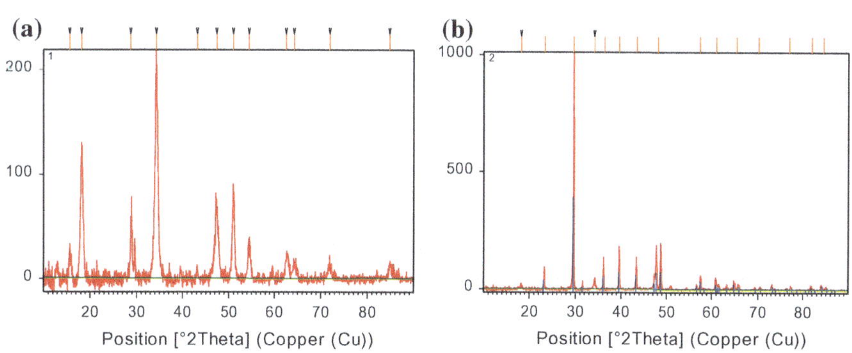

Fig. 5 XRD pattern of CaO from **a** coprecipitation of calcium nitrate and **b** thermal decomposition of eggshells

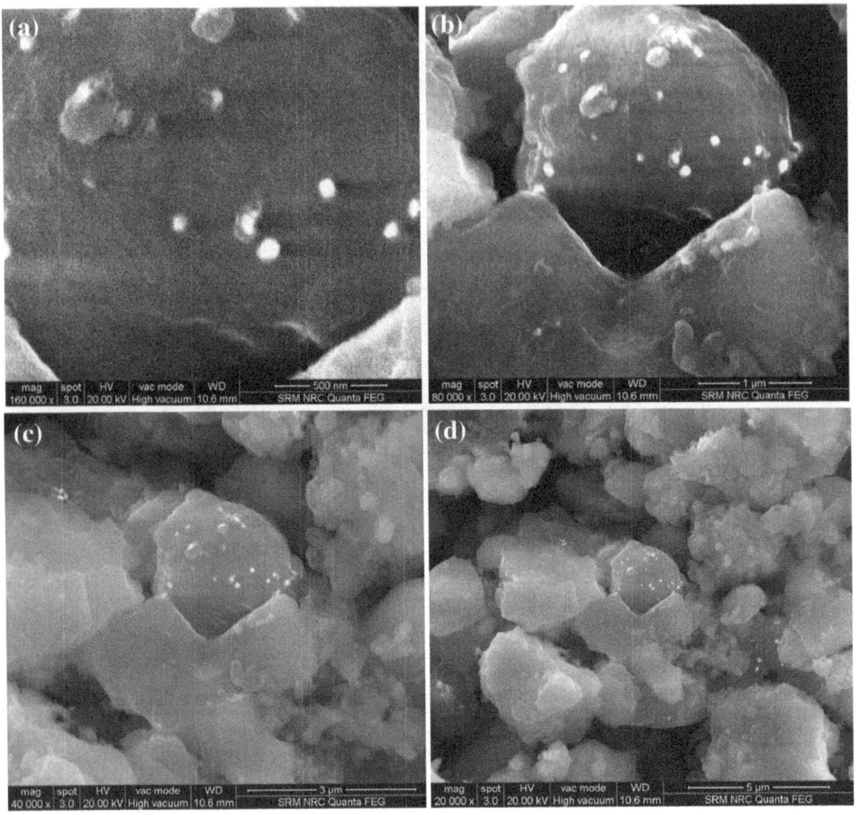

Fig. 6 SEM images of CaO nanoparticles by coprecipitation method at different scale of **a** 500 nm **b** 1 μm **c** 3 μm **d** 5 μm

calcium oxide nanoparticles were obtained with particle size 14–21 nm. Agglomeration showed that the particles are porous in nature (Figs. 6 and 7).

3.4 Filtration Studies Using Gas Analyser

The powdered product obtained was fabricated into membranes and thin films. The nanofibrous membrane was formed using electrospinning technique, and the thin films were formed using a conventional method using PVA. The fabricated membranes and films were used for filtration of vehicle gas exhaust. The CaO particles in membrane and film, when exposed to the vehicle exhaust, absorb carbon dioxide and hydrocarbon and carbon monoxide.

Fig. 7 SEM images of CaO by thermal decomposition of eggshells at different scales of **a** 500 nm **b** 1 μm **c** 3 μm **d** 5 μm

3.4.1 Filtration Studies Using Thin Films of Coprecipitated CaO

The thin films were fabricated with varying weight percentage of powdered CaO. The filtration potential of thin films of coprecipitated CaO and CaO from egg shells were depicted in Tables 1 and 2. In order to check the exact potential of calcium oxide, we need to neglect the absorption value of the added polymer (PVA). The PVA film was also used for filtration, and it showed that it can also reduce few amounts of carbon dioxide and HC and CO content. The composition of gas constituent without the filtration aid was also determined using gas analyser to observe the amount of reduction and filtration of CaO membrane. With the reduction in the quantity of CO, HC, CO_2 by reaction with CaO, there is an increase in the level of oxygen this is due to the breakdown of HC and its reaction with PVA. It is found that the filtration efficiency increases with increasing concentration of calcium oxide nanoparticles. Calcium oxide from egg shells showed better

Table 1 Different filtration potential of coprecipitated CaO

Content	CO (%)	HC (%)	CO_2 (%)	O_2 (%)
Without filtration	3.69	1405	12.40	1.63
PVA blank	2.97	460	12.00	1.03
I (0.5 g)	2.92	340	10.10	4.11
II (1.0 g)	2.68	273	9.30	5.65
III (1.5 g)	1.21	395	10.80	4.39
IV (2.0 g)	0.02	44	0.10	4.11

Table 2 Different filtration potential of CaO from egg shells

Content	CO (%)	HC (%)	CO_2 (%)	O_2 (%)
Without filtration	3.69	1405	12.40	1.63
PVA blank	2.97	460	12.00	1.03
I (0.5 g)	3.69	410	11.70	0.84
II (1.0 g)	0.71	108	2.30	16.26
III (1.5 g)	0.37	113	1.30	17.98
IV (2.0 g)	0.05	48	0.20	19.88

filtration than coprecipitated CaO. The films with 2.0 g can reduce CO_2, HC, CO to the maximum extent. This could be nearer to the optimal value for complete filtration of vehicle gas exhaust.

3.4.2 Filtration Studies With CaO From Egg Shell

See Table 2.

3.4.3 Filtration Studies Using Electrospun Membrane

From Table 3, membranes were formed using only PVA to set as a blank and only the CaO from egg shell formed was fabricated into fibrous membrane but the coprecipitated membrane did not form fibrous membrane but resulted in bead formation at all varying parameters. Since it contains only 0.1 g of CaO it showed very low filtration potential. The filtration of PVA blank of nanofilm and fibrous membrane is almost similar. But the nanoporous membrane does not have as much potential as nanofilms.

Thus, this work showed that CaO can be a better candidate in the purification of vehicle gas exhaust. This could be a cheapest and efficient way for pollution control. Though it has a great potential in purification it has a limitation in its

Table 3 Different filtration potential of electrospun CaO membrane

Content	CO (%)	HC (%)	CO_2 (%)	O_2 (%)
Without filtration	3.69	1405	12.40	1.63
PVA blank	3.08	885	11.40	1.81
Eggshell membrane (0.1 g)	2.31	834	9.40	5.74

durability because of absorption of CO_2 will soon result in a carbonate formation. Further works were going on to enhance the efficiency and durability of the film.

4 Conclusion

Calcium oxide nanoparticles synthesised from coprecipitation and thermal decomposition of egg shells were fabricated into nanofibrous membranes and thin films. Both the coprecipitated CaO and CaO from egg shells were characterised using FT-IR, XRD, SEM. The size of coprecipitated CaO was ~21.33 nm and CaO from eggshells was in ~14.6 nm FT-IR results showed the presence of Ca–O bond at 875.524 and 711.604 cm^{-1}. The nanoporous electrospun membrane has low filtration capacity. The other two CaO nanoparticles with varying weight percentage showed great potential towards purifying vehicle gas exhaust. It reduces the quantity of HC, CO, CO_2 in high rates. It is also proved that increase in the concentration of CaO increased the efficiency of filtration.

References

1. Pimenta MF (2010) Flue gas desulphurization through wet limestone process—adding acids and bases to the limestone suspension
2. Prabhavathi SP, Ranjith SR, Maruthamuthu KR (2014) Sol Gel synthesis of MgO and CaO nanoparticles and their characterization. Wjpr 3(7):362–370
3. Mastin J, Aranda A, Meyer J (2011) New synthesis method for CaO-based synthetic sorbents with enhanced properties for high-temperature CO_2-capture. Energy Procedia 4:1184–1191
4. Darčanova O, Beganskiene A, Kareiva A (2015) Sol-gel synthesis of calcium nanomaterial for paper conservation. Chemija 26(1):25–31
5. Tangboriboon N, Kunanuruksapong R, Sirivat A (2012) Preparation and properties of calcium oxide from eggshells via calcination. Mater Sci-Pol 30(4):313–322
6. Roy A, Gauri SS, Bhattacharya M, Bhattacharya J (2013) Antimicrobial activity of CaO nanoparticles. J Biomed Nanotechnol 9(9):1570–1578
7. Alavi MA, Morsali A (2010) Ultrasonic-assisted synthesis of $Ca(OH)_2$ and CaO nanostructures. J Exp Nanosci 5(2):93–105
8. Bhavya C, Yogendra K (2015) A study on the synthesis, characterization and photocatalytic activity of CaO nanoparticle against some selected azo dyes. Indian J Appl Res, pp 361–365
9. Sadeghi M, Husseini MH (2013) A novel method for the synthesis of CaO nanoparticle for the decomposition of sulfurous pollutant. J Appl Chem Res 7(4):39–49

10. Munchow EA, Pankajakshan D, Albuquerque MTP, Kamocki K, Piva E, Gregory RL, Bottino MC (2016) Synthesis and characterization of CaO-loaded electrospun matrices for bone tissue engineering. Clin Oral Invest 20(8):1921–1933
11. Singh IHNB (2016) Green synthesis of nanoparticles and its potential application. Biotech Lett 38(4):545–560
12. Joya MR, Bautista Ruiz JH, Raba AM (2016) Quicklime as an alternative in the photodegradation of contaminants. J Phys: Conf Ser 687:12044
13. Kim SH, Jeon SY, Yoo PJ, Pu LS, Lee JY (2013) Metal oxide/polymer hybrid nanofiber as flexible moisture absorbent. Fibers Polym 14(12):1975–1980
14. Weinstein B (n.d.) Reaction of acid gases with solid alkali flowing in a duct: SO_2 reaction with lime

Exhaust Heat Recovery Using Thermoelectric Generators: A Review

Sarthak Nag, Atul Dhar and Arpan Gupta

Abstract With the major concern to increase the efficiency of internal combustion (IC) engines, various technologies and innovations have been implemented to improvise efficiency and reduction of emissions. Since 60–70% of the energy produced during combustion is rejected as heat through exhaust and coolant channels, it is important to recover that waste heat. Numerous technologies have been invented and applied to the diesel engine unit to harness the waste heat. One such is the use of solid-state device thermoelectric generator (TEG). In the late 1980s, many automobile manufacturers implemented automotive exhaust thermoelectric generators (AETEGs) in their respective vehicles, and since then, the work on AETEGs has picked at gradual pace. Advantages of using TEG are its noise-free operation, low failure rate and lack of moving components. However, it is not a very popular solution due to the low energy conversion efficiency (~6–8%) of thermoelectric modules and the incompetence to produce high power at low-temperature gradient. Engineers and researchers are basically working for improving the conversion efficiency of TEG modules by developing and doping semiconductors and optimization of the AETEG system to utilize and recover maximum heat available from the exhaust line by designing efficient heat exchanger systems, thus trying to improve its feasibility. This chapter covers the wide spectrum of feasibility of application of TEG modules in diesel engines with possible ways to utilize the generated power.

Keywords Thermoelectric generator (TEG) · Exhaust heat recovery (EHR) IC engines

S. Nag · A. Dhar (✉) · A. Gupta
School of Engineering, IIT Mandi, 175005 Mandi, India
e-mail: add@iitmandi.ac.in

D. K. Srivastava et al. (eds.), *Advances in Internal Combustion Engine Research*, Energy, Environment, and Sustainability,
https://doi.org/10.1007/978-981-10-7575-9_10

1 Introduction

There is growing concern for degrading environment, depleting natural resources and incorporating efficient and environment-friendly technologies. India Energy Outlook states that net oil consumption for 2014 was 3.8 million barrel per day out of which 40% is consumed by the transport sector [1]. Diesel is the primary fuel with its utilization being 70% of the net consumption [1]. It is expected that the consumption of oil for transport sector will increase from 43 million tonnes of equivalent oil (MTOE) in 2014 to 132 MTOE in 2040 [1]. The number of the new vehicles on the road is increasing at the rate of 19% every year, which adds up to the already existing vehicles on the road. Figure 1 shows the recent data for the vehicles registered from 2006 to 2013. Therefore, the energy demand for the transport sector, majorly the road transport, is increasing at a gradual pace which puts pressure to look for more sustainable technologies.

Most of the vehicles of the present day run on conventional fuels. Fuel is supplied to the engine, where combustion takes place and the energy released is used for the mobility. It is important to note here that only 25–30% energy is utilized for mobility and the major part of remaining energy either is lost to the coolant or goes out in the form of heat from exhaust [2]. Figure 2 shows the net fuel energy distribution in an engine. Hence, it is important to recover and utilize the energy lost in the form of heat. Heat is lost both from coolant and exhaust, but the latter is more viable option when it comes to heat recovery due to high temperature and high flow rate. Exhaust heat recovery (EHR) simply converts the waste heat of the exhaust into some useful energy. The recovered energy can be used to either produce electric energy for storage in batteries or can be reintroduced as mechanical energy into the engine.

EHR methods can be broadly classified in two approaches: (1) fluid-based EHR and (2) solid-state-based EHR [2]. Fluid-based EHR basically uses simple Rankine

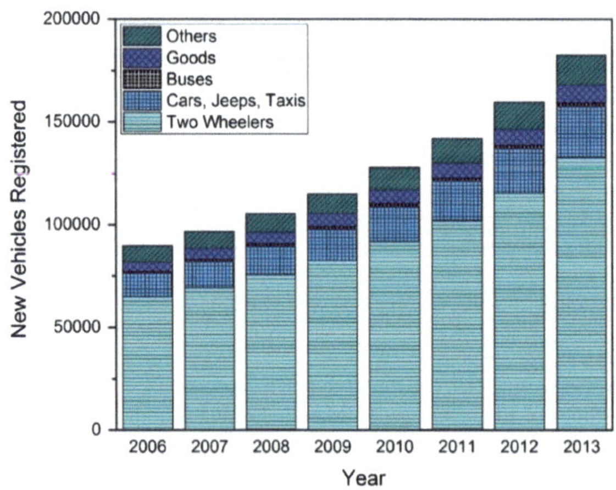

Fig. 1 New vehicles registered in India [1]

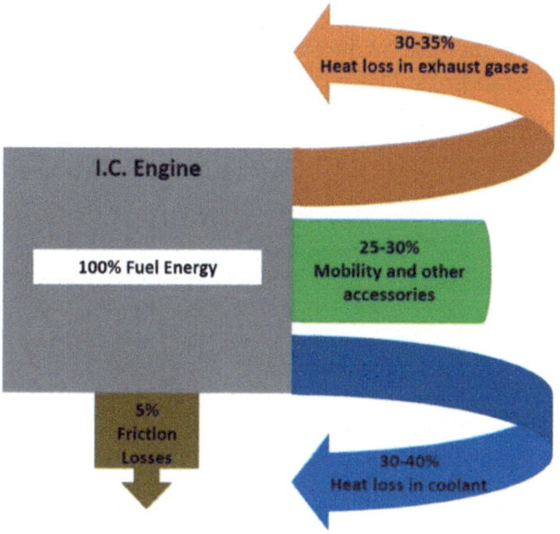

Fig. 2 Energy distribution in a typical engine

cycle for exhaust heat recovery. The working fluid is circulated to extract heat from the exhaust. Rankine cycle can be further divided into steam Rankine (SRC) and organic Rankine cycles (ORC). The major difference is in the type of working fluid used. SRC uses water as working fluid, whereas ORC uses organic fluids. Solid-state-based EHR is another approach which uses thermoelectric generator (TEG) to recover heat from the exhaust to reduce fuel consumption. TEG works on the principle of Seebeck effect simply converting temperature difference into electricity. This chapter deals with the solid-state EHR, its advantages, challenges and further possibilities.

2 Thermoelectric Generation

Thermoelectric generator is a solid-state device based on thermoelectricity, which converts heat directly into electricity. The phenomenon of thermoelectricity was first discovered by Thomas Seebeck in 1821 [3]. TEG works on Seebeck effect which states that when there is a temperature difference between two dissimilar electrical conductors or semiconductors, a voltage difference is produced between the two materials. Electrons may be considered as the working fluid in TEGs. Since electrons are the only moving species, TEGs enjoys great simplicity due to the lack of moving components [4]. Other added advantages of using TEGs are the small size, low failure rates, robustness and fast response [5]. Along with these, they are silent in operation, have high reliability and dos not cause pollution [6]. However, to utilize thermoelectricity to the fullest, there is still a need to improve thermoelectric materials and the way of utilization of the TE modules.

Figure 3 shows a single thermocouple of TE module. Two dissimilar semiconductors, a p-type and an n-type, are joined with temperature difference at their ends. This results in the flow of electrons from p-type to n-type semiconductor, leading to

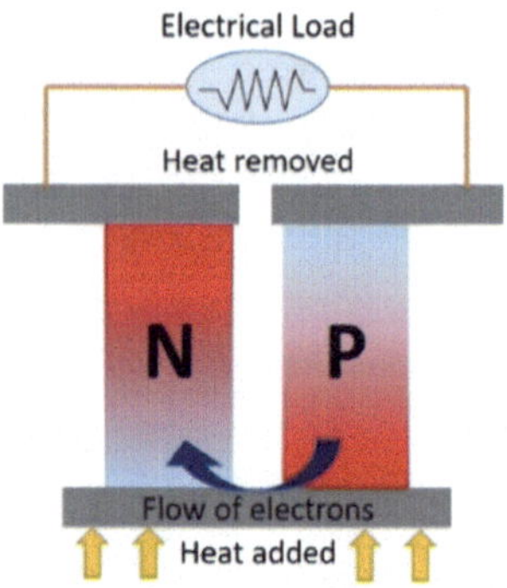

Fig. 3 Single thermocouple of TE module for power generation

the flow of electricity in direction opposite to the flow of electrons. The magnitude of current or, more precisely, the electromotive force (emf) generated depends on the magnitude of temperature difference. The net emf may be defined as:

$$V_o = S\Delta T$$

where V_o is the emf, S is the Seebeck coefficient, and ΔT is the temperature difference.

The properties of materials, which are beneficial to TE modules, are high electrical conductivity (σ) and low thermal conductivity (κ) of the material [7]. Low thermal conductivity helps in maintaining the temperature difference between the cold and the hot side, whereas high electrical conductivity helps in ensuring the rapid flow of electrons through the material. Efficiency of TE modules is set using a dimensional parameter known as figure of merit (ZT) [8]. ZT depends on Seebeck coefficient (S), thermal conductivity (κ), electrical conductivity (σ) and average operating temperature (T) and is given by [7–9]:

$$\mathrm{ZT} = \frac{S^2 \sigma T}{\kappa}$$

ZT values of various TEG materials and their variation with temperature are shown in Table 1. Although Bi_2Te_3-based TEGs are low in cost, they have a low ZT above 400 K. Therefore, based on the temperature range and the ZT value, lead-based TEGs are best suited for automotive application. Si-Ge TEGs also find their application in automobiles at high temperatures (up to 1200 K). Bi_2Te_3 super-lattices have high ZT values but are functional at low temperatures. Lead-based quantum dots have the best ZT, but their high cost is a major concern for deployment.

Thermocouples, the building unit of a thermoelectric module, are electrically in series but thermally in parallel [10]. Parallel thermal connection ensures equal heat transfer through each thermocouple. Figure 4 shows the conduction paths in a TE module.

Table 1 ZT values of various TEG materials [6]

Material of TEG	Temperature (K)	ZT Value
Bi_2Te_3	200	0.4
	400	0.95
	600	0.45
PbTe	400	0.6
	600	0.86
	800	0.77
Si-Ge	400	0.17
	600	0.31
	800	0.52
	1000	0.78
	1200	0.93
$AgPb_{18}SbTe_{20}$	400	0.73
	600	1.4
	800	2
$Yb_{0.19}Co_4Sb_{12}$ (n-type)	200	0.16
	400	0.47
	600	0.87
$La_{0.9}CoFe_3Sb_{12}$ (p-type)	400	0.28
	600	0.7
	800	0.96
Bi_2Te_3/Sb_2Te_3 super-lattices	200	1.7
	400 K	2.8
$PbSe_xTe_{1-x}/PbTe$ quantum dots	400	2.3
	600	3.4

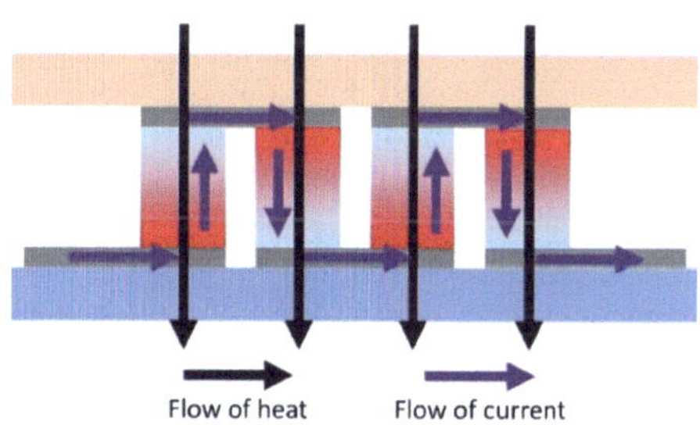

Fig. 4 Conduction paths in TE module [10]

3 Basic Components of AETEG

For the recovery of waste heat from engine's exhaust, the major components of AETEG unit are (see Fig. 5):

1. **Heat exchanger**: to extract heat from the engine exhaust.

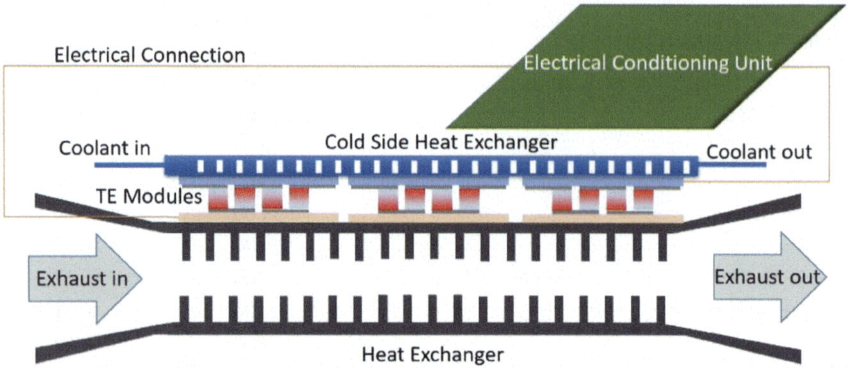

Fig. 5 Schematic diagram for basic components of AETEG

2. **TE modules**: to convert the heat energy to electrical energy. The TE modules are the most important component. They must be able to withstand exhaust temperatures with failures and solder melting. Lead-based TE modules are best suited for exhaust applications.
3. **Cold side heat exchanger**: to maintain the temperature at the cold side of the TE module. It needs to be efficient to provide a good temperature gradient.
4. **Power unit**: to match the output with automobile's electrical unit. It is basically a DC-to-DC convertor, which matches TEG's output voltage to automobile's bus voltage.
5. **Auxiliaries**: temperature sensors, electronic unit, coolant pump.

Figure 6 shows the typical heat recovery system using AETEG.

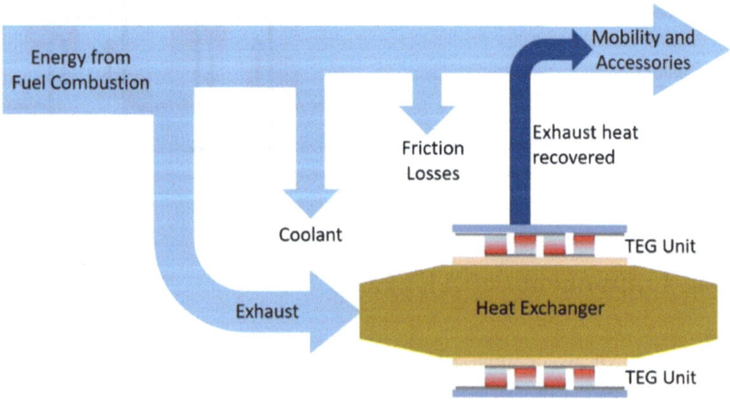

Fig. 6 Schematic of typical waste heat recovery system

4 AETEG Positioning

The exhaust line of any automobile consists of emission control components such as catalytic converter and particulate filter and sound attenuation components such as silencers or mufflers, depending on the type of vehicle. The exhaust is hottest at just after the exhaust manifold, and the temperature gradually decreases as the exhaust moves further. Hence, AETEG unit can be installed at various positions depending on the TEG material and the required output. Some of the TEG mounting positions are [11, 12]:

1. Engine manifold: exhaust is hottest in this region. Temperature may range up to 1000 °C. Preferable TEG material for application at this region is lead-based alloys. Si-Ge alloys can also be used.
2. Just before catalytic converter: the exhaust temperature at this region is between 750 and 400 °C. Lead telluride TE modules can be used at these regions.
3. Just before exhaust muffler: the exhaust temperature at this region is between 400 and 150 °C. Bismuth telluride semiconductors can be used at this region due to their high ZT at this temperature range.

5 Major Milestones in AETEG Development

The application of thermoelectric generators in vehicles goes back to the second decade of the twentieth century when Leigh E Hale [13] filed a US patent on thermoelectric battery for motor vehicles where a thermoelectric battery was attached to the motor of the vehicle to generate electricity to assist lighting systems and ignition. In 1962, another patent was filed based on the same theory of application of TEG for assisting auxiliary electrical units in an automobile [14]. Neild reported thermoelectric-based automobile waste heat recovery system (WHR) in 1963 [15] where he reviewed the major thermal design aspects of AETEG units and their application in specialized military equipment with portable thermoelectric generators. This provided the initial work on the concept of application of TEG in automobile WHR.

Bauer studied the feasibility to recover heat from engine coolant and found it to be an unviable option [16]. Anthony also carried a feasibility study of replacing generator with TEG unit with heat recovered from exhaust gases and reported that the minimum power requirement for electrical auxiliaries was met only at speeds above 20 mph [17]. Parallel to their work, Birkholz et al. [18] designed a AETEG unit using $FeSi_2$ thermoelectric material and tested it on Porche 944. They used 90 thermoelectric modules and reported a peak power of 58 W at full power.

In the 1990s, many organizations like Hi-Z, Nissan entered in the field of WHR using thermoelectric modules. Hi-Z studied various heat sources in an automobile and the thermoelectric material compatible with these sources. They concluded the best option being the exhaust gas for the heat recovery, whereas the best material being Bi_2Te_3 [19]. They designed a AETEG unit and tested it on 14 L Cummins engine and got a maximum output of over 1000 W [20]. Nissan worked on Si-Ge-type thermoelectric modules where they designed AETEG unit with water cooling and produced a maximum output of 36 W on a 3 L road vehicle at 60 kmph uphill drive [21].

In 2004, US Department of Energy sponsored a project on Thermoelectric WHR for passenger vehicles [22–24] for BSST, BMW, Ford, Faurecia, Visteon, Caltech, Virginia tech, National Renewable Energy Laboratory (NREL). They designed AETEG unit for BMW X6 vehicle and Ford fusion. The system produced 500–750 W of electricity, thus increasing the fuel economy by 2–5% by assisting in car's electrical components. The reduction in fuel usage directly leads to reduction in emissions. This was a 7-year programme with great outcomes in the field of WHR using thermoelectrics. US DOE has targeted it for vehicles in 2018 models.

A similar project was initiated by General Motors in 2012 with an objective of improving fuel economy by 5% and developing low-cost TEG units. AETEG unit was installed on GM Suburban Vehicle, and an average power output of 50 W was achieved. The cooling system was integrated with engine cooling unit. A notable fuel economy gain of 2% has been achieved so far in the testing phase with much scope of improvements.

In 2014, researchers from Wuhan University developed AETEG unit with power output of 944 W [25]. Their TEG's output is said to meet the automotive electrical requirement. Their AETEG is called "four-TEG system" and is made from 240 thermocouples [25]. It was installed on prototype vehicle "Warrior", a military purpose vehicle. It was then tested on various speeds, and the results showed the feasibility of the replacement of alternator with AETEG unit.

6 Challenges in TEG Implementation

Since the introduction of TEGs, they have made a huge progress in terms of materials and cost cutting. TEGs are seen as a technology which can help in solving climate issues. TEGs are known to increase the fuel efficiency of vehicles by assisting the power unit with electrical power. However, there are various challenges that stand in the way for large-scale TEG implementation.

Figure of merit (ZT) is the measure of efficiency of TEGs. TEGs have a low ZT value. The problem is on the materials end: physics and chemistry associated with materials [7]. ZT value of best-known TEGs is near to 1 only. The thermal efficiency of TEG is also low, less than 4% [26]. Other known EHR technologies have

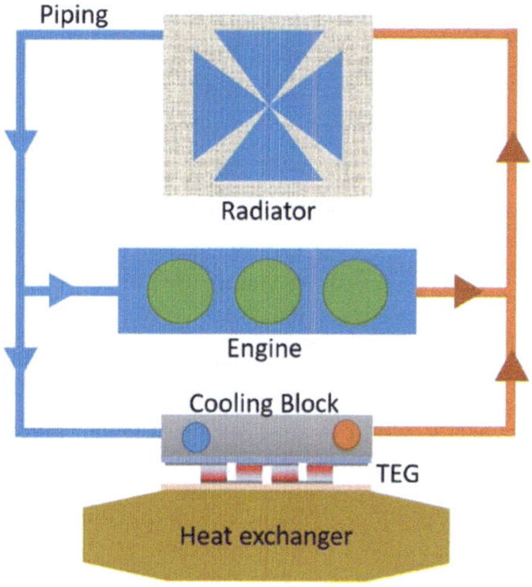

Fig. 7 Schematic diagram of cooling of TEG from radiator

much higher thermal efficiency values. Since ZT values are dependent on temperature, material's ZT values peak at nearly the melting point of that material. This causes degradation of TEG material and defeats its goal. The decrease in the temperature decreases the ZT values by significant amount resulting in lower efficiencies. Moreover, TEG materials are hard and brittle; thus, they may develop cracks owing to vibrations and thermal loading. These cracks decrease the electrical conductivity, resulting in decrease in ZT values and the life of TEGs.

Other major challenge is the efficient cooling on the cold side of the TEG. The cooling unit for TEGs can be merged with existing engine cooling unit. This leads to the need for bigger radiator for efficient cooling. With this, the pipes carrying the cooling fluid need to be extended till the exhaust unit of the automotive, thus adding to the cost and making the system more complex. Figure 7 shows the schematic diagram for complete radiator circuit for cooling of TEG. The viable solution to this problem is the use of nanofluids, which tends to enhance the heat transfer characteristics of the cooling fluids. Researchers have reported up to 30% enhancement of heat transfer coefficient in radiator of automobile using nanofluids [27]. Moreover, heat transfer can be improved by efficient circulation of cooling fluid through proper channelized copper cooling blocks.

7 Broad Advantages of AETEG

Presently, the use of AETEG assists the requirement of both users and the government. AETEG can be solution to both energy crisis and environmental degradation. The users want their vehicles to be more fuel efficient, whereas the governments are pushing towards stringent emission norms.

1. **Fuel economy improvement**: Fuel economy of a vehicle can be enhanced by using AETEG unit. Since the power produced by engine also provides support to the electrical components such as headlights, AC, electronics, AETEG can directly charge a battery and can provide assistance to aforementioned components. This will decrease the load on the engine as all the power is now available for drive train and hence improves fuel economy. General Motors have reported a fuel economy improvement up to 5% with the assistance of AETEG unit [28]. Freedom car, an initiative by US Government, aims at improving fuel economy by 10% [7].
2. **Limiting exhaust emissions**: Less fuel burnt leads to less exhaust emissions. Governments across the world are moving to stringent emission norms. Since AETEG unit supplies some power back to the battery, the decreased load on the engine leads to less fuel burning which directly leads to lower emissions.

8 Possible AETEG Applications

The heat recovered by AETEG systems can be used to power various electrical components. One such critical system where this power can be used is air conditioning system of automobiles. An air conditioning unit, in general, uses 2–3 l/100 km of fuel. But fuel consumption may increase depending on the load on the unit. If a vehicle covers a distance of 10,000 km annually, the fuel consumption by the air conditioner may be between 200 and 300 l. This much fuel gives around 460–700 kg of CO_2 emissions. All of this emission can be prevented by using the heat recovered by AETEG unit to power air conditioning systems. Feasibility of powering AC unit with AETEG is low since conventional AC system requires around 180–2000 W depending on conditions; however, improvements on AETEG unit may help us in running air conditioning units. Researchers have successfully developed AETEG-assisted temperature-controlled car seat to increase the comfort level while driving [29].

AETEG can also assist other car electrical equipment such as headlamps, parking lights, wipers. Conventionally, these units are powered by batteries which are charged by alternators. Alternators use car's energy to recharge battery. However, it is very difficult to replace alternator due to its ability to provide high power. Moreover, AETEG needs some initial warm-up time to start producing electricity and the amount of electricity produced may vary with driving patterns. But once the efficiency is improved from materials' field, it can definitely replace alternator unit of car.

9 Some Case Studies on AETEGs

9.1 BMW X6 and Lincoln MKT Vehicles

AETEG unit was installed on BMW X6 and Lincoln MKT [30]. This project was the outcome of US Department of Energy's 7-year programme on thermoelectric waste heat recovery. Fuel economy improvement was estimated by researchers. They built liquid/liquid AETEG unit in their second phase of development, which produced power in access of 500 W. They also attempted gas/liquid AETEG unit. The role of pressure application to reduce contact resistance for efficient heat transfer is important. They were able to harness a power of 100 W. After viewing the shortcomings of flat plat design, they developed a cylindrical AETEG unit. A provision of bypass was kept for excessive pressure drop across AETEG unit. They were able to achieve more than 200 W of power with this configuration. The cooling unit was integrated with engine radiator with an additional pump for the flow of coolant to the TEG unit. The electrical power improved with vehicle speed from 100 W at 50 kmph to around 600 W at 125 kmph. The increase in the fuel efficiency was also calculated, and it improved from 0.7% at 50 kmph to 1.2% at 130 kmph. The cold start performance of TEG was also studied for Lincoln MKT. The AETEG unit required 800 s to produce any substantial power. At a speed of 120 kmph, the AETEG unit produced a power of 225 W. At the city driving conditions, a power output of 80 W was observed, which is low pertaining to the electrical requirements. The major conclusions of this work were: improvements in ZT value will greatly affect the scope of use of AETEG values in a positive manner; critical management of the thermal expansion coefficients of TEG material; TEGs require a warm-up time before efficient functioning.

9.2 1995 Kawasaki Ninja 250 R

Schlichting et al. tested the feasibility of AETEG unit on a motorbike [31]. They used 1995 Kawasaki Ninja 250 R and mounted a TEG unit on its exhaust pipe. Since motorcycle does not have catalytic converter, the positioning of AETEG unit was not a big concern. The only deciding factor was the maximum allowable operational temperature of TE modules. The alternator of the bike produced 14 V at 17 A. They aimed at replacing the alternator with AETEG unit. They ran various test trials at different speeds and driving conditions. They obtained an average output power of 0.47 W from the module. They estimated the use of 570 modules, 10 modules in series and 57 combinations in parallel with matching the output of alternator. The conclusion was that the feasibility to replace alternator was low. There is still a major need to improve TEG materials.

9.3 1999 GMC Pickup Truck

Clarkson University attempted the development of AETEG unit for 1999 GMC Pickup Truck. The model for power measurement, exhaust temperature and back pressure was also developed [32]. They aimed at developing a 330 W AETEG unit using 16 HZ-20 modules. They also designed a power conditioning unit (PCU) for power management and DC/DC conversion. Their experiments were able to reach a power of 150 W at 70 mph at coolant temperature of 80 °C, with improvement to 255 W at 25 °C coolant temperature. They also developed a PCU with 88% efficiency. They also concluded that the cooling load on engine radiator was not very high. They reported an increased fuel efficiency of 2% [33].

9.4 GMZ 1000 W AETEG for US Military

GMZ Energy developed a 1000 W AETEG for US military diesel vehicles. It was developed by the integration of 5200 W AETEG units. They used the half-Heusler material for the TE module development [34]. The major advantage of this unit was the continuous power production. It was tested at 15 l V8 engine. GMZ is a leader in TEG materials with best ZT values. In the past, GMZ has also produced AETEG units with 3.45% improvement in fuel consumption and removal of belts and alternator from the engine.

10 Outlook

There is a large potential of waste heat recovery technologies. Thermoelectric technology has been progressive since its advent and has gained a lot from where it started. With the introduction of new materials and nanostructures, ZT values are already increasing. Some cases have also shown a possibility of replacement of alternator with AETEG units to save on fuel consumption and exhaust emissions. However, there is still a need to improve the ZT value for further replacement of mechanical systems. The design of heat exchangers to extract exhaust heat and to cool the colds side of TE modules is also critical, while designing AETEG unit since temperature gradient is beneficial to its operation. There has been an extra challenge to integrate the TEG cooling with automobile's radiator system and the requirement of bigger radiator, but with the use of nanofluids, this problem can be solved very easily. The recovered exhaust heat can be used in various ways and can be stored in batteries too. In a nutshell, analysing the current scenario of TEG and the improvement it does on fuel economy, there are systems which are more efficient than TEG unit, but the fullest potential of TEG will be unveiled once there is an improvement on materials side.

References

1. Birol F (2015) India energy outlook. International Energy Agency, Paris
2. Tripathi G, Dhar A (2017) Exhaust heat recovery options for diesel locomotives. In: Locomotives and rail road transportation. Springer Singapore, pp 27–40
3. Kasap S (2001) Thermoelectric effects in metals: thermocouples. Department of Electrical Engineering University of Saskatchewan, Canada
4. Kim TY, Negash AA, Cho G (2016) Waste heat recovery of a diesel engine using a thermoelectric generator equipped with customized thermoelectric modules. Energy Convers Manag 124:280–286
5. Bell LE (2008) Cooling, heating, generating power, and recovering waste heat with thermoelectric systems. Science 321(5895):1457–1461
6. Yang J, Stabler FR (2009) Automotive applications of thermoelectric materials. J Electron Mater 38(7):1245–1251
7. Vining CB (2009) An inconvenient truth about thermoelectrics. Nat Mater 8(2):83–85
8. Behnia K (2015) Fundamentals of thermoelectricity. OUP Oxford
9. (2005) Sorrell CC, Nowotny J, Sugihara S (eds) Materials for energy conversion devices. Elsevier
10. Armstead JR, Miers SA (2014) Review of waste heat recovery mechanisms for internal combustion engines. J Therm Sci Eng Appl 6(1):014001
11. Jumade SR, Khond VW (2012) A survey on waste heat recovery from internal combustion engine using thermoelectric technology. Int J Eng Res Technol IJERT 1(10)
12. Nag S, Thangavelu SB, Tripathi G, Dhar A, Gupta A (2017) Studies on temperature variation in automotive exhaust thermoelectric generator with exhaust pipe length. J Energy Environ Sustain 3:53–58
13. Hale LE (1915) Thermo-electric battery for motor-vehicles. U.S. Patent 1,134,452
14. Carl KE, Kiekhaefer Corp (1965) Thermoelectric generator for internal combustion engine. U.S. Patent 3,217,696
15. Neild AB (1963) Portable thermoelectric generators (No. 630019). SAE Technical Paper
16. Bauer RH (1963) Auxiliary electric power for an automobile through the utilization of a thermoelectric generator: a critical examination. Department of Mechanical Engineering, Clarkson College of Technology, Potsdam, New York
17. Tomarchio AJ (1964) A feasibility study of replacing an electrical generator of a standard American automobile with a thermoelectric generator. Department of Mechanical Engineering, Clarkson College of Technology, Potsdam, New York
18. Birkholz U, GroB E, Strohrer U, Voss K, Gruden DO, Wurster W (1988 March) Conversion of waste exhaust heat in automobiles using $FeSi_2$-thermoelements. In: Proceeding of 7th international conference on thermoelectric energy conversion, pp 124–128
19. Bass JC, Elsner NB, Slone R (1991 April) Design study and experience with thermoelectric generators for diesel engines. In: Proceedings of the annual automotive technology development contractors. SAE P-243
20. Bass JC, Elsner NB, Leavitt FA (1995) Performance of the 1 kW thermoelectric generator. In: Proceedings of 13th International AIP Conference on Thermoelectrics (ICT'94), vol 316, p 295
21. Ikoma K, Munekiyo M, Furuya K, Kobayashi MAKM, Izumi TAIT, Shinohara KASK (1998 May) Thermoelectric module and generator for gasoline engine vehicles. In Thermoelectrics, 1998. Proceedings ICT 98. XVII International Conference on (pp 464–467). IEEE
22. Crane DT, Lagrandeur JW (2010) Progress report on BSST-Led US department of energy automotive waste heat recovery program. J Electron Mater 39(9):2142–2148
23. Crane D, LaGrandeur J, Jovovic V, Ranalli M, Adldinger M, Poliquin E, Dean J, Kossakovski D, Mazar B, Maranville C (2013) TEG on-vehicle performance and model validation and what it means for further TEG development. J Electron Mater, 1–10

24. Department of Energy. Could TEG improve your cars efficiency? Available at: https://energy.gov/articles/could-teg-improve-your-cars-efficiency. Accessed 2 June 2017
25. Liu X, Deng YD, Li Z, Su CQ (2015) Performance analysis of a waste heat recovery thermoelectric generation system for automotive application. Energy Convers Manag 90:121–127
26. Saidur R, Rezaei M, Muzammil WK, Hassan MH, Paria S, Hasanuzzaman M (2012) Technologies to recover exhaust heat from internal combustion engines. Renew Sustain Energy Rev 16(8):5649–5659
27. Selvam C, Lal DM, Harish, S (2017) Enhanced heat transfer performance of an automobile radiator with graphene based suspensions. Appl Therm Eng
28. Meisner G (2012 March). Skutterudite thermoelectric generator for automotive waste heat recovery. In: 3rd Thermoelectrics Applications Workshop
29. Du H, Wang YP, Yuan XH, Deng YD, Su CQ (2016) Experimental investigation of a temperature-controlled car seat powered by an exhaust thermoelectric generator. J Electron Mater 45(3):1529
30. Crane D, LaGrandeur J, Jovovic V, Ranalli M, Adldinger M, Poliquin E, Dean J, Kossakovski D, Mazar B, Maranville C (2013) TEG on-vehicle performance and model validation and what it means for further TEG development. J Electron Mater, 1–10
31. Schlichting AD, Anton SR, Inman DJ (2008 March) Motorcycle waste heat energy harvesting. In: The 15th international symposium on: smart structures and materials & nondestructive evaluation and health monitoring (pp 69300B-69300B). International Society for Optics and Photonics
32. Karri MA (2005) Modeling of an automotive exhaust thermoelectric generator. Master's thesis, Clarkson University, Department of Mechanical and Aeronautical Engineering, Potsdam, NY
33. Zhang Y, Cleary M, Wang X, Kempf N, Schoensee L, Yang J, Joshi G, Meda L (2015) High-temperature and high-power-density nanostructured thermoelectric generator for automotive waste heat recovery. Energy Convers Manag 105:946–950
34. Thacher EF, Helenbrook BT, Karri MA, Richter CJ (2007) Testing of an automobile exhaust thermoelectric generator in a light truck. Proc Inst Mech Eng Part D: J Automobile Eng 221 (1):95–107

Part IV
Numerical/Simulation

Chemical Kinetic Simulation of Syngas-Fueled HCCI Engine

Rakesh Kumar Maurya, Mohit Raj Saxena, Akshay Rathore and Rahul Yadav

Abstract Energy safety concern and depletion of fossil fuel resources lead towards the investigation of an efficient and clean alternative combustion strategy as well as renewable biofuels. Homogeneous charge compression ignition (HCCI) engine has demonstrated the potential for higher thermal efficiency along with simultaneous reduction of NO_x and PM emissions to ultra-low level. Syngas is a potential alternative fuel. Syngas-fueled HCCI engine combines the advantages of advanced combustion strategy and biofuels. This chapter provides the overview of HCCI combustion and its chemical kinetic simulation using stochastic reactor model (SRM). This chapter also presents the comparative analysis of performance of various syngas reaction mechanisms in the HCCI engine at different inlet temperature and equivalence ratio using stochastic reactor model. For validating the reaction mechanisms, experimental in-cylinder pressure data is compared with the numerically simulated data. Syngas reaction mechanism CRECK-2014 (consisting of 32 species and 173 reactions) is found suitable for syngas-fueled HCCI combustion simulation.

Keywords HCCI · Combustion · Reaction mechanism · Syngas
Compression ignition · Diesel engine

1 Introduction

The advancement in modern society tremendously depends on the utilization of hydrocarbon fossil fuels. Combustion of these fossil fuels generates the energy for human requirement like electricity generation, transportation. However, combustion of these fuels emits harmful emissions in the significant amount. Diesel engines are the most preferred combustion engine for heavy-duty transportation and power

R. K. Maurya (✉) · M. R. Saxena · A. Rathore · R. Yadav
Advanced Engine and Fuel Research Laboratory, Department of Mechanical Engineering, Indian Institute of Technology Ropar, Rupnagar 140001, Punjab, India
e-mail: rakesh.maurya@iitrpr.ac.in

D. K. Srivastava et al. (eds.), *Advances in Internal Combustion Engine Research*, Energy, Environment, and Sustainability,
https://doi.org/10.1007/978-981-10-7575-9_11

generation due to its higher fuel conversion efficiency. Additionally, diesel engine also emits nitrogen oxides (NO_x) and particulate matter (PM) in relatively higher concentrations. To meet the energy demand and overcome the challenges of conventional diesel engines, researchers have suggested two strategies, either to replace the fossil fuel with renewable fuel or employ premixed low-temperature combustion (LTC) strategies instead of conventional non-premixed combustion strategy. Premixed LTC strategies have a relatively higher thermal efficiency along with lower NO_x and soot emissions. In recent decades, various LTC strategies were proposed such as homogeneous charge compression ignition (HCCI), premixed charge compression ignition (PCCI), partially premixed combustion (PPC), reactivity-controlled compression ignition (RCCI). In all the LTC strategies, significant fraction of fuel is premixed, which wards off the diffusion combustion phase, and higher fraction of charge is burnt in premixed combustion phase. Combustion of premixed homogeneous charge leads to reduction in the PM emission. In this type of premixed combustion, typically fuel–air mixture used for combustion is leaner, which results in lower combustion temperature leading to lower NO_x emissions. Figure 1 illustrates the operating range of conventional diesel engine, spark ignition (SI) engine, and advanced LTC engines with respect to soot-NO_x map with local equivalence ratio and local combustion temperature.

Figure 1 depicts that conventional diesel engine operates at relatively higher temperature region and higher (fuel-rich) local equivalence ratio. Therefore, diesel engines generate higher levels of NO_x and PM from the combustion chamber. For cleaner engine operation, it is recommended to operate the engine with local combustion temperature in the range of 1400–2000 K (approximately) and local equivalence ratio less than two (Fig. 1). In LTC engine, lower combustion temperature is kept by using charge dilution (either by using EGR or excess oxygen) and lower local equivalence ratio, which is achieved by premixing of charge.

Among all the premixed LTC strategies, homogeneous charge compression ignition (HCCI) is the most investigated engine combustion strategy by the researchers because of its potential of higher thermal efficiency along with ultra-low

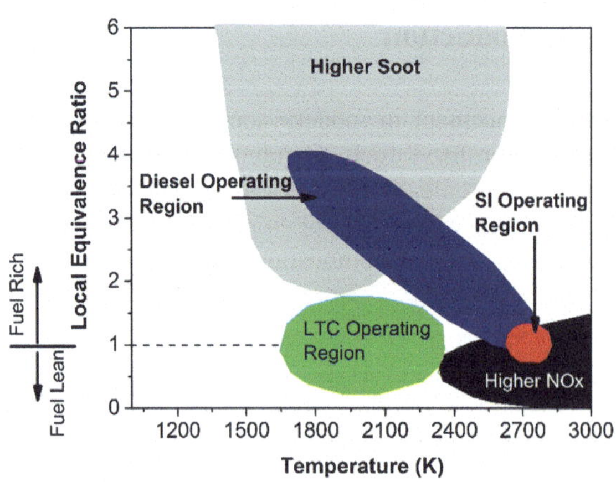

Fig. 1 Operating range of different combustion strategies. Adapted from [37]

NO_x and soot emissions [1, 2]. The HCCI engine has a higher fuel conversion efficiency as compared to conventional engines of same size [3]. However, the HCCI combustion strategy has demerit of higher hydrocarbon (HC) and carbon monoxide (CO) emission due to lower combustion temperature. The HCCI engine also lacks the direct control on the combustion phasing; however, conventional engine has direct control on combustion phasing by spark or fuel injection timings.

To meet the future energy demand and decline in the supply of fossil fuels, renewable fuels (such as biofuels) are presently investigated as one of the options for internal combustion engines. The HCCI combustion strategy has a potential to utilize any such fuels which can be autoignited in the combustion chamber using suitable control strategy. Several studies [4–6] have demonstrated that HCCI engine can be operated with various fuels such as methanol, butanol, ethanol, and hydrogen.

In the past few years, syngas has attracted the attention of researchers for application as fuel in internal combustion engines [7, 8]. Syngas is produced by the gasification process of biomass or coal. Composition of syngas (mainly consist of CO, N_2, H_2, and CO_2) depends on the type of biomass used, process of gasification, etc. Syngas can also be utilized in HCCI combustion engine, which combines the advantage of alternative fuels as well as alternative combustion strategy. To reduce the cost and time of detailed experimental investigation, numerical studies are quite helpful. This chapter presents insight into chemical kinetic modeling of syngas-fueled HCCI engine. Different reaction mechanisms developed for syngas combustion [9–15] are compared for utilization in HCCI engine simulation. Before discussing details on chemical kinetic model of syngas HCCI, fundamentals of HCCI combustion and syngas application in combustion engines are discussed in the following subsections.

1.1 HCCI Combustion Engine

The HCCI engines have received significant attention in recent decades because of their potential of lower NO_x and PM emission along with higher fuel conversion efficiency. The HCCI combustion is an alternative engine operating mode. HCCI combustion relies on the independent ignition of a premixed charge rather than on an external source of ignition. As the name suggests, HCCI engine uses well-mixed or premixed fuel–air mixture. Typically in HCCI engine, fuel is injected into the intake manifold to prepare close to homogeneous charge (air–fuel mixture) towards the end of compression stroke. During intake stroke, fuel–air mixture is typically inducted into the cylinder. After the intake stroke, the piston starts moving from the bottom dead center (BDC) to top dead center (TDC) in the compression stroke. During compression stroke, in-cylinder pressure and temperature increase. Towards the end of compression stroke (when the piston reaches close to TDC position), fuel–air mixture autoignites when charge temperature reaches the autoignition temperature. During autoignition process, the fuel starts oxidizing and releasing chemical energy of fuel that further increases the pressure and temperature of the cylinder. Power is produced by the engine in expansion stroke after the combustion. In HCCI engine, typically autoignition occurs simultaneously throughout the combustion chamber at different locations leading to very high heat release rate (HRR). High HRR in the engine cylinder leads to very high peak pressure rise rate (PPRR) in comparison with conventional diesel combustion engines. However, the peak in-cylinder combustion temperature in HCCI combustion is relatively lower due to lean engine operation. In HCCI engine, the combustion is chemical kinetically controlled as combustion is initiated by autoignition reactions of premixed charge. The chemical kinetics is mainly controlled by charge temperature and concentration of fuel species. At particular equivalence ratio of fuel–air mixture, kinetics of reaction is mainly controlled by pressure and temperature of the charge (air–fuel mixture) at the beginning of the compression stroke (at intake valve closing position). Typically, temperature of charge at the end of intake valve closing is indirectly controlled by either preheating of air or by using exhaust gas recirculation (EGR).

The HCCI combustion strategy has very short combustion duration (i.e., 6–15 CAD) as compared to conventional diesel engine (i.e., 30–90 CAD) [16]. Combustion duration has significant effect on the HCCI combustion, and it limits the HCCI engine operating range in terms of engine load. The HCCI combustion is limited by high and lower operating load limits. The high load limit is constrained by very high heat release rate, leading to ringing operation. High load limit occurs at richer mixture and higher inlet temperature engine operation. The HCCI engine load limit is typically characterized by a threshold value of ringing intensity or peak pressure rise rate. To reduce the combustion rate, HCCI engine is typically operated at leaner charge, which increases the combustion duration and thus lower HRR is achieved. At very lean engine operation, misfire or partial burn cycle occurs due to lower combustion temperature. The lower engine load is limited by misfire or partial burn. Generally, lower engine load limit is characterized by cyclic variation of IMEP, which is defined by acceptable value coefficient of variation in IMEP.

The HCCI combustion can be achieved by using gasoline as well as diesel like fuels. Depending on the autoignition characteristics of fuel, adequate control strategies need to be selected to achieve combustion/autoignition in the cylinder. Diesel is relatively less volatile fuel, and therefore, it is directly injected into the cylinder at higher pressure and temperature condition for premixed charge preparation. For diesel fuel, typically port injection is not preferred for well-mixed charge preparation. However, for gasoline like fuels, well-mixed charge preparation is done by port injection of fuel due to high volatility. However, diesel has relatively lower autoignition temperature than gasoline. To achieve the autoignition using gasoline fuel in cylinder, higher charge temperature is required. Higher charge temperature requirement is typically fulfilled by intake preheating or by trapping internal residual gases.

Combustion process in HCCI engine can be summarized as

- Preparation of well-mixed highly diluted charge to achieve acceptable combustion rate by extending the combustion duration;

Table 1 Summary of important characteristics of different combustion concepts

Spark ignition	Compression ignition	HCCI
Spark ignition	Autoignition	Autoignition
Homogeneous charge (stoichiometric)	Heterogeneous charge (over all lean)	Homogeneous charge (lean)
Premixed combustion	Premixed and diffusive combustion	Premixed kinetically controlled combustion
Throttled engine operation	Un-throttled engine operation	Un-throttled engine operation
Port injection	Direct injection with swirl	Port or direct injection depending on fuel quality

- Achieve the autoignition of homogeneous charge toward the end of compression stroke by rise in charge temperature by compression;
- Precise control of HRR and combustion phasing for better performance [17].

Table 1 presents the summary of important characteristics of conventional engines (spark ignition and compression ignition) along with HCCI combustion engine. In HCCI combustion, charge preparation is similar to spark ignition engine and combustion is similar to compression ignition engine.

In HCCI engines, the air–fuel mixture is well premixed similar to spark ignition engines. Combustion of well-premixed charge leads to lower particulate matter emission. Effective charge preparation and avoiding fuel–wall interactions are crucial issues in HCCI engine for having high fuel conversion efficiency, lower unburned hydrocarbon and particulate matter emissions, and lubricating oil dilution prevention. The HCCI fuel–air mixture preparation strategies can be divided into three categories: port injection (external mixture preparation), in-cylinder injection (internal mixture preparation), and strategies using both port and in-cylinder injections (in case of dual fuel injection). This method consists of mixing fuel with air prior to entering the cylinder. In port fuel injection method, fuel is injected upstream of the intake valves and mixture is inducted into the engine cylinder during intake stroke. In port fuel injection, fuel has relatively more and sufficient time for evaporation and mixing. Fuel is injected into intake manifold at relatively lower fuel injection pressure. However, in case of direct injection, fuel is directly injected into the cylinder at higher fuel injection pressure. Fuel injection is kept in such a way that fuel has sufficient time for evaporation and mixing to prepare a premixed charge in the cylinder. Too early direct injection timings can lead to cylinder wall wetting, while too late injection timing can result in insufficient mixing of air and fuel.

Although term “homogeneous charge”, is used for HCCI combustion engine. However, mixture inhomogeneity or temperature inhomogeneity always exits to a certain extent in the combustion chamber. In real engine, it is impossible to create a perfectly homogeneous fuel-air mixture and purely homogeneous temperature distribution in the cylinder. The inhomogeneity in fuel and temperature distribution

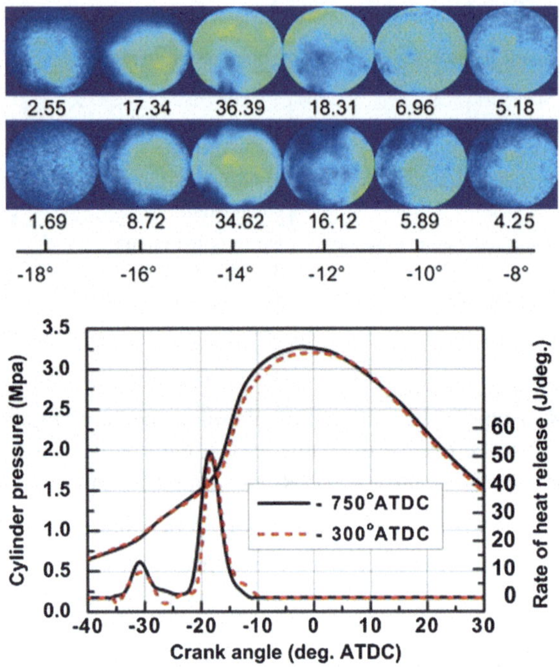

Fig. 2 Combustion images, pressure, and ROHR under different injection strategies as start of injection (SOI) = −750° ATDC (first) and −300° ATDC (last) [18]

exists due to imperfect mixing of fuel, air, and residual gases along with heat transfer through combustion chamber surfaces [18]. Figure 2 illustrates the combustion in HCCI combustion using DME (dimethyl ether). Two different port fuel injection timings during closed and open intake valve condition are used at intake temperature of 125 °C. Injection timings have relatively less difference at this temperature, and difference is more visible at lower intake temperatures [18]. However, at particular injection timing, chemiluminescence images show sufficient inhomogeneity in the combustion (Fig. 2). Study summarized that the port fuel injection timing, different intake and coolant temperatures, and turbulence produced by higher engine speed can affect HCCI combustion processes [18]. These factors affect temperature distribution in the combustion chamber during HCCI combustion. Relatively larger stratification of local charge temperature in the combustion chamber can reduce the rate of pressure rise through smoothing the reaction rates and extend the HCCI engine operating range [19].

1.1.1 Advantages of HCCI Engine

The potential advantages of HCCI combustion, which make this combustion strategy attractive that can replace the conventional engine combustion modes in the future, are as follows [19, 20].

- HCCI engine has a relatively higher fuel conversion efficiency due to lower heat transfer loss, higher compression ratio engine operation, and leaner mixture operation (higher ratio of specific heat). In addition, this strategy has no throttling loss at part load engine operation as it occurs in spark ignition engine.
- The HCCI combustion simultaneously reduces the NO_x and PM emissions to an ultra-low level.
- The HCCI engine operation has a flexibility to use both gasoline and diesel like fuels.
- Higher fuel economy can be achieved in HCCI engines due to higher fuel conversion efficiency.

1.1.2 HCCI Engine Challenges

Even though HCCI engine has benefits over conventional diesel combustion strategy, there are several challenges which must be resolved before commercialization of HCCI combustion engine for automobiles. The HCCI engine challenges are as follows [19, 20].

- Control of ignition timing and combustion rate over wide engine operating range;
- Higher level of CO and HC emissions particularly at lower engine loads;
- Higher level of combustion noise particularly at higher engine loads;
- Lower engine load operating range;
- Cold-start problems.

1.2 Syngas Application in Combustion Engines

Syngas mainly consists of CO, N_2, H_2, and CO_2. Syngas composition depends on the type of biomass used and process of gasification. Common examples of syngas production processes are gasification of coal and steam reforming of natural gas (NG) to produce hydrogen. The energy density of syngas is typically 50% energy density of the NG, and syngas has a potential to be used as engine fuel.

Application of syngas in internal combustion engine can possibly reduces the emissions (as compared to conventional fuels) and could be a good replacement for conventional fuels. Several researchers investigated the application of syngas in internal combustion engines [21–26]. Yamasaki and Kaneko [21] investigated the impact of mock syngas composition on autoignition and combustion characteristics of HCCI engine. Main composition of mock syngas is H_2, CO, CO_2, N_2, and small fraction of methane. Their results reveal that ignition timing depends on the autoignition temperature (i.e., 1100 K, which is almost similar to hydrocarbon fuels) and in-cylinder gas temperature. Additionally, rate of combustion can be estimated by CO_2 and H_2 contents. Their results also indicated that conversion time of H_2 into final product is longer as compared to CO. Combustion duration can be roughly determined by the ratio of H_2 to CO_2 in the fuel. Przybyla et al. [22] experimentally investigated the producer gas-fueled SI and HCCI engines of similar displacement volume under similar load condition. Results indicate that HCCI engine has higher CO emissions as compared to SI engine. In addition, SI engine has higher indicated thermal efficiency as compared to HCCI engine. Study also reveals that combustion duration in HCCI engine is four times shorter as compared to SI engine. Sahoo et al. [23] investigated the impact of H_2 and CO fraction (i.e., 100% H_2, 75% H_2, and 50% H_2) in syngas on the performance and emissions characteristics of dual fuel diesel engine. Study showed that at higher load condition, brake thermal efficiency increases with increase in H_2 fraction but at lower load condition, the efficiency reduces for all the H_2 and CO fractions. Hydrogen has higher energy content and flame speed, and thus, peak pressure and in-cylinder temperature are higher with 100% H_2 which leads to NO_x formation. In syngas combustion engines, CO emission depends on the concentration of CO fraction in syngas and increases with increase in CO fraction.

2 Simulation Methodology

The HCCI combustion engine is numerically investigated by different simulation and modeling strategies including 0-D stochastic reactor model (SRM), single and multi-zone models, and 2-D computational fluid dynamics (CFD) models. To compare the simulation ability of different syngas reaction mechanisms, SRM based simulation approach is used and results are presented in Sect. 3. This section briefly discusses the SRM approach. The syngas HCCI engine is numerically investigated using stochastic reactor model. Numerical investigation is conducted during closed valve conditions. Various syngas reaction mechanisms are validated by comparing the in-cylinder pressure data from previously published study [27].

A single-cylinder four-stroke direct injection diesel engine was used in original study. Detailed specifications of the engine used for the study are specified in Table 2. Intake manifold of the engine was modified for port fuel injection of syngas (mixture of H_2 and CO) during the intake stroke. Details of the experimental setup are provided in the original study [27]. All the simulations were conducted using zero-dimensional SRM model. This model replaced the homogeneity inside the cylinder and real fluid particles with statistical homogeneity and virtual stochastic particles, respectively. All these virtual stochastic particles have its own chemical composition, mass, and temperature. In addition to this, these virtual particles have a capability to mix with other virtual particles and they can also exchange heat from the cylinder wall. In this study, SRM engine suite software is used which is an in-cylinder engine combustion simulator.

The details of the governing equations used in numerical investigation (simulation) by SRM are discussed in detail in the previous studies [28–30]. SRM model

Table 2 Engine specifications

Engine characteristics	Specifications
Piston shape	Reentrant bowl
Bore (mm)	86
Stroke (mm)	75
Compression ratio	19.0
Engine speed (RPM)	1825 ± 25
Swept volume (L)	0.435
Injection system	Direct injected (pump–line–nozzle)
Crank radius	3.22
Connecting rod	120.75 mm

is based on probability density function, which predicts better results close by 3D-CFD analysis. In SRM, statistical homogeneity is assumed throughout the combustion chamber which means that probability density function is not changed throughout the combustion chamber. SRM determines the progression of N_s chemical species, their temperature and mass fraction as a function of time. The 'N_{s+1}' random scalar variables are combined into the vector $\psi = (\psi_1, \psi_2, \psi_3, \cdots, \psi_{N_s}, \psi_{N_{s+1}})$ whose probability density function is 'f' [28]. In internal combustion engines, the density of charge (air–fuel mixture) is varying in combustion cycles. Thus, the mass density function was used in place of probability density function in SRM [31]. The mass density function (F) is related to probability density function (f) and represented as

$$F(\psi, t) = \rho(\psi) f(\psi; t) \tag{1}$$

where ρ denotes the mass density.

The variation in mass density function with time is represented as follows [31]

$$\begin{aligned}\frac{\partial}{\partial t} F(\psi; t) = & -\sum_{j=1}^{N_s+1} \frac{\partial}{\partial \psi_j}\left[G_j(\psi) F(\psi; t)\right] + \sum_{j=1}^{N_s+1} \frac{\partial}{\partial \psi_j}[A(\psi) F(\psi; t)] \\ & - \frac{1}{V}\frac{\mathrm{d}V}{\mathrm{d}t} F(\psi; t) - \frac{\partial}{\partial \psi_{N_{s+1}}}\left[U(\psi_{N+1}) F(\psi; t)\right] \\ & + \frac{F_c(\psi; t)}{\tau_{\text{crev}}} - \frac{F(\psi; t)}{\tau_{\text{cycle}}} + \frac{F_f(\psi; t)}{\tau_f}\end{aligned} \tag{2}$$

where

the term $-\sum_{j=1}^{N_s+1} \frac{\partial}{\partial \psi_j}\left[G_j(\psi) F(\psi; t)\right]$ describes the in-cylinder chemistry process,

the term $\sum_{j=1}^{N_s+1} \frac{\partial}{\partial \psi_j}[A(\psi) F(\psi; t)]$ describes the turbulent mixing,

the term $\frac{1}{V}\frac{\mathrm{d}V}{\mathrm{d}t} F(\psi; t)$ describes the piston movement,

the term $-\frac{\partial}{\partial \psi_{N_{s+1}}}\left[U(\psi_{N+1}) F(\psi; t)\right]$ describes the convective heat transfer,

the term $\frac{F_c(\psi;t)}{\tau_{\text{crev}}} - \frac{F(\psi;t)}{\tau_{\text{cycle}}}$ describes the crevice flow,
and the term $\frac{F_f(\psi;t)}{\tau_f}$ describes the fuel injection [28].

This multi-dimensional mass density function is then computed by using Monte Carlo particle method with a second-order operator splitting algorithm [30].

The chemical reactions have significant effect of the temperature and composition of the species in the combustion chamber. The effect is described by a function $G(\psi)$ and represented as

$$G_j(\psi) = \frac{M_j\dot{\omega}_j}{\rho} \quad j = 1, 2, 3, \ldots, N_s \tag{3}$$

$$G_{N_s+1}(\psi) = -\frac{1}{c_V\rho}\sum_{j=1}^{N_s} e_j m_j \dot{\omega}_j - \frac{p}{c_V\rho}\frac{\mathrm{d}V}{\mathrm{d}t} \tag{4}$$

where

$\dot{\omega}_j$ denotes the production rate,
M_j denotes molecular mass,
e_j denotes specific internal energy of species j,
c_V denotes specific heat capacity at constant volume, and
m and V are the total mass and volume of cylinder.

A more completed discussion of SRM model and relevant equations can be found in the study [5].

By SRM model, thirteen syngas combustion mechanisms developed by researchers are used to simulate HCCI combustion engine. Table 3 presents the details about the reaction mechanisms used in this study for comparison. A study conducted by Olm et al. [32] compared various syngas reaction mechanisms for general combustion. Most of the reaction mechanism used in this study is used to compare in the present study. In few reaction mechanisms, updated version of reaction mechanism is used in the present study. A recently developed mechanism ELTE-2016 [33] is also used for comparison in the present study, which is not compared in earlier study [32].

3 Comparison of Different Syngas Reaction Mechanisms

This section presents the performance of several syngas oxidation reaction mechanisms for predicting combustion characteristics of HCCI engine. Combustion characteristics are analyzed by heat release analysis of cylinder pressure data. To experimentally validate the different syngas reaction mechanisms, measured in-cylinder pressure data of the syngas HCCI engine from a published study [27] is used. Experimental results of the syngas HCCI engine are validated by comparing

Table 3 Details of syngas combustion reaction mechanisms used for comparison

No.	Mechanism	No. of species	No. of reactions	References
1	Ahmed-2007	14	37	[38]
2	Davis-2005	14	38	[11]
3	ELTE-2016	15	44	[33]
4	GRI 3.0-1999	15	48	[35]
5	Keromnes-2013	15	49	[10]
6	Li-2015	14	37	[12]
7	NUIG-NGM-2010	15	41	[9]
8	Ramussen-2008	15	59	[39]
9	Starik-2009	16	44	[15]
10	Sun-2007	15	48	[40]
11	USC-II-2007	14	48	[13]
12	Zseley-2005	13	44	[36]
13	CRECK-2014	32	173	[14]

numerically simulated data using various syngas combustion reaction mechanisms at different operating conditions [such as different inlet valve closing temperature (T_{ivc}) and equivalence ratio (ϕ)]. A combustion reaction mechanism (CRECK-2014) [14] consisting of 32 species and 173 reactions is best matched with experimental results of the study [27]. Figure 3 shows that SRM is able to simulate syngas fueled HCCI engine accurately using CRECK-2014 reaction mechanism.

Figure 3 shows the comparison of predicted in-cylinder pressure with experimental data using different reaction mechanisms at different T_{ivc} and ϕ. Figure 3 indicates that for both the equivalence ratios, with an increase in the T_{ivc}, start of combustion (SOC) is slightly advanced. That means during compression stroke combustion chamber temperature increases and reaches to an autoignition temperature relatively early, which results in the advanced SOC at higher inlet valve closing temperature. A study reported that SOC mainly depends on the T_{ivc} and has lower dependency on ϕ [34]. However, at same operating conditions, some of the syngas reaction mechanisms do not show even the autoignition of the charge and some mechanism has predicted higher peak pressure during HCCI combustion. Detailed comparison reaction mechanism in predicting different combustion parameters is presented in Figs. 4, 5, 6, 7, and 8. For combustion process, other than HCCI engine, a detailed comparison of syngas reaction mechanism can be found in the study [32].

To determine the optimal syngas reaction mechanism for HCCI engine, error in maximum cylinder pressure (P_{max}), SOC (CA_{10}), crank angle position for 50% heat release (CA_{50}), crank angle position for 90% heat release (CA_{90}), and combustion duration (CA_{90-10}) are calculated for different T_{ivc} and ϕ. The error is calculated with respect to experimental data or best-matched syngas reaction mechanism CRECK-2014 [14], which predicted experimental data accurately. Figure 4

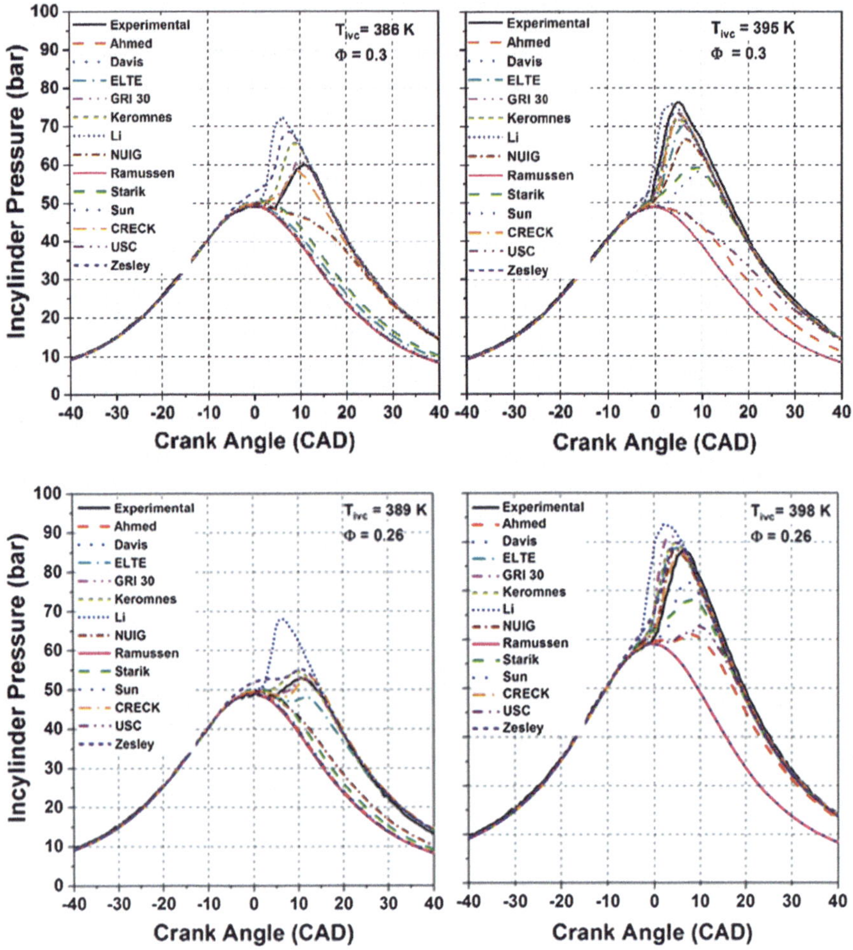

Fig. 3 Comparison of predicted in-cylinder pressure with experimental data using different reaction mechanisms in HCCI engine

indicates the variations in the error percentage of P_{max} for various syngas reaction mechanisms relative to experimental results at different T_{ivc} and ϕ.

Figure 4 reveals that some of the syngas reaction mechanisms such as GRI 3.0-1999 [35], Keromnes-2013 [10], and Zseley-2005 [36] have small variations at lower equivalence ratio (ϕ = 0.26) for both T_{ivc}. In addition to this, the variation of P_{max} is higher for higher equivalence ratio (ϕ = 0.3) for all the test mechanism. Combustion of rich mixture (higher ϕ) leads to increase the P_{max}, and reactions show slightly higher variations in prediction of P_{max} at higher fuel concentration (ϕ).

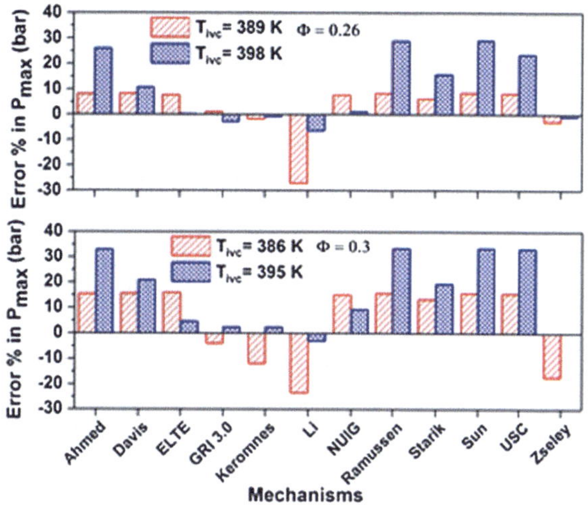

Fig. 4 Error in predicting P_{max} for different syngas reaction mechanisms in HCCI engine

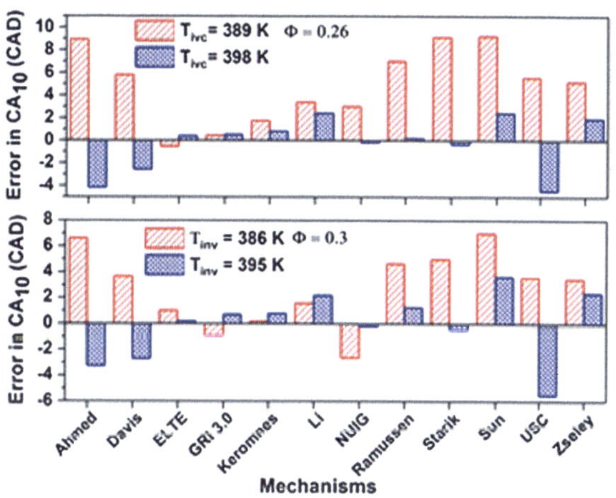

Fig. 5 Error in predicting CA10 using different syngas reaction mechanisms in HCCI engine

In HCCI combustion strategy, SOC mainly depends on the chemical kinetics, along with initial pressure and temperature of the combustion chamber. However, SOC is an essential parameter for performance of HCCI engine. With an increase in the air intake temperature [for any equivalence ratio (ϕ)], SOC advances because of increased reaction rates at higher air intake temperature, which leads to prior autoignition. Figure 5 shows the variations in predicting SOC for various syngas reaction mechanisms relative to experimental results at different T_{ivc} and ϕ. It can be depicted from Fig. 5 that reaction mechanisms ELTE-2016 [33], GRI 3.0-1999 [35], and Keromnes-2013 [10] have lower variations at both the equivalence ratio and T_{ivc}. However, Starik-2009 [15] reaction mechanism has higher variations in CA_{10} for both the T_{ivc} and ϕ.

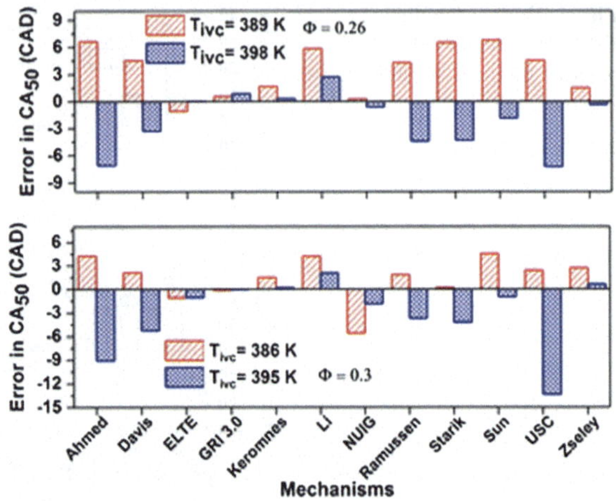

Fig. 6 Error in predicting CA50 for different syngas reaction mechanisms in HCCI engine

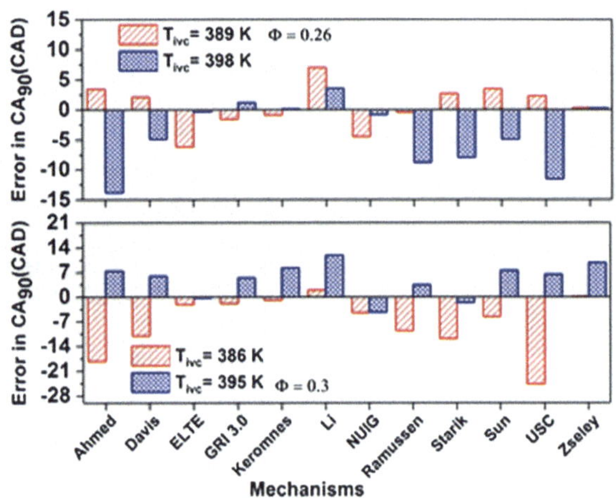

Fig. 7 Error in predicting CA90 for different syngas reaction mechanisms in HCCI engine

The overall combustion phasing during an engine combustion cycle is characterized by CA_{50}. A study reported that early or late combustion phasing directly affects the efficiency of HCCI engine [5]. With the variation in the air intake temperature, ignition timing strongly affects the combustion phasing (CA_{50}). In HCCI combustion strategy, with an increase in the air intake temperature, combustion phasing advances. Additionally, combustion phasing is very sensitive to air intake temperature in HCCI engine. Therefore, prediction of CA_{50} by syngas reaction mechanism is essential. It has been observed that with the reaction mechanisms ELTE-2016 [33], GRI 3.0-1999 [35], Zseley-2005 [36], and Keromnes-2013 [10] have small variations at both the equivalence ratio and T_{ivc} (as shown in Fig. 6). Results also reveal that variations in end of combustion (CA_{90}) are also lower for same mechanisms as for CA_{90} (as shown in Fig. 7).

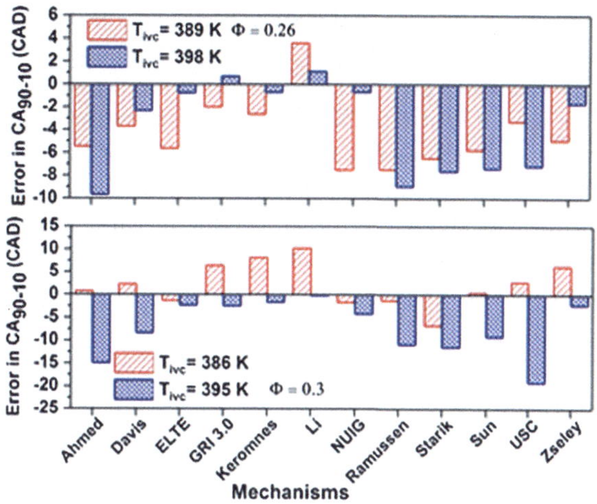

Fig. 8 Error in predicting combustion duration (CA_{90-10}) for different syngas reaction mechanisms in HCCI engine

Combustion duration is defined in terms of difference in crank angle position between 90% and 10% rate of heat release during combustion, i.e., CA_{90-10}. In an HCCI engine, with the retarded start of combustion timing, combustion duration increases rapidly due to reduction in reaction rate at lower combustion temperature. Figure 8 represents the variations in combustion duration for various syngas reaction mechanisms at different T_{ivc} and ϕ. It is observed that with an increase in air intake temperature, combustion duration decreases for both the equivalence ratios. Figure 8 also reveals that for both the equivalence ratio, the reaction mechanisms ELTE-2016 [33], GRI 3.0-1999 [35], and Keromnes-2013 [10] have lower variations for higher T_{ivc} as compared to lower T_{ivc}. However, with some of the reaction mechanism higher reductions in combustion duration are observed.

From the above discussion, it can be summarized that CRECK-2014 is most suitable mechanism for the syngas HCCI engine simulation. However, GRI 3.0-1999 and Keromnes-2013 are the other most preferred mechanism for syngas simulation with minimum error in combustion characteristic prediction of HCCI engine.

4 Summary

This chapter presented the overview of HCCI combustion engine and its simulation using stochastic reactor model. This model is able to accurately simulate the experimental cylinder pressure data using CRECK-2014 syngas oxidation mechanism. The simulation ability of thirteen syngas reaction mechanisms in HCCI engine using the SRM is also compared at different T_{ivc} and equivalence ratio (ϕ). Ability to predict the syngas HCCI combustion characteristics is analyzed using error in prediction of $P_{\max}$, SOC (CA_{10}), combustion phasing (CA_{50}), end of combustion (CA_{90}), and combustion duration (CA_{90-10}) for different T_{ivc} and ϕ. It is concluded that CRECK-2014 reaction mechanism (consisting of 32 species and 173 reactions) provides best accurate prediction of experimental in-cylinder pressure data. It is also found that GRI 3.0-1999 and Keromnes-2013 are the other most preferred mechanism for syngas simulation with minimum error in combustion characteristic prediction of HCCI engine. However, other syngas reaction mechanisms have higher error at certain or all the test conditions, and thus, these mechanisms are not suitable for chemical kinetic modeling of syngas HCCI combustion.

References

1. Maurya RK, Agarwal AK (2015) Experimental investigations of particulate size and number distribution in an ethanol and methanol fueled HCCI engine. J Energy Res Technol 137 (1):012201
2. Maurya RK, Agarwal AK (2014) Particulate morphology and toxicity of an alcohol fuelled HCCI engine. SAE Int J Fuels Lubr **7**(2014-01-9076):323–336
3. Yao M, Zheng Z, Liu H (2009) Progress and recent trends in homogeneous charge compression ignition (HCCI) engines. Prog Energy Combust Sci 35(5):398–437
4. Komninos NP, Rakopoulos CD (2012) Modeling HCCI combustion of biofuels: a review. Renew Sustain Energy Rev 16(3):1588–1610
5. Maurya RK, Akhil N (2016) Numerical investigation of ethanol fuelled HCCI engine using stochastic reactor model. Part 1: development of a new reduced ethanol oxidation mechanism. Energy Convers Manag 118:44–54
6. Maurya RK, Akhil N (2017) Development of a new reduced hydrogen combustion mechanism with NO_x and parametric study of hydrogen HCCI combustion using stochastic reactor model. Energy Convers Manag 132:65–81
7. Veetil JE, Rajith CV, Velamati RK (2016) Numerical simulations of steady perforated-plate stabilized Syngas air pre-mixed flames. Int J Hydrog Energy 41(31):13747–13757
8. Bhaduri S, Berger B, Pochet M, Jeanmart H, Contino F (2017) HCCI engine operated with unscrubbed biomass syngas. Fuel Process Technol 157:52–58
9. Healy D, Kalitan DM, Aul CJ, Petersen EL, Bourque G, Curran HJ (2010) Energy Fuel 24:1521–1528
10. Kéromnès A, Metcalfe WK, Heufer KA, Donohoe N, Das AK, Sung C-J, Herzler J, Naumann C, Griebel P, Mathieu O, Krejci MC, Petersen EL, Pitz WJ, Curran HJ (2013) Combust Flame 160:995–1011
11. Davis SG, Joshi AV, Wang H, Egolfopoulos F (2005) Proc Combust Inst 30:1283–1292
12. Li X, You X, Wu F, Law CK (2015) Proc Combust Inst 35. https://doi.org/10.1016/j.proci.2014.07.047
13. Wang H, You X, Joshi AV, Davis SG, Laskin A, Egolfopoulos F, Law CK. USC Mech version II. High-temperature combustion reaction model of H_2/CO/C1–C4 Compounds. http://ignis.usc.edu/USC_Mech_II.htm/
14. CRECK modeling group hydrogen/CO mechanism version 2014. http://creckmodeling.chem.polimi.it/kinetic.html/
15. Starik AM, Titova NS, Sharipov AS, Kozlov VE (2010) Combust Explos Shock Waves 46:491–506
16. Anders H, Christensen M, Johansson B, Franke A, Richter M, Aldén M (1999) A study of the homogeneous charge compression ignition combustion process by chemiluminescence imaging (No. 1999-01-3680). SAE Technical Paper

17. Maurya RK, Agarwal AK (2009) Experimental investigation of the effect of the intake air temperature and mixture quality on the combustion of a methanol-and gasoline-fuelled homogeneous charge compression ignition engine. Proc Inst Mech Eng Part D J Automob Eng 223(11):1445–1458
18. Liu H, Zheng Z, Yao M, Zhang P, Zheng Z, He B, Qi Y (2012) Influence of temperature and mixture stratification on HCCI combustion using chemiluminescence images and CFD analysis. Appl Therm Eng 33:135–143
19. Maurya RK (2018) Characteristics and control of low temperature combustion engines: employing gasoline, ethanol and methanol. Springer. ISBN 978-3-319-68507-6
20. Agarwal AK, Singh AP, Maurya RK (2017) Evolution, challenges and path forward for low temperature combustion engines. Prog Energy Combust Sci 61:1–56
21. Yamasaki Y, Kaneko S (2014) Prediction of ignition and combustion development in an HCCI engine fueled by syngas (No. 2014-32-0002). SAE Technical Paper
22. Przybyla G, Szlek A, Haggith D, Sobiesiak A (2016) Fuelling of spark ignition and homogenous charge compression ignition engines with low calorific value producer gas. Energy 116:1464–1478
23. Sahoo BB, Sahoo N, Saha UK (2012) Effect of H_2:CO ratio in syngas on the performance of a dual fuel diesel engine operation. Appl Therm Eng 49:139–146
24. Sahoo BB, Saha UK, Sahoo N (2011) Effect of load level on the performance of a dual fuel compression ignition engine operating on syngas fuels with varying H_2/CO content. J Eng Gas Turbines Power 133(12):122802
25. Boehman AL, Corre OL (2008) Combustion of syngas in internal combustion engines. Combust Sci Technol 180(6):1193–1206
26. Martínez JD, Mahkamov K, Andrade RV, Lora EES (2012) Syngas production in downdraft biomass gasifiers and its application using internal combustion engines. Renew Energy 38 (1):1–9
27. Bika AS (2010) Synthesis gas use in internal combustion engines, PhD thesis, University Of Minnesota, USA
28. CMCL innovations, kinetics & SRM Engine SuiteTM, version 8.5.0. (user manual). URL: http://www.cmclinnovations.com (17 Nov 2015)
29. Bhave A, Balthasar M, Kraft M, Mauss F (2004) Analysis of a natural gas fuelled homogeneous charge compression ignition engine with exhaust gas recirculation using a stochastic reactor model. Int J Engine Res 5(1):93–104
30. Bhave A, Kraft M (2004) Partially stirred reactor model: analytical solutions and numerical convergence study of a PDF/Monte Carlo method. SIAM J Sci Comput 25(1):1798–1823
31. Bernard G, Scaife M, Bhave A, Ooi D, Dizy J (2016) Application of the SRM engine suite over the entire load-speed operation of a US EPA Tier 4 capable IC engine (No. 2016-01-0571). SAE Technical Paper
32. Olm C, Zsély IG, Varga T, Curran HJ, Turányi T (2015) Comparison of the performance of several recent syngas combustion mechanisms. Combust Flame 162(5):1793–1812
33. Varga T, Olm C, Nagy T, Zsély IG, Valkó É, Pálvölgyi R, Curran H, Turányi T (2016) Development of a joint hydrogen and syngas combustion mechanism based on an optimization approach. Int J Chem Kinet 48(8):407–422. https://doi.org/10.1002/kin.21006
34. Maurya RK, Akhil N (2016) Numerical investigation of ethanol fuelled HCCI engine using stochastic reactor model. Part 2: parametric study of performance and emissions characteristics using new reduced ethanol oxidation mechanism. Energy Convers Manag 121:55–70

35. Smith GP, Golden DM, Frenklach M, Moriary NW, Eiteneer B, Goldenberg M, Bowman CT, Hanson RK, Song S, Gardiner WC, Lissianski VV, Qin Z. GRI-Mech 3.0. http://www.me.berkeley.edu/gri_mech/
36. Zsély IG, Zádor J, Turányi T (2005) Proc Combust Inst 30:1273–1281
37. Singh G (2010) Overview of the DOE advanced combustion engine R&D. DOE hydrogen program and vehicle technologies program, annual merit review, Washington, DC
38. Ahmed SS, Mauß F, Moréac G, Zeuch T (2007) Phys Chem Chem Phys 9:1107–1126
39. Rasmussen CL, Hansen J, Marshall P, Glarborg P (2008) Int J Chem Kinet 40:454–480
40. Sun H, Yang SI, Jomaas G, Law CK (2007) Proc Combust Inst 31:439–446

Gasoline Compression Ignition—A Simulation-Based Perspective

Janardhan Kodavasal and Sibendu Som

Abstract Gasoline compression ignition (GCI) is an advanced combustion concept for internal combustion engines, where gasoline is ignited purely through compression, without the use of a spark. Combustion is the result of a sequence of autoignition events based on reactivity stratification within the charge. In recent years, GCI has garnered significant interest owing to its potential to deliver diesel-like efficiency with much lower engine-out soot and nitrogen oxides (NO_x) emissions. In this work, we present results from a series of computational fluid dynamics (CFD) simulation studies performed by us to understand the impact of design features and operating conditions on GCI, focusing on idle to low loads, where igniting gasoline purely through compression is challenging. These simulations are based on experiments performed at Argonne National Laboratory (Argonne) on a four-cylinder diesel engine modified to run in GCI mode. We studied the impact of factors like injector nozzle inclusion angle, injection timing, injection pressure, boost level, and swirl ratio. The preignition reaction space from the results was analyzed to understand the interplay between these factors and the overall reactivity. We also delve into the impact of uncertainties in CFD model inputs such as model parameters and initial and boundary conditions on simulation results by performing a global sensitivity analysis (GSA), based on thousands of CFD calculations run on a supercomputer at Argonne.

1 Introduction

With increasingly stringent emissions and fuel economy regulations, there is sustained interest in advanced internal combustion (IC) engine concepts that simultaneously reduce emissions and improve fuel economy over conventional engines. Low-temperature combustion (LTC) encompasses a wide array of combustion modes for IC engines that keep in-cylinder temperatures low primarily through the

J. Kodavasal (✉) · S. Som
Center for Transportation Research, Argonne National Laboratory, Argonne, USA
e-mail: jkodavasal@anl.gov

D. K. Srivastava et al. (eds.), *Advances in Internal Combustion Engine Research*, Energy, Environment, and Sustainability,
https://doi.org/10.1007/978-981-10-7575-9_12

use of techniques like dilution and flame-less combustion. Gasoline compression ignition (GCI) is one such LTC mode, wherein gasoline is used in a compression ignition engine, without the use of a spark plug or throttle [1–16]. GCI can be thought of as the spiritual successor to homogeneous charge compression ignition (HCCI) which garnered significant research interest in the 1980s through the late 2000s. The main challenge with HCCI, where port fuel injection or early direct injection is used for fuel delivery, is that controlling ignition timing is quite challenging, since ignition in HCCI is primarily controlled by chemical kinetics, and we do not really have any control over the charge after intake valve closing (IVC). This control challenge is the main reason why HCCI has been primarily limited to research projects and has not found its way into production vehicles. On the other hand, GCI presents an additional fluid mechanics-based dimension of control that is more closely coupled in time to the ignition event (than IVC)—direct injection of fuel during the compression stroke. Through manipulating injection system design (injector number, geometry, and location) and injection strategy (start of injection or SOI timing, injection pressure, multiple injections, etc.), one can achieve a greater degree of control over the ignition event and burn duration, than is possible with traditional HCCI. Combustion occurs because of sequential autoignition without significant flame propagation [1]. Operation is overall lean, with the absence of significant flame-based combustion, which keeps in-cylinder temperatures low, resulting in reduced heat losses and improved thermodynamic efficiency over traditional spark-ignited gasoline engines, and significantly lower engine-out nitrogen oxides (NO_x) emissions compared to diesel engines. Further, since gasoline is more volatile compared to diesel, fuel–air mixing is improved, resulting in much lower engine-out soot emissions compared to diesel engines [1]. Combustion is the result of sequential autoignition of the fuel–air charge within the cylinder based on reactivity stratification in the charge. This is typically achieved through higher compression ratios (diesel-like) compared to SI engines. Since there is no need to sustain a propagating premixed flame front (as in SI engines), GCI engines do not need throttling at low loads, which results in significant gains in terms of reducing pumping losses compared to SI engines.

Among the major bottlenecks associated with realizing GCI is ensuring stable idle and low-load operation, with brake mean effective pressures (BMEPs) in the range of 0–2 bar. This is because it is difficult to ignite gasoline through compression alone as it has a relatively high octane number, which makes it resistant to autoignition. High octane numbers are desirable in traditional SI combustion, where we do not want the end-gas to autoignite and cause knock; however for GCI, we depend on autoignition for combustion, as there is no flame front. Charge preparation is critical to ensure that the gasoline–air mixture within the cylinder maintains a high level of reactivity under these conditions. One way to ensure this is through optimization of the injection system as well as injection strategy, which can be expedited through computer simulation. This forms one of the big motivations for our simulation work. Through our simulation work [13], we discovered that there is an optimum start of injection (SOI) timing that facilitates reactivity under low-load conditions for a diesel engine that we are running in GCI mode at

Argonne [9]. This optimum injection timing is advanced enough compared to diesel, to provide enough chemical residence time to autoignite gasoline, while at the same time not being too advanced where the fuel spray gets into the squish region, resulting in reactivity losses.

Thus, computer simulations play a critical role in providing insights into the complex interactions between the injection process, spray–geometry interactions, thermal and compositional stratification, and chemical kinetics—insights that simply cannot be obtained from metal engine experiments alone. These insights are vital in designing injection systems and operating strategies that can realize the emissions and efficiency benefits of GCI while maintaining stable ignition and combustion. Computational fluid dynamics (CFD) techniques and computing hardware have evolved significantly over the last decade, enabling higher fidelity of simulation. Other researchers have also performed simulation studies to better understand and optimize GCI [6–9], but there has not been an extensive simulation-driven analysis of this concept unlike traditional gasoline and diesel concepts, which is one of the motivations for our work.

Additionally, another important consideration for simulation-based studies is understanding the sensitivity of model predictions to uncertainties in model inputs. Our group has performed such studies using a technique called global sensitivity analysis (GSA) for diesel spray [17] and diesel engine [18] simulations. Using GSA in conjunction with CFD simulations, we can understand the interactions between different inputs to the model and their relative importance on target predictions from the computer model. In a GSA, a set of model inputs is chosen to be perturbed, and uncertainty ranges are assigned to the inputs about their baseline values. Then, multiple CFD simulations are performed (realizations), where each of the inputs has a random value within its uncertainty range. Further, all inputs are perturbed simultaneously for each realization, resulting in a Monte Carlo sampling of the uncertainty hyperspace. This is different from brute force sensitivity analysis, where each input is perturbed individually. The strength of the GSA approach arises from two factors—(1) interactions between input perturbations are implicitly accounted for and (2) fewer simulations (realizations) are needed to characterize the uncertainty space with the same level of fidelity as the brute force method.

This chapter describes the computational simulation efforts underway at Argonne to inform GCI combustion system design and provide a fundamental understanding of the GCI combustion process. We start off by describing the simulation model used and the engine that forms the basis for this model. We then describe results from our studies investigating the effects of injector design and operating strategy on GCI combustion. This work is primarily focused on the idle to low-load portions of the operating map, since one of the biggest challenges associated with GCI on this engine is achieving gasoline autoignition at lower reactivity (lower fueling) conditions. Then, we explore the sensitivity of this model to uncertainties in model constants as well as experimental measurements through a technique called global sensitivity analysis (GSA).

2 Computer Model and Engine Specifications

The engine that forms the basis for this work is a production four-cylinder 1.9 L turbocharged diesel engine, with an 1800-bar capable common rail diesel injection system. This engine was manufactured by General Motors (GM) for offerings in the European market. At Argonne, this engine has been run—without any functional modifications—in the GCI mode by running 87 anti-knock index (AKI) gasolines with a lubricity enhancer. The stock common rail injection system is employed; however, injection pressures on the order of 250–500 bar are used with gasoline. The key specifications of this engine are listed in Table 1. Note that there are two values of injector inclusion angle shown—148° represents the stock diesel injector for this engine, on which initial experimental testing was performed. Based on the experiments and simulation studies, it was concluded that a smaller, 120° inclusion angle would perform better in the context of GCI (with earlier injection timings compared to conventional diesel), and this is the injector that is currently fitted onto the engine.

Simulations were performed with a 1/7th sector mesh which represents one cylinder of the four cylinders. The mesh is shown in Fig. 1. Only the closed portion of the cycle is simulated going from 132° (intake valve closing—IVC) through

Table 1 Engine specifications

Cylinders	4
Geometric compression ratio	17.8:1
Bore (mm)	82
Stroke (mm)	90.4
Connecting rod length (mm)	145.4
IVC (°bTDC)	132
EVO (°aTDC)	116
Number of injector nozzle holes	7
Nozzle hole diameter (μm)	141
Injector inclusion angle (°)	148/120
Injection pressure (bar)	250–500

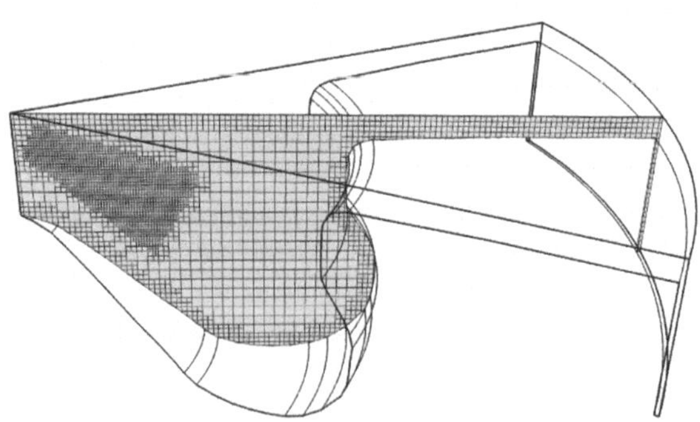

Fig. 1 Sector mesh for CFD model of the engine

116° (exhaust valve opening—EVO), with combustion top dead center (TDC) at 0°. A fuel surrogate comprised of 87% isooctane and 13% *n*-heptane (by mass) was used to represent 87 AKI gasolines. A primary reference fuel (PRF) mechanism by Liu et al. [19] consisting of 48 species and 152 reactions was used to model chemistry. The simulations used a lower compression ratio of 17.5 (relative to measured) to account for potential blow-by-effects and better match motoring pressure.

A base mesh size of 0.7 mm was used with two levels of adaptive mesh refinement (AMR) based on temperature and velocity gradients. Two levels of fixed embedding were activated in the nozzle area and walls. The minimum cell size with these settings is 0.175 mm. A multi-zone model was used for chemistry with bin sizes of 0.5 in equivalence ratio and 5 K in temperature. All simulations performed used Reynolds-averaged Navier–Stokes (RANS) with the renormalization group (RNG) k-e turbulence model and other standard models within the CFD software CONVERGE v2.1 [20]. We have validated this model against experimental data at low load for this engine in our previous work [13], where we also showed that the simulation was converged with respect to grid resolution. More details on models and settings are provided in [21]. The operating conditions studied representing low load and idle are listed in Table 2.

Table 2 Operating conditions for the simulations

Parameter	Low load	Idle
Commanded SOI (°aTDC)	−30	−18.0
SOI actual (°aTDC)	−27.0	−16.2
Fuel (mg/cycle/cylinder)	10.0	4.0
Engine speed (RPM)	1500	850
Load (bar BMEP)	2	0
T_{IVC} (K)	400	400
P_{IVC} (bar)	1.14	1.14
Equivalence ratio	0.35	0.14
EGR (% mass due to leakage)	5	5
Estimated internal residual (% mass)	5	5
Number of injector nozzle holes	7	7
Nozzle hole diameter (μm)	141	141
Baseline nozzle inclusion angle (°)	120	120
Baseline injection pressure (bar)	250	250
Baseline swirl ratio	1.7	1.7
Nozzle inclusion angle (°)	70–160	70–160
Injection pressure (bar)	100–500	100–500
Swirl ratio	0.7–2.0	0.7–2.0

3 Effect of Design and Operating Parameters

With the baseline settings listed in Table 2, we performed a simulation sweep of nozzle inclusion angle for both idle and low load. The sweep ranged from 70° to 160° total inclusion angle. The stock injection system for this engine uses a 148° inclusion angle injector since this is a diesel engine, with a conventional bowl in piston design, with SOI being near top dead center (TDC) when operated in diesel mode. However, relatively early injection timings are needed with gasoline compared to diesel to achieve GCI. This is because gasoline is less chemically reactive than diesel (it has a high octane number, or resistance to autoignition), and thus, it needs more chemical residence time to autoignite, and hence earlier SOI timings. With these early SOI timings, and the larger inclusion angle of 148°, we found that a significant portion of the injected fuel gets into the squish region resulting in reduced reactivity and difficulty in autoignition. A smaller inclusion angle is thus desirable when early SOI timings are used to retain all the injected fuel in the more reactive bowl region of the piston. Based on performing a sweep of various inclusion angles ranging from 70° to 160°, we found that an inclusion angle of 120° is optimum for this particular engine for both low-load and idle conditions with the injection timings used here. In our previous work [12], we showed that inclusion angles larger than 120° tend to spray fuel into the squish, while angles smaller than 120° result in the fuel spray going tangentially across the bowl and also ending up near the head and squish regions due to momentum. Fuel in the squish suffers from higher rates of heat losses, which results in retarded ignition. This is illustrated in Fig. 2.

We also evaluated the effect of injection pressure (ranging from 100 to 500 bar) on charge preparation and ignitability under both these operating conditions. In general, we find that varying injection pressure does not have as strong an impact as varying inclusion angle. The idle condition is more affected by variations in injection pressure than the low-load condition. As injection pressure is reduced, we found that there is a higher degree of local reactivity being retained under idle

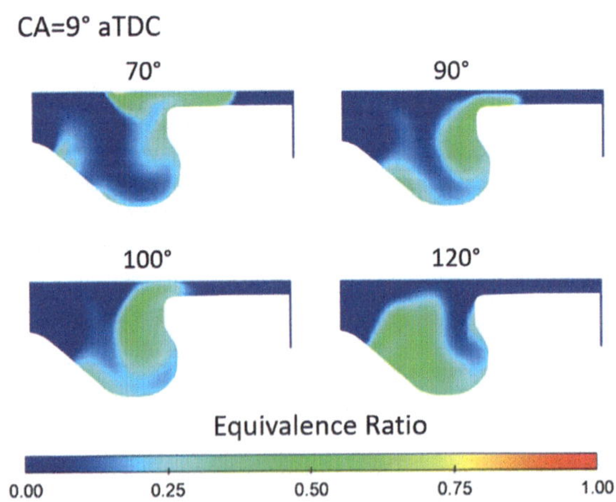

Fig. 2 Fuel vapor distribution for smaller inclusion angles

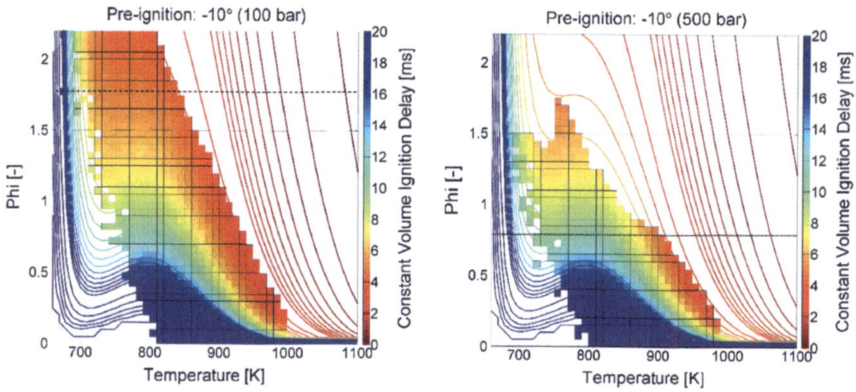

Fig. 3 Effect of injection pressure (100 bar on the left and 500 bar on the right) on preignition reaction (−10° aTDC) space for the idle condition (non-reacting simulations). The horizontal black-dashed line indicates the fuel-mass-weighted equivalence ratio for both cases. The colors and isolines indicate constant volume ignition delays computed using the Goldsborough [22] correlation in every CFD cell for each of these simulations

conditions (where reactivity is at a premium) enhancing autoignition, as lower injection pressure avoids over-mixing and leaning out of the fuel–air mixture. The preignition reaction space under idle load for injection pressures of 100 and 500 bar is shown in Fig. 3. We note that at the 100 bar condition, there is more stratification in the fuel as compared to the 500 bar condition, resulting in higher overall fuel-mass-weighted equivalence ratio (1.75 at 100 bar vs. 0.75 at 500 bar) at the lower injection pressure. In the figure, the horizontal black-dashed line indicates the fuel-mass-weighted equivalence ratio for both cases. The colors and isolines indicate constant volume ignition delays computed using the Goldsborough [22] correlation for isooctane in every CFD cell for each of these simulations. This kind of visualization of the Φ temperature space gives us a better idea of the actual reactivity of various parts of this space from a *qualitative* standpoint, as the real fuel chemistry (gasoline) is more complicated, and a constant volume ignition delay does not account for thermodynamic history of the charge through compression. Nevertheless, this is a useful metric to characterize the reaction space. We note that the 100 bar injection pressure has a larger portion of the fuel distributed in the high-reactivity region (as evidenced by lower constant volume ignition delays) of the Φ temperature space.

Under low-load conditions, reactivity is not at as much of a premium as under idle conditions, and lower injection pressure does not help in this case. In fact, the lowest injection pressure of 100 bar resulted in an elongated injection duration, thus reducing the chemical residence time available for the later portions of the fuel injected, reducing autoignition propensity.

Both the idle and low-load conditions showed increased autoignition propensity and advanced ignition with increasing levels of boost. This is expected as GCI is primarily kinetically controlled and higher pressures increase mixture chemical reactivity. This is

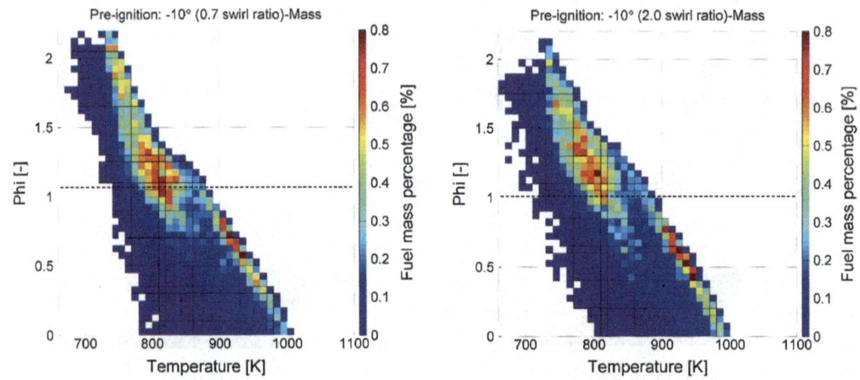

Fig. 4 Effect of swirl ratio on preignition reaction space for the idle condition (non-reacting simulations). The horizontal black line indicates the fuel-mass-averaged equivalence ratio for both cases

in spite of the fact that with higher boost there is a higher level of dilution and overall lower global equivalence ratio for both cases, as the fueling was kept fixed.

Another parameter that we varied was the swirl ratio, which we swept from 0.7 to 2.0. Note that the engine used is a conventional diesel engine and hence swirl is a relevant parameter here. The baseline value is 1.7 for both operating conditions. We found that increasing the swirl ratio tends to retard ignition under both conditions. Our initial hypothesis was that higher swirl ratio values would result in an increase in the mixing of the fuel with the rest of the charge and over-leaning of the charge. However, upon examining the preignition reaction space (at $-10°$ aTDC) for non-reacting simulations corresponding to the reacting CFD simulations, we found that the swirl ratio does not have an appreciable impact on the stratification in the reaction space. As shown for the idle load case as an example (Fig. 4), the distribution of fuel in the Φ temperature space does not appear too different, and the fuel-mass-weighted equivalence ratio is roughly the same for both cases. This indicates that the swirl ratio does not alter the fuel mixing under these conditions.

From Fig. 5, examining the in-cylinder temperatures from non-reacting simulations based on the reacting simulations, we find that with an increased level of swirl ratio, there is a reduction in the in-cylinder temperature near TDC. This is due to increased heat loss with the higher levels of swirl. Since GCI is primarily kinetics-driven, the lower in-cylinder temperatures for the higher swirl ratio cases result in later ignition for those cases, compared to those with lower swirl ratios.

4 Sensitivity to Uncertainty in Model Inputs

Another focus of our efforts was to determine the impact of uncertainty in CFD model inputs (including model parameters, initial and boundary conditions) on the simulation predictions. To do this, we adopted a technique called global sensitivity

Fig. 5 In-cylinder temperatures for the low- and high-swirl cases under idle load (non-reacting)

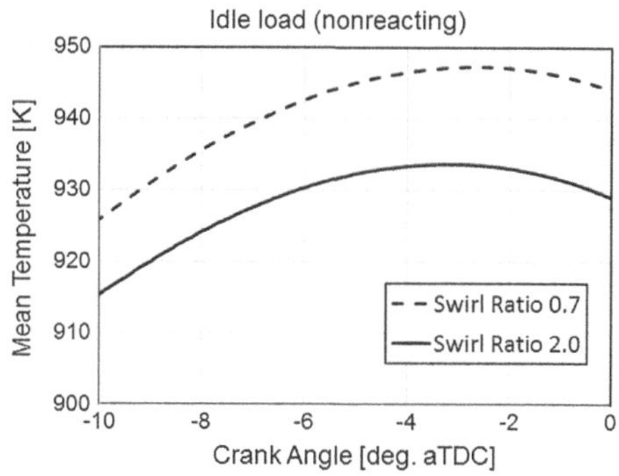

analysis (GSA) wherein we randomly perturb a selected set of model parameters and inputs simultaneously, within their prescribed uncertainty ranges, to generate multiple realizations (on the order of 100–1000) of the baseline CFD simulation. Using GSA, nonlinear relationships between multiple inputs and a given target can be accounted for. Within the chemical kinetics community, GSA is typically used to understand the effects of uncertainties in reaction rates on targets such as ignition delay [23–25]. Based on the predicted values of the targets of interest (such as emissions and combustion characteristics) from all these realizations, we can calculate the sensitivity of these simulation targets to each of the perturbed inputs and also rank these sensitivities. To calculate the sensitivities, we first build a regression model of each of the targets as a function of all the perturbed inputs. This regression model includes a constant term, linear terms, and square terms, and the model parameters (coefficients) are calculated analytically using the normal equation method, which gives us coefficients that minimize the residual sum squares error of the fit with respect to the data. This analytical model is then used to compute individual sensitivity indices. Suppose there are 'N' input variables (34 perturbed inputs in this study) to the model with each of these variables having a *uniform* probability distribution within their respective uncertainty ranges, then we define the following:

$$v_i \in \left[v_i^L v_i^H\right], \quad i = 1, 2, \ldots, N \tag{1}$$

Here, v_i represents the ith input variable, and v_i^L and v_i^H are the lower and upper limits of its uncertainty, respectively. For example, one of the N inputs perturbed in our study is the intake valve closing temperature (T_{IVC}), and this can take values anywhere between 386.5 K ($T_{\min}$ or $v_{\mathrm{TIVC}}^{\mathrm{L}}$) and 392.5 K ($T_{\max}$ or $v_{\mathrm{TIVC}}^{\mathrm{H}}$). Similarly, we would have ranges of values for other inputs such as intake valve closing pressure (P_{IVC}) and nozzle inclusion angle (θ_{noz}). To illustrate how the simulations performed are distributed in the input space, Fig. 6 shows what the distribution of the CFD simulations performed to generate data for the sensitivity analysis would look like if only two inputs were perturbed (N = 2), and Fig. 7 shows the distribution if only three inputs were perturbed (N = 3). Of course, we cannot visually

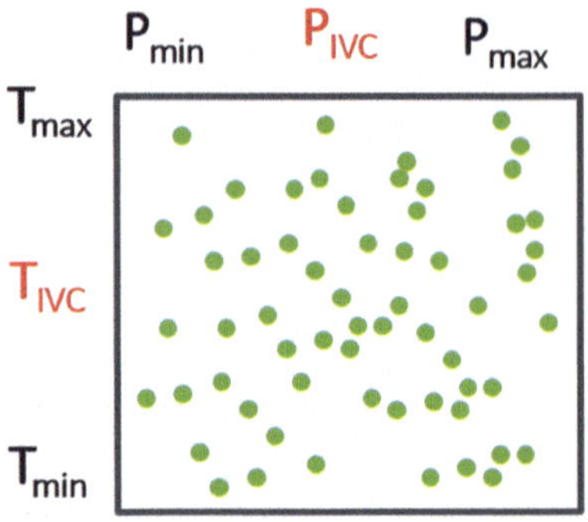

Fig. 6 Illustration showing how CFD simulations would be distributed if only two variables were perturbed—each green dot represents a unique CFD simulation that has a certain value of T_{IVC} and a certain value of P_{IVC}, between their minimum and maximum allowed values based on the uncertainty ranges prescribed to them

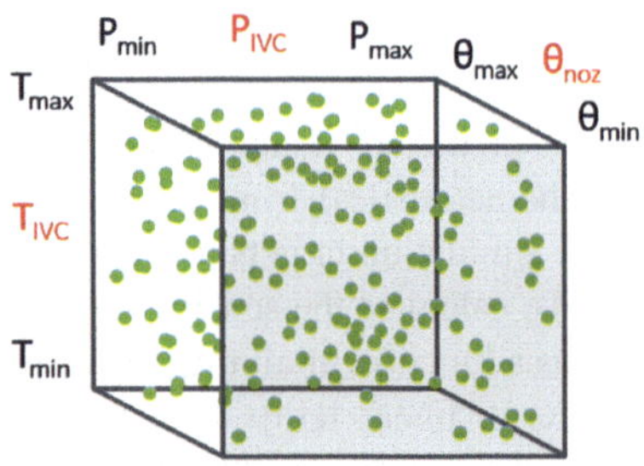

Fig. 7 Illustration showing how CFD simulations would be distributed if only three variables were perturbed—each green dot represents a unique CFD simulation that has a certain value of T_{IVC}, a certain value of P_{IVC}, and a certain value of θ_{noz}, between their minimum and maximum allowed values based on the uncertainty ranges prescribed to them

represent a hypercube that has more than three dimensions. In our current work, the hypercube has $N = 34$ dimensions, and we sample randomly and uniformly from this 34-dimensional hypercube.

To generalize and represent mathematically, the probability distribution of each of the N variables is assumed to be uniform within the N-dimensional uncertainty hypercube [25] given by Eq. (2).

$$\Omega = \left[v_1^L \quad v_1^H\right] \otimes \cdots \otimes \left[v_i^L \quad v_i^H\right] \otimes \cdots \otimes \left[v_N^L \quad v_N^H\right] \tag{2}$$

To obtain the values of the input variables within the uncertainty hypercube, the Monte Carlo method is used to generate a large number, 'M' sets, of values for the variables independently and randomly selected from their ranges. In the present study, 1024 simulations are used for each operating conditions, i.e., $M = 1024$. Each of the targets (outputs) obtained from the simulations based on the randomly sampled variables can be written as:

$$\mathbf{f} = \mathbf{f}(\mathbf{v}), \tag{3}$$

where **f** is a regression function that predicts the target, based on the perturbed inputs represented by the vector v.

$$f(\mathbf{v}_i) = \sum_{i=1}^{N}\sum_{k=0}^{n} a_{ik}v_i^k, \tag{4}$$

where n is the order of the expansion or the degree of the polynomial we use to fit our regression curve; here, n = 2. Note this is a diagonal expression; i.e., there are no cross-terms. The effect of cross-terms, which require larger numbers of simulations, will be a topic of future studies.

For a specific example illustration of what Eq. (4) looks like, let us say we are trying to represent the output CA50 (crank angle of 50% mass percent burned), as a function of the perturbed inputs. Let us assume that we have only perturbed two inputs, T_{IVC} (v_1) and P_{IVC} (v_2) for ease of explanation. Thus, here, N = 2 (two inputs), and n = 2 (polynomial of degree = 2), and we do not have cross-terms. Then, we would like to represent CA50 as follows:

$$\mathrm{CA50} = f(v) = a_0 + a_1T_{\mathrm{IVC}} + a_2P_{\mathrm{IVC}} + a_3T_{\mathrm{IVC}}^2 + a_4P_{\mathrm{IVC}}^2 \tag{5}$$

Here, we have essentially fit a quadratic polynomial model, without cross-terms (we do not have the term $T_{\mathrm{IVC}} * P_{\mathrm{IVC}}$) to predict CA50, given T_{IVC} and P_{IVC}. Here, f is the function represented by the right-hand side of Eq. (5). We can treat this as a linear regression problem and determine the model parameters ($a_0 \ldots a_4$), by considering T_{IVC}^2 and P_{IVC}^2 as new, distinct inputs for the model. We can then use the normal equation method (linear regression) to find what values of $a_0 \ldots a_4$ give us the best fit, or in other words the smallest sum of the squares of the errors of our regression model's prediction compared to the actual value of CA50 from the CFD simulations we have performed.

The sensitivity coefficient (S_i) for each variable can then be calculated based on the ratio of the partial variance and the total variance as shown below:

$$s_i = \sigma_i^2/\sigma_t^2, \tag{6a}$$

$$\sigma_t^2 = \langle target^2 \rangle - \langle target \rangle^2, \tag{6b}$$

$$\sigma_i^2 = \mathrm{var}(E(f|v_i)) = E\left(E(f|v_i)^2\right) - (E(E(f|v_i)))^2 \tag{6c}$$

Here, σ_t^2 represents the total variance of the target (say CA50) over the whole dataset, and the brackets "$\langle\ \rangle$" indicate the mean value over the M simulations, and Eq. (6b) just represents the formula for computing variance for a set of observations. We compute this quantity *directly* from the CFD predictions of that target (total M observations) and do not use the regression model to calculate σ_t^2. The quantity σ_i^2 stands for the partial variance of the regression function f in Eq. (4) over the ith input, which is calculated as the variance of the expected value of the regression function (f) as a function of $\mathrm{v_i}$ alone over the range of values that $\mathrm{v_i}$ can take. The notation $E(\,)$ represents "expected value" of the quantity within the parentheses. Note that the variance of any quantity x, var(x), is equal to $E(x^2)$ minus $[E(x)]^2$, and this is how we calculate $\mathrm{var}(E(f|v_i))$.

Since the mathematical notation can be quite confusing, we illustrate here by reverting to our simple example from Eq. (5). Let us assume we want to evaluate the partial variance of CA50 (our target) over the input T_{IVC} only (σ^2_{TIVC}). First, we want to find $E(\mathrm{CA50}|T_{\mathrm{IVC}})$. To do this, we have to take our expression for f in Eq. (4) and integrate over the other inputs (in this case, only one other input, P_{IVC}), and apply limits of integration as $P_{\min}$ and $P_{\max}$ to find the expected (in other words average over the whole P_{IVC} dimension) value of the f as a function of T_{IVC} only.

$$E(CA50|T_{\mathrm{IVC}}) = \int_{P_{\min}}^{P_{\max}} f(T_{\mathrm{IVC}}, P_{\mathrm{IVC}}) \cdot \mathrm{PDF}(P_{\mathrm{IVC}})\mathrm{d}P_{\mathrm{IVC}} \tag{7}$$

To simplify evaluation of this integral, let us assume that all our inputs were normalized between 0 and 1, where 0 represents the minimum value of the input and 1 represents the maximum value of the input. Further, since we have uniformly sampled our inputs between their minimum and maximum (normalized to 0 and 1), the PDF or probability density function of every input is equal to 1, i.e., PDF (T_{IVC}) = PDF(P_{IVC}) = 1 in the interval [0,1] and 0 everywhere else. Thus, the expression for $E(\mathrm{CA50}|T_{\mathrm{IVC}})$ simplifies to:

$$E(\mathrm{CA50}|T_{\mathrm{IVC}}) = \int_{P_{\mathrm{IVC}}=0}^{P_{\mathrm{IVC}}=1} f(T_{\mathrm{IVC}}, P_{\mathrm{IVC}})\mathrm{d}P_{\mathrm{IVC}} \tag{8}$$

$$\begin{aligned}
\sigma^2_{T_{\mathrm{IVC}}} &= \mathrm{var}(E(\mathrm{CA50}|T_{\mathrm{IVC}})) \\
&= E\left(E(\mathrm{CA50}|T_{\mathrm{IVC}})^2\right) - (E(\mathrm{CA50}|T_{\mathrm{IVC}}))^2 \\
&= \int_{T_{\mathrm{IVC}}=0}^{T_{\mathrm{IVC}}=1} (E(\mathrm{CA50}|T_{\mathrm{IVC}}) \cdot \mathrm{PDF}(T_{\mathrm{IVC}}))^2 \mathrm{d}T_{\mathrm{IVC}} \\
&\quad - \left(\int_{T_{\mathrm{IVC}}=0}^{T_{\mathrm{IVC}}=1} E(\mathrm{CA50}|T_{\mathrm{IVC}}) \cdot \mathrm{PDF}(T_{\mathrm{IVC}})\mathrm{d}T_{\mathrm{IVC}}\right)^2 \\
&= \int_{T_{\mathrm{IVC}}=0}^{T_{\mathrm{IVC}}=1} (E(\mathrm{CA50}|T_{\mathrm{IVC}}))^2 \mathrm{d}T_{\mathrm{IVC}} \\
&\quad - \left(\int_{T_{\mathrm{IVC}}=0}^{T_{\mathrm{IVC}}=1} E(\mathrm{CA50}|T_{\mathrm{IVC}})\mathrm{d}T_{\mathrm{IVC}}\right)^2
\end{aligned} \tag{9}$$

The sensitivity coefficient can provide the sensitivity of each input variable to the target function over the entire input space. Therefore, the sensitivity coefficient can provide the sensitivity of each input variable to the target function over the entire input space. A more detailed derivation is given in our prior paper [26]. The input variables that were selected for this study are listed in Tables 3, 4, and 5. In total, 34 variables representing initial or boundary conditions from experiments, fuel properties, and CFD model parameters were chosen to be perturbed in this GSA, for each of the low-load and the idle operating conditions. Each variable is assigned an uncertainty range (i.e., a minimum and maximum value), which is a combination of measurement uncertainty and cyclic variations. For the CFD model parameters, the range was based on the range of values typically used in engine simulations. For the experimental boundary conditions, the ranges were based on typically observed/expected variation in these quantities. A total of 1024 simulations were run for each operating condition on an IBM Blue Gene/Q supercomputer. The reader is advised that changing the ranges of these uncertainties can result in a different ordering of the sensitivities. The baseline setup values for these variables are also listed in the table.

Table 3 Experimental and fuel property inputs perturbed for the low-load case

Variable	Units	Description	Baseline	Min	Max
T_{piston}	K	Piston temperature	400	385	415
$T_{cylinder}$	K	Cylinder temperature	380	370	390
T_{head}	K	Cylinder head temperature	400	385	415
RPM	–	Engine speed	1500	1495	1505
SR	–	Swirl ratio	2.2	2.0	2.4
T_{tke}	m^2/s^2	Initial turbulence	1.6	1.2	2.0
L_i	mm	Initial turbulence length scale	8.2	3.0	13.0
EGR fraction	–	Exhaust gas recirculation	0.1	0.08	0.12
T_{IVC}	K	Temperature at IVC	389.5	386.5	392.5
P_{IVC}	bar	Pressure at IVC	1.38	1.35	1.40
$T_{f,crit}$	K	Fuel critical temperature	540	530	550
ρ_f	–	Fuel density	1.00	0.95	1.05
HOV_f	–	Fuel heat of vaporization	1.0	0.9	1.1
VPf	–	Fuel vapor pressure	1.0	0.9	1.1
μ_f	–	Fuel viscosity	1.0	0.7	1.3
D_{noz}	μm	Nozzle diameter	141	139	143
θ_{noz}	°	Nozzle inclusion angle	148	144	152
C_d	–	Discharge coefficient	0.80	0.75	0.85
SOI	°CA	Start of injection	−21.0	−21.2	−20.8
DOI	°CA	Duration of injection	9.0	8.8	9.2
M_{inj}	mg	Fuel mass injected	9.68	9.20	10.16
T_{inj}	K	Injected fuel temperature	353	348	358

Table 4 Experimental inputs perturbed for the idle case (only the inputs that have different ranges compared to those listed in Table 3 are listed here)

Variable	Units	Description	Baseline	Min	Max
RPM	–	Engine speed	850	845	855
SR	–	Swirl ratio	1.7	1.5	1.9
T_{IVC}	K	Temperature at IVC	400	397	403
P_{IVC}	bar	Pressure at IVC	1.14	1.10	1.18
θ_{noz}	°	Nozzle inclusion angle	120	116	124
C_d	–	Discharge coefficient	0.70	0.65	0.75
SOI	°CA	Start of injection	−16.2	−16.4	−16.0
DOI	°CA	Duration of injection	2.15	2.10	2.20
M_{inj}	mg	Fuel mass injected	4.0	3.8	4.2
T_{inj}	K	Injected fuel temperature	340	335	345

Table 5 CFD model parameters perturbed and their ranges for both the low-load and idle conditions

Variable	Description	Baseline	Min	Max
ceps1	Turbulence dissipation const	1.42	1.35	1.5
ceps2	Turbulence dissipation const	1.68	1.6	1.76
c_s	Turbulent spray source const	0	0	1.5
c_ps	Drop turbulent dispersion const	0.03	0	0.16
schmidt	Schmidt number	0.78	0.6	0.9
prandtl	Prandtl number	0.9	0.85	0.95
Balpha	KH breakup model size const	0.6	0.55	0.65
kh_cnst2	KH breakup model time const	7	5	9
rt_cnst2b	RT breakup model time const	1	0.5	2
cnst3rt	RT breakup model size const	0.1	0.1	1
cone_noz	Spray cone angle	9	7	12
new parcel cut off	Fraction of injected mass per parcel used to determine when to create a child parcel in the KH breakup model	0.05	0.03	0.07

Figures 8 and 9 show the spread of in-cylinder pressure and NO_x evolution, respectively, for the 1024 cases of the low-load operating condition. Figures 10 and 11 show the spread of in-cylinder pressure and NO_x evolution, respectively, for the 1024 cases run of the idle operating condition. It may be seen from these figures that there is a significant spread in combustion and emission predictions under both operating conditions based on the perturbed inputs. The reason for this variation is that the inputs going into each of the 1024 simulations are different, based on our imposed perturbations. Some inputs such as T_{IVC} can have a large impact on the ignition timing, combustion characteristics, and consequently emissions, as GCI is a kinetically controlled combustion mode, where in-cylinder temperature largely determined ignition timing.

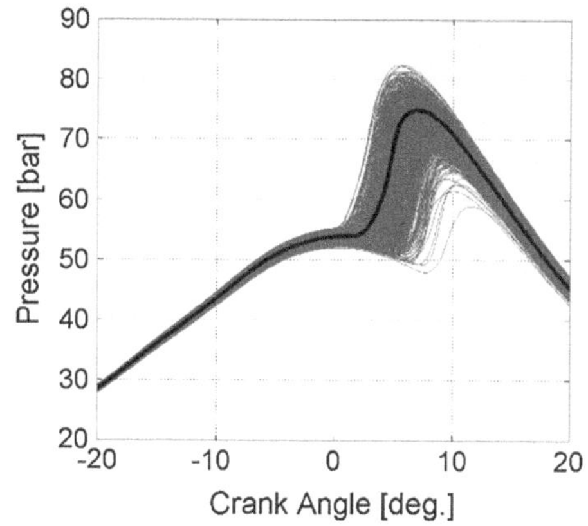

Fig. 8 Spread in in-cylinder pressure traces from all 1024 cases simulated at the low-load condition—baseline pressure trace shown as solid black line

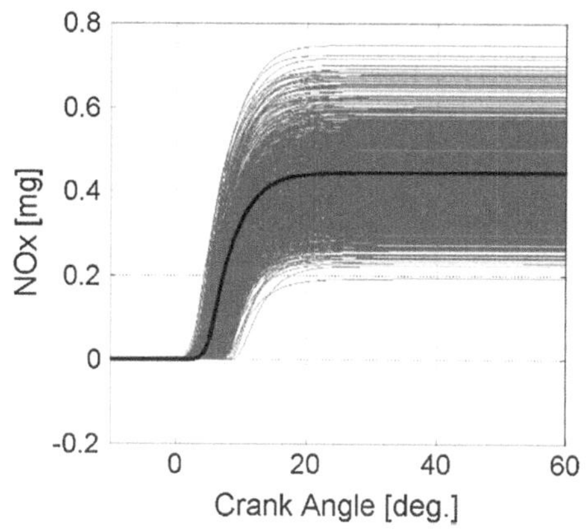

Fig. 9 Spread in NO_x emissions from all 1024 cases simulated at the low-load condition—baseline NO_x shown as solid black line

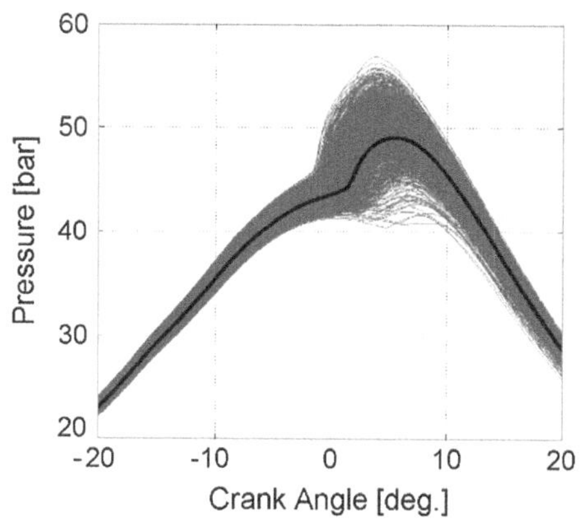

Fig. 10 Spread in in-cylinder pressure traces from all 1024 cases simulated at the idle condition—baseline pressure trace shown as solid black line

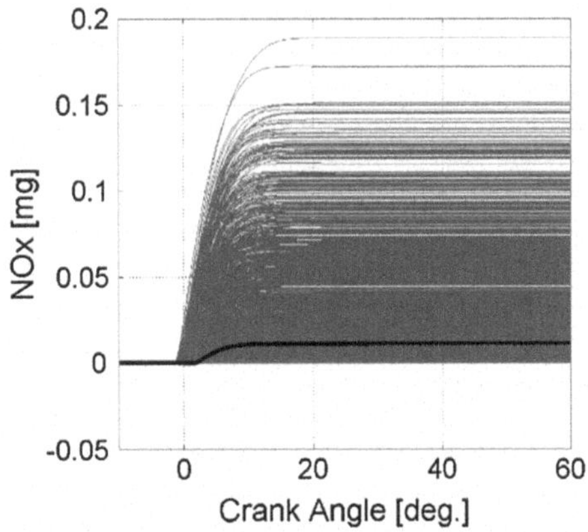

Fig. 11 Spread in NO_x emissions from all 1024 cases simulated at the idle condition—baseline NO_x shown as solid black line

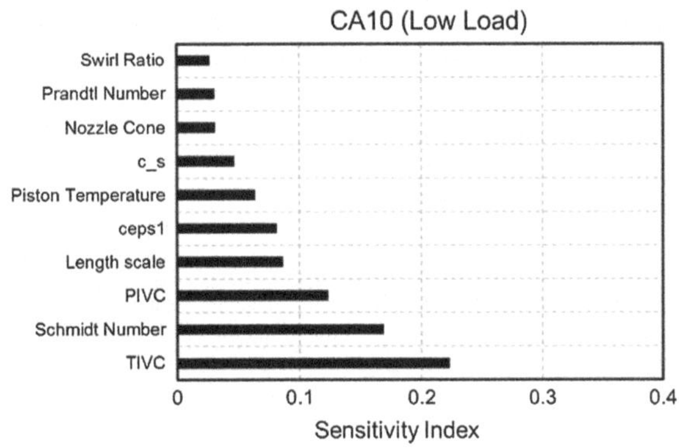

Fig. 12 Sensitivity of ignition timing (CA10) to various inputs at the low-load condition

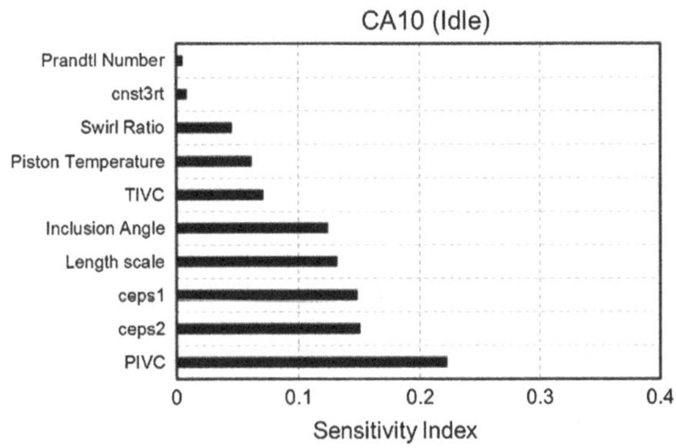

Fig. 13 Sensitivity of ignition timing (CA10) to various inputs at the idle condition

Figures 12, 13, 14, 15, 16, and 17 show the rankings of the sensitivities of selected targets to the perturbed inputs, under both operating conditions. The top ten ranking inputs have been shown in every figure. The sensitivity index is normalized

Fig. 14 Sensitivity of combustion phasing (CA50) to various inputs at the low-load condition

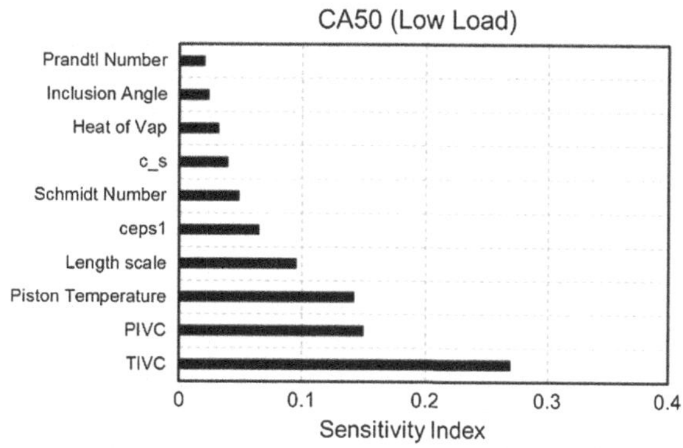

Fig. 15 Sensitivity of combustion phasing (CA50) to various inputs at the idle condition

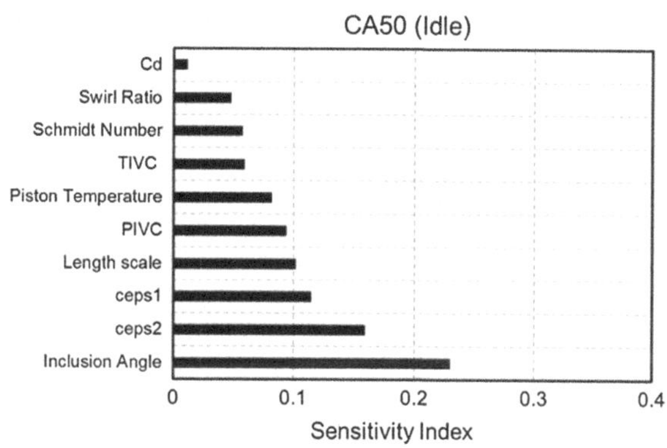

Fig. 16 Sensitivity of NO_x to various inputs at the low-load condition

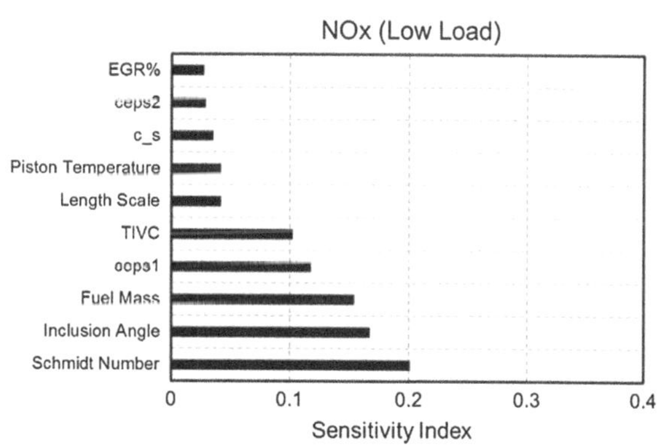

based on the sum of all the sensitivities of all the 34 inputs perturbed. One key observation from these figures is that both uncertainties in boundary conditions and fuel properties, as well as uncertainties in CFD model parameters, have a similar order effect on the targets of interest; i.e., they both show up in the rankings of the

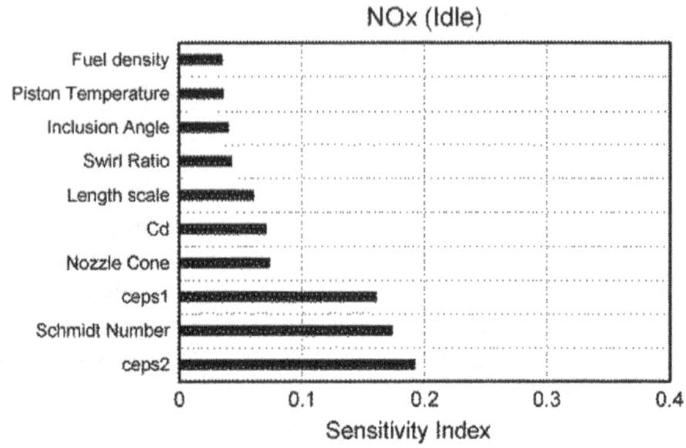

Fig. 17 Sensitivity of NO_x to various inputs at the idle condition

top ten parameters for various targets, without one form of uncertainty completely dominating any given target. It must be noted that in our previous work [27], we studied the effect of only perturbing experimental boundary conditions and fuel properties.

Figures 12 (low load) and 13 (idle) show the impact of uncertainties in input parameters for the CA10 ignition timing target. From Fig. 12, we note that T_{IVC} and P_{IVC} seem to have a significant impact on CA10 under low load, whereas for the idle condition (Fig. 13), T_{IVC} has a relatively smaller impact. Does this perhaps indicate that at idle conditions, reactivity enhancement by means of boosting might have a bigger impact than elevating T_{IVC} through some means? Of course, this cannot be conclusively said as this is a sensitivity study on uncertainties, and not an actual parametric variation. The range of perturbation of these values is much smaller, than if the objective were to examine the ability of an input to control a target. Another important note here is that the Schmidt number seems to have a significant effect on CA10 under low-load conditions, but does not feature in the ranking for this target at idle. It is interesting to note that the imposed uncertainties in the turbulence model constants, ceps1 and ceps2, have a significant impact on CA10 for the idle condition, but not as much for the low-load condition. The sensitivities of the CA5 ignition timing target have been omitted for brevity.

Figures 14 (low load) and 15 (idle) show the impact of uncertainties in input parameters for the CA50 combustion phasing target. As with CA10, we note that T_{IVC} and P_{IVC} seem to have a significant impact on CA50 for low load, whereas for the idle condition, these have a relatively smaller impact. We further note that CA50 is significantly impacted by the nozzle inclusion angle for the idle condition, while this is not the case for the low-load condition. It must be noted that the low-load condition has a baseline nozzle inclusion angle of 148°, while the idle condition has a baseline nozzle inclusion angle of 120°. Again, as with CA10 ignition timing, the turbulence model constants, ceps1 and ceps2, seem to have a significant impact on CA50 combustion phasing for the idle condition, while this is not the case for the low-load condition.

Figures 16 and 17 show the sensitivity of NO_x emissions to various inputs at low load and idle, respectively. From these figures, it seems that the Schmidt number is an important input for NO_x predictions under both low-load and idle conditions. The turbulence constant ceps2 seems to dominate NO_x for the idle condition, but does not have a large impact on the low-load condition. The turbulence constant ceps1 has a notable impact on NO_x at both low-load and idle conditions. The inclusion angle has a significant impact on NO_x for the low-load condition, but not for the idle condition. The nozzle cone angle and discharge coefficient (C_d) have an impact on NO_x for the idle condition, but do not show up in the top ten inputs for the low-load condition. The exhaust gas recirculation (EGR) percentage has somewhat of an impact on NO_x for the low-load condition, but not on the idle condition. This is understandable, as both conditions have quite low overall equivalence ratio, and EGR is practically fresh air.

Additionally, we examined the impact on HC and CO emission predictions. The rankings for these are not shown for brevity. We found that perturbations in charge temperature (affected by T_{IVC} and T_{piston}) have the biggest impact on HC emission predictions. Additionally, we found that fuel mass and swirl ratio have a significant impact on HC predictions at low load, but this is not the case at idle. The nozzle inclusion angle and the turbulence constant, ceps1, have a notable impact on this target for the idle case, while this target does not seem as sensitive to these inputs for the low-load condition. We found that the perturbation in the nozzle inclusion angle has a significant impact on the predictions of CO, for both operating conditions. The input T_{IVC} also has a marked impact on this target for both conditions. The turbulence constant ceps1 has a significant impact on CO for the idle condition, but a minor impact on the low-load condition. Fuel mass has a high impact on CO for the low-load condition, but this is not the case for the idle condition.

In order to understand the impact of variation in fuel properties alone, we performed another GSA—only for the low-load condition as an example—where we perturbed just the fuel properties ($T_{f,crit}$, ρ_f, HOV_f, VP_f, and μ_f) in the same Monte Carlo fashion based on the ranges provided in Table 3. We performed 128 simulations (realizations) for this analysis. Figure 18 shows the sensitivity rankings for various fuel properties in terms of the impact of their uncertainties on the target predictions CA50, CO, NO_x, and HC.

We note from Fig. 18 that clearly the uncertainty in HOV dominates target predictions. The critical temperature also has a major impact on all the targets. Viscosity has a not insignificant impact on the emissions, but its impact on CA50 is negligible. Uncertainties in density and vapor pressure do not seem to be a major factor in terms of these target predictions.

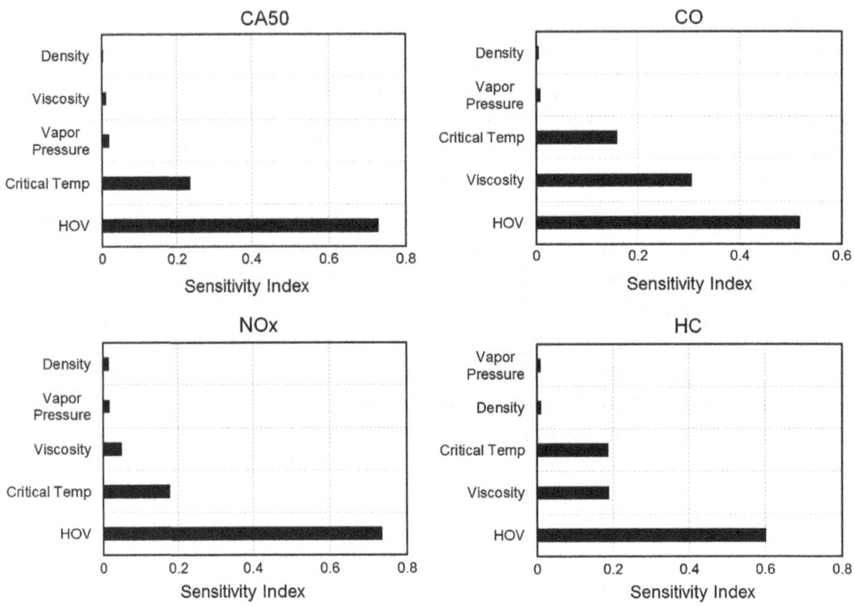

Fig. 18 Sensitivity CA50, CO, NO_x, and HC to uncertainties in fuel properties at the low-load condition

5 Summary

In this work, we have presented a simulation perspective on GCI. We investigated the effect of various operating and design parameters on achieving stable idle to low-load operation with GCI. In general, parameters that enhance reactivity of the charge within the combustion chamber are desirable under these conditions in order to facilitate compression ignition of gasoline without a spark. We found that for this engine, under these conditions there is an optimum nozzle inclusion angle, which is 120°. Inclusion angles larger or smaller than this value both tend to deposit fuel in the squish region, resulting in reduced reactivity and retarding of ignition. Injection pressure does not play a significant role under low-load conditions (for the range of injection pressures studied here—100 to 500 bar). However, under idle conditions, lower injection pressures reducc the possibility of over-leaning of the charge, retaining a higher level of reactivity, resulting in earlier ignition. Higher levels of boost aid GCI under these idle to low-load conditions, as ignition in GCI is primarily driven by kinetics. Higher in-cylinder pressures result in higher rates of reactions leading to autoignition, and thus earlier combustion phasing even when overall dilution is increased with boost. Swirl ratio does not have a significant impact on the charge stratification in the reaction space; however, higher values of swirl ratios tend to increase heat losses, resulting in lower in-cylinder temperatures and retarded ignition.

In order to understand the impact of uncertainties in model inputs on model predictions of key combustion and emissions targets, we performed global sensitivity analysis based on this GCI CFD simulation. The analysis indicates that there can be a significant spread in the predicted combustion and emissions based on the uncertainties in these model inputs. In particular, the uncertainties in temperature and pressure at IVC, inclusion angle, Schmidt number, and piston temperature have the greatest impact on the combustion and emission targets. Uncertainty in inputs such as initial turbulence, RPM, SOI, DOE, nozzle diameter, injection temperature, head temperature, and liner temperature did not have a significant impact on the predicted targets. Based on the stand-alone GSA on fuel properties at the low-load condition, we note that the HOV has the highest impact on combustion and emission targets. The fuel critical temperature is also a significant factor. The viscosity impacts emissions, but not combustion phasing. Uncertainties in density and vapor pressure do not have a major impact on any of the targets.

GCI is a viable alternative to current gasoline and diesel engines, by simultaneously addressing the efficiency issues associated with gasoline spark-ignited engines and emission issues associated with diesel engines. Simulations play a vital role in developing the GCI concept, as demonstrated in this chapter, by providing valuable fundamental insights into this novel combustion mode.

Disclaimer and Funding Acknowledgement The submitted manuscript has been created by UChicago Argonne, LLC, Operator of Argonne National Laboratory (Argonne). Argonne, a US Department of Energy Office (DOE) of Science Laboratory, is operated under Contract No. DE-AC02-06CH11357. The US government retains for itself, and others acting on its behalf, a paid-up non-exclusive, irrevocable worldwide license in said article to reproduce, prepare derivative works, distribute copies to the public, and perform publicly and display publicly, by or on behalf of the government. This research was funded by US DOE Office of Vehicle Technologies, Office of Energy Efficiency and Renewable Energy under Contract No. DE-AC02-06CH11357. The authors wish to thank Gurpreet Singh and Leo Breton, program managers at DOE, for their support. This research used resources of the Argonne Leadership Computing Facility (ALCF), which is a DOE Office of Science User Facility supported under Contract DE-AC02-06CH11357. We gratefully acknowledge the computing resources provided by the ASCR Leadership Computing Challenge (ALCC) Award of 60 million core-hours on Mira supercomputer at the ALCF at Argonne National Laboratory. We would also like to acknowledge computational resources provided by the Argonne Laboratory Computing Resource Center. We wish to thank Drs. Stephen Ciatti and Christopher Kolodziej for experimental data and valuable discussions on GCI. We also wish to acknowledge Drs. Albert Wagner, Noah Van Dam, Yuanjiang Pei, and Michael Davis for advice on performing the GSA. We thank Kevin Harms, Dr. Marta Garcia, and Dr. Ketan Maheshwari for helping us run the GSA on ALCF systems. We acknowledge Dr. Peter Kelly Senecal of Convergent Science for providing us with CONVERGE licenses and support, and for valuable discussions.

References

1. Ciatti S (2015) FY2015 annual progress report—advanced combustion engines research and development. Sec. II-15. US Department of Energy
2. Kalghatgi GT, Risberg P, Ångström HE (2006) Advantages of fuels with high resistance to auto-ignition in late-injection, low-temperature, compression ignition combustion. SAE Technical Paper 2006-01-3385
3. Manente V, Johansson B, Tunestal P (2009) Partially premixed combustion at high load using gasoline and ethanol, a comparison with diesel. SAE Technical Paper 2009-01-0944
4. Sellnau M, Sinnamon J, Hoyer K, Husted H (2011) Gasoline direct injection compression ignition (GDCI)—diesel-like efficiency with low CO_2 emissions. SAE Technical Paper 2011-01-1386
5. Opat R, Ra Y, Gonzalez MA, Krieger R, Reitz RD Foster DE Durrett RP, Siewert RM (2007) Investigation of mixing and temperature effects on HC/CO emissions for highly dilute low temperature combustion in a light duty diesel engine. SAE Technical Paper 2007-01-0193
6. Ra Y, Yun JE, Reitz RD (2009) Numerical parametric study of diesel engine operation with gasoline. Combust Sci Tech 181(2):350–378
7. Adhikary BD, Ra Y, Reitz R, Ciatti S (2012) Numerical optimization of a light-duty compression ignition engine fuelled with low-octane gasoline. SAE Technical Paper 2012-01-1336
8. Badra JA, Sim J, Elwardany A, Jaasim M, Viollet Y, Chang J, Amer A, Im HG (2016) Numerical simulations of hollow-cone injection and gasoline compression ignition combustion with naphtha fuels. J Energy Resour Technol 138(5):052202
9. Ciatti S, Johnson M, Adhikary BD, Reitz R, Knock A (2013) Efficiency and emissions performance of multizone stratified compression ignition using different octane fuels. SAE Technical Paper 2013-01-0263
10. Hanson R, Splitter D, Reitz R (2009) Operating a heavy-duty direct-injection compression-ignition engine with gasoline for low emissions. SAE Technical Paper 2009-01-1442
11. Kolodziej C, Ciatti S, Vuilleumier D, Adhikary BD, Reitz RD (2014) Extension of the lower load limit of gasoline compression ignition with 87 AKI gasoline by injection timing and pressure. SAE Technical Paper 2014-01-1302
12. Kolodziej CP, Kodavasal J, Ciatti S, Som S, Shidore N, Delhom J (2015) Achieving stable engine operation of gasoline compression ignition using 87 AKI gasoline down to idle. SAE Technical Paper 2015-01-0832
13. Kodavasal J, Kolodziej C, Ciatti S, Som S (2015) Computational fluid dynamics simulation of gasoline compression ignition. J Energy Res Technol 137(3):032212
14. Benajes J, Novella R, De Lima D, Tribotté P (2015) Analysis of combustion concepts in a newly designed two-stroke high-speed direct injection compression ignition engine. Int J Engine Res 16(1):52–67
15. Benajes J, Molina S, Novella R, De Lima D (2014) Implementation of the partially premixed combustion concept in a 2-stroke HSDI diesel engine fueled with gasoline. Appl Energy 122:94–111
16. Yang B, Yao M, Cheng WK, Zheng Z, Yue L (2014) Regulated and unregulated emissions from a compression ignition engine under low temperature combustion fuelled with gasoline and n-butanol/gasoline blends. Fuel 120:163–170
17. Pei Y, Davis MJ, Pickett LM, Som S (2015) Engine combustion network (ECN): global sensitivity analysis of Spray A for different combustion vessels. Combust Flame 162 (6):2337–2347
18. Pei Y, Shan R, Som S Lu T, Longman D, Davis MJ (2014) Global sensitivity analysis of a diesel engine simulation with multi-target functions. SAE paper 2014-01-1117

19. Liu Y-D, Jia M, Xie M-Z, Pang B (2014) Enhancement on a skeletal kinetic model for primary reference fuel oxidation by using a semi-decoupling methodology. Energy Fuels 26 (12):7069–7083
20. Richards KJ, Senecal PK, Pomraning E (2013) CONVERGE (version 2.1.0) manual. Convergent Science Inc, Middleton, WI
21. Kodavasal J, Kolodziej C, Ciatti S, Som S (2016) Effect of injection parameters, boost, and swirl ratio on gasoline compression ignition operation at idle and low-load conditions. Int J Engine Res (in press). https://doi.org/10.1177/1468087416675709
22. Goldsborough SS (2009) A chemical kinetically based ignition delay correlation for iso-octane covering a wide range of conditions including the NTC region. Combust Flame 156(6):1248–1262
23. Scire JJ Jr, Dryer FL, Yetter RA (2011) Comparison of global and local sensitivity techniques for rate constants determined using complex reaction mechanisms. Int J Chem Kinet 33:784–802
24. Zador J, Ig Zseley, Turanyi T (2006) Local and global uncertainty analysis of complex chemical kinetic systems. Reliab Eng Syst Saf 91:1232–1240
25. Zhou DY, Davis MJ, Skodje RT (2013) Multitarget global sensitivity analysis of n-butanol combustion. J Phys Chem A, pp 3569–3584
26. Kodavasal J, Van Dam N, Pei Y, Harms K, Maheshwari A, Wagner A, Garcia M, Ciatti S, Senecal PK, Som S (2016) Sensitivity analysis on key CFD model inputs for gasoline compression ignition on an IBM Blue Gene/Q Supercomputer. In: THIESEL 2016 Conference on thermo- and fluid dynamic processes in direct injection engines
27. Kodavasal J, Pei Y, Harms K, Ciatti S, Wagner A, Senecal P, García M, Som S (2016) Global sensitivity analysis of a gasoline compression ignition engine simulation with multiple targets on an IBM Blue Gene/Q Supercomputer. SAE paper 2016-01-0602

Application of CFD for Analysis and Design of IC Engines

Vijayashree and V. Ganesan

Abstract Most of us are not familiar with the concept that an internal combustion (IC) engine is working on a four-stroke six-event principle. The six events are suction, compression, combustion, expansion, blow-down and exhaust. However, the expansion and blow-down happen in the power stroke and they can be clubbed together. Therefore, we can say that an engine, whether diesel or gasoline, is working on four-stroke five-event principle. The purpose of this chapter is to make the reader to familiarize with the complexities involved in the working of a four-stroke engine. The five events which are completed in four strokes are: suction, compression, combustion, expansion and exhaust. Application of computational fluid dynamics (CFD) principles for each process mentioned above is a challenging job. The difficulty in understanding the working of an IC engine is due to the fact that we cannot see what is happening inside the cylinder piston arrangement. All that is described in textbooks is based on our knowledge gained over a period of time by conducting experiments. There is no doubt in saying that "seeing is believing". As it is next to impossible to have a complete experimental flow visualization, nowadays CFD comes in handy to have theoretical flow visualization. Well-developed software is available for the simulation in 3D geometries. In this chapter, we explain step-by-step the details required for the CFD simulation of various processes in an IC engine. Extensive results obtained over a period of 20 years of research by the application of CFD in analysing the flow in engines are comprehensively presented and discussed. CFD can be very well applied for analysing any particular process. It can also be used for the modification of the existing engine design or can also be employed for a new design of an engine. It is hoped that readers may be benefitted in understanding the application of CFD for fluid flow analysis and engine design by reading this chapter. Therefore, the main aim of this chapter is to make the reader appreciate how exactly CFD can be applied for

Vijayashree · V. Ganesan (✉)
Department of Mechanical Engineering, Indian Institute of Technology Madras, Chennai 600 036, India
e-mail: vganesan.iitm@gmail.com

Vijayashree
e-mail: vijaya.shree.iitm@gmail.com

D. K. Srivastava et al. (eds.), *Advances in Internal Combustion Engine Research*, Energy, Environment, and Sustainability,
https://doi.org/10.1007/978-981-10-7575-9_13

design of an engine. As it is application oriented, we are not going deep into the equations, modelling, etc. A number of case studies are presented and discussed.

1 Introduction

Engineers always look for improvement in the design and manufacture of internal combustion (IC) engines because there is tremendous competition and pressure in achieving higher performance with lower and lower emissions. The next generation of engines needs to be compact, flexible, light, and powerful. At the same time, they should discharge minimum amount of pollutants and use less fuel. This is a conflicting requirement. To achieve this, innovative engine design is a must to meet these competing requirements. The ability to accurately and rapidly analyse the performance of multiple engine designs is very critical. This chapter explains in detail the various processes in an IC engine, and the results are presented as case studies at appropriate places. Details regarding these aspects are organized into the following sections in this chapter.

2 Engine Performance

The performance of an IC engine depends upon complex interactions between chemical, electronic, fluid, and mechanical systems. However, the major challenge in design is the complex fluid dynamics of turbulent reacting flows interacting with moving parts through the intake/exhaust manifolds, valves, cylinder, and piston. The timescales of the intake air flow, fuel injection, vaporization of fuel, turbulent mixing process, species transport, chemistry, and pollutant formation all overlap and need to be considered simultaneously.

Computational fluid dynamics (CFD) has emerged as a useful and powerful tool in understanding the fluid dynamics of IC engines for design purposes. This is due to the fact that, compared to analytical, experimental, or lower dimensional computational methods, multidimensional CFD modelling allows designers to simulate and visualize the complex fluid dynamics involved. This has become possible due to the availability of superfast digital computers in solving the governing physical laws for mass, momentum, and energy transport on a 3D geometry, along with submodels for critical phenomena such as turbulence and fuel chemistry. Insight provided by CFD analysis helps in design of appropriate geometry, such as ports, valves, and pistons, as well as engine parameters such as valve timing and fuel injection.

It should be noted that engine analysis using CFD software has always been hampered due to the inherent complexity in:

- creating a computational mesh in both the moving and non-moving portions of the domain;
- decomposing the geometry into a topology which should exactly duplicate the piston motion;
- specifying the motion of the valves;
- solving the unsteady equations for flow, turbulence, energy and chemistry;
- post-processing of results for extracting useful information from very large number of data sets.

Those who are contemplating to venture on CFD studies on engines should be aware that this is a time-consuming and error-prone process. Further, it can create a significant impediment to rapid engine analysis and design feedback. The solution to this problem is an integrated approach and environment specifically tailored to the needs of modelling the IC engine. The environment requirements should have:

- minimal inputs from the user;
- the necessary tools to automatically perform a problem set-up;
- the ability to transfer information very rapidly between the different stages of the CFD analysis;
- the capability to significantly compress the set-up and analysis process;
- minimal loss in the accuracy;
- minimal errors.

The IC Engine Analysis System should provide such an integrated environment with the capabilities of integrating the setup of most IC engine designs. Further, it should include:

- bidirectional CAD connectivity to mainstream CAD systems;
- powerful geometry modelling tools in design mode;
- flexible meshing;
- solution using appropriate software;
- powerful post-processing;
- persistent parameterization and design exploration;
- provision to allow users to modify geometry or problem set-up parameters; and
- provision to automatically regenerate analysis results.

Over a period of last one decade, the time taken for geometry adaptation, mesh generation, and solution convergence has been reduced considerably from several hours of work to minutes, with minimum errors. The user should specify the engine parameters and geometry at the beginning of setting up the problem, instead of at the solution stage.

The next few sections we will introduce the tools in the IC Engine Design and Analysis System. A deeper examination of the fluid dynamics issues in IC engines and the CFD modelling process will be carried out first, followed by details of the IC engine port flow analysis, using STAR-CD platform.

3 Engine Design

Efficient design of an IC engine requires a number of critical decisions to establish the impact of fluid dynamics on the overall performance of the engine. The major decisions on the design involve in taking care of the following:

- the type of engine and its specifications;
- peak power to be developed at the design speed;
- the number of cylinders;
- the type of fuel and its emissions characteristics;
- the cubic capacity of the engine;
- overall "packaging" of the engine including all the subsystems.

In future, electronics may play an important role. However, at present, mechanical systems, especially the port design using different cam configurations and analysing the flow through them, are quite important. For this, there may be further specifications, such as the output of the engine at idling speed or at low RPM.

These design decisions are vital for the evaluation of air–fuel ratio needed by the engine at that particular operating conditions. This will lead to a number of design decisions in order to achieve the maximum overall efficiency of the engine. The power output in terms of various efficiencies is given by the following equation:

$$p_b = \frac{\eta_i \eta_c \eta_m \eta_v \rho_i V_d N}{nFQ} \tag{1}$$

where

η_i is the indicated thermal efficiency
η_c is the combustion efficiency
η_m is the mechanical efficiency
η_v is the volumetric efficiency of the engine
ρ_i is the density of the air at the intake
V_d is the engine displacement volume
N is the rotational speed
N is the number of revolution per power stroke (n = 2 for 4-stroke and 1 for 2-stroke engine)
F is the fuel air ratio
Q is the calorific value of the fuel per unit mass

- The important point to be taken into account in port design is to achieve maximum value for each of the efficiency mentioned above. This is required to obtain maximum power from the minimum fuel input. Of the various efficiencies mentioned in Eq. (1), fluid dynamics plays a vital role in the volumetric and combustion efficiency.

- The secondary goal of engine design is to meet emissions standards. These are always imposed by regulations. The major pollutants include three oxides, viz. CO_x, NO_x and SO_x, where CO_x, refers to oxides of carbon [CO, CO_2], NO_x refers to oxides of nitrogen [NO, NO_2], and SO_x, refers to oxides of sulphur [SO_2, SO_3]. Further there are unburned hydrocarbons (HC), and polyaromatic hydrocarbons (PAH or “soot”), which are all products of combustion.
- Pollutants are formed inside the engine due to intense interactions of the mechanical and chemical processes. They are intimately associated with the fluid dynamics inside the combustion chamber. With after treatment, the pollutants in the exhaust stream can be considerably reduced. However, it becomes very costly. Therefore, it is desirable to minimize the pollutant formation at the source itself. In this, fluid dynamics in the port plays a vital role.

4 Fluid Dynamics During the Four Strokes

Achieving the maximum volumetric efficiency is one of the most important factors in engine design. It depends on several fluid dynamic phenomena in the intake and exhaust tracts leading to the combustion chamber. This is the subject matter of this chapter.

- When the air is sucked into the cylinder during the intake cycle, it passes through the gap between the intake valve and the valve seat. As it squeezes through the gap, the flow separation takes place from the walls of the port and valve surfaces, forming a tangential jet. The jet coming out of the gap may impinge on the cylinder walls and tumble into the space between the valves and the piston. This jet imparts angular momentum known as swirl and tumble, to the fresh charge.
- Swirl is usually defined by a normalized angular momentum value about the vertical axis through which the piston motion is constrained. Swirl can be imparted to the incoming air by the proper design of the intake port. This is called suction swirl.
- The gross motions of the fresh charge in the cylinder form recirculation regions that enhance mixing. If there is strong swirl, the flow may develop stratification with regions of high and low velocities. The intake port should be designed to impart additional angular momentum to the air.
- Squish happens during the last part of the compression stroke where there is a narrow region between the piston and the cylinder head, and the air may be squeezed (or “squished”) from the sides of the cylinder into the combustion chamber. This converts the energy in the swirl into turbulence. Further, it is to be noted that when the piston travels towards the top during the compression stroke, most of the energy contained in the tumble (or angular momentum

orthogonal to the swirl axis) of the jet is converted to turbulence as the available space in the vertical direction is reduced significantly. The swirl will become stronger as the air is squeezed out to the side. Flow phenomena that affect volumetric efficiency are:

- flow separation, jet formation and reattachment on the cylinder wall/head;
- mixing due to swirl and tumble in the cylinder volume;
- turbulence production during the compression of air due to squish;
- flow stratification in the cylinder.

Engines that utilize carburetion or port fuel injection (PFI) are known as premixed engines. If an engine uses a carburettor, then the air and fuel are mixed in the carburettor and the fuel–air mixture enters the intake manifold. In this type of engine, at least from the fluid mechanical point of view, a mixture of fresh air and fuel is inducted into the engine through the intake port. For a port fuel-injected engine, the fuel is sprayed into the ports. Usually, spraying takes place onto the back of the intake valve, where it vaporizes and mixes with the intake air.

In all modern direct injection diesel (DI diesel) engines, high-pressure fuel is injected directly into the combustion chamber as the piston nears the end of the compression stroke. The liquid fuel spray breaks up into finer droplets and vaporizes into the surrounding air taking the latent heat of vapourization from air. High pressure of the fuel spray enhances break-up, and high levels of turbulence in the cylinder enhance the mixing. In any engine, proper charge motion at the start of the combustion process is essential for efficient burning of the mixture. However, often some compromises are needed when an engine operates in the range of speeds as in the case of automotive engines.

In a spark ignition (SI) engine, a flame front is formed which travels outwards from the ignition point, consuming the available fuel air mixture. Turbulence again plays an important role in flame propagation, since the flame travels at the turbulent flame speed. Hence, if the turbulence levels are high, then the flame front will move very fast to all parts of the combustion chamber. In case of SI engines, the rapid flame propagation avoids knock due to autoignition of fuel–air mixture ahead of the flame. The flame speed depends on the air–fuel ratio of the mixture. If the mixture is outside of the flammability limits, usually between equivalence ratios of less than 0.5 and greater than 4, the flame will not propagate. The engine starts misfiring. Similarly, if regions exist inside the cylinder that are outside of the flammability limits, these regions will not burn. The unburnt mixture will most likely be pushed out through the exhaust and into the atmosphere.

In compression ignition (CI) engines, due to higher compression ratio, air is compressed to a high temperature and pressure. Fuel is injected directly into the combustion chamber. After sometime (physical delay period), spray break-up, mixing and low-temperature chemical reactions occur. The mixed air and fuel in the spray plume ignites and starts burning, usually forming a stratified or diffusion flame. CI engines have no knock limit. However, they are limited by the amount of mixing in the cylinder and the material strength of the components.

Combustion increases pressure and temperature as the energy contained in the fuel is liberated and the chemical reaction is completed. The combustion of fuel produces a spike in pressure and temperature due to sudden energy release taking place with the production of exhaust gases. Some of the energy is radiated and convected to the cylinder walls, cylinder head, piston and the valves, and some of it is lost. One-third of the energy goes into the power stroke, where the exhaust gases expand under high pressure and push the piston down towards bottom centre position. A thermodynamic energy balance will show that the energy produced due to combustion is used for work done due to expansion. The thermal losses include heat losses through the walls and the enthalpy loss in the exhaust gases at high temperature.

During the exhaust stroke, the exhaust gases are pushed out through the exhaust valves, which start opening towards the end of the expansion stroke. Like intake process, this process also involves formation of a high-speed, high-temperature jet in the gap between the exhaust valves and ports.

During the process of combustion, the fuel, which is a long chain hydrocarbon, breaks up into smaller molecules. The carbon and hydrogen contained in these molecules combine with the oxygen in the air in exothermic reactions. If the fuel–air ratio is stoichiometric at each location in the combustion chamber, carbon dioxide and water vapour will be formed. However, at some regions, where the fuel–air ratio is rich due to inadequate mixing, there will be insufficient oxygen molecules and the combustion will not be complete. Because of this, carbon monoxide (CO) and unburnt hydrocarbon molecules will be produced.

Some of the unburnt hydrocarbons will be polycyclic or polyaromatic as the carbon chains wrap around each other and form solid particles. This is called soot. If the carbon monoxide and unburnt hydrocarbons are then transported to a region with adequate oxygen, then there is a possibility of combustion to reach completion. Otherwise, they will leave with the exhaust gases and cause loss of energy. Due to high temperatures in cylinder, the nitrogen molecules contained in the air break up under intense heat. The nitrogen ions combine with the available oxygen radicals to form nitrous oxides or NO_x. If the fuel contains nitrogen or sulphur atoms, they will also form NO_x and sulphur oxides (SO_x).

Now, it must be clear that the combustion efficiency of the engine and pollutant formation depends upon:

1. fluid dynamics of swirl;
2. tumble;
3. mixing;
4. turbulence production during the intake and compression strokes.

Energy loss is mainly due to incomplete combustion, the heat transfer losses to the wall, and the exhaust losses. Engine emissions include CO_x, NO_x, SO_x, soot, unburnt hydrocarbons and water.

5 Design of Engines with High Efficiency

The design of engines with high efficiency has to take into account the complex fluid dynamics that occurs in the manifolds and cylinders. Several design issues come to the forefront here:

- port flow design;
- combustion chamber and piston shape;
- squish;
- compression ratio;
- design for low speed and idle;
- spark and injection timing.

5.1 *Port Flow Design*

The air flow rate through the intake manifold ports depends on the pressure difference between the cylinder and the manifold. Further, it also depends on the throttle position in case of SI engines. A critical factor here is the packaging. That is to say, the engine and its supporting systems have to fit in a certain amount of space. Further, it should allow easy access for future maintenance. This means that the intake manifolds and engine ports might be routed around other parts. This introduces an additional resistance to the air flow and affects the swirl and tumble in the cylinder. Port flow design to achieve a given air flow rate and desired levels of swirl/tumble within a certain packaging layout to maximize volumetric efficiency thus becomes a critical fluid dynamics design problem. This chapter is devoted to understanding the fluid dynamics of port flow right from the straight port and ports with bend angle

5.2 *Combustion Chamber and Piston Shape*

Apart from port flow, another critical design issue is the size and shape of the combustion chamber, the piston crown shape, and the layout of the valves. Here, the chamber can be flat, a hemispheric dome or a pent roof, while the piston crown can be flat, domed, inclined or a bowl.

The valves can be positioned as "straight", i.e. the valves are aligned with the cylinder axis as shown in Fig. 1, or they can be "canted", i.e. they are at an angle to the cylinder axis and normal to the surfaces of the combustion chamber (Fig. 2).

It is a known fact that the volumetric efficiency and the amount of air intake into the cylinder are dependent on the ratio of the intake valve area relative to the cross-sectional area of the cylinder. Hence, it is desirable to have the intake valves

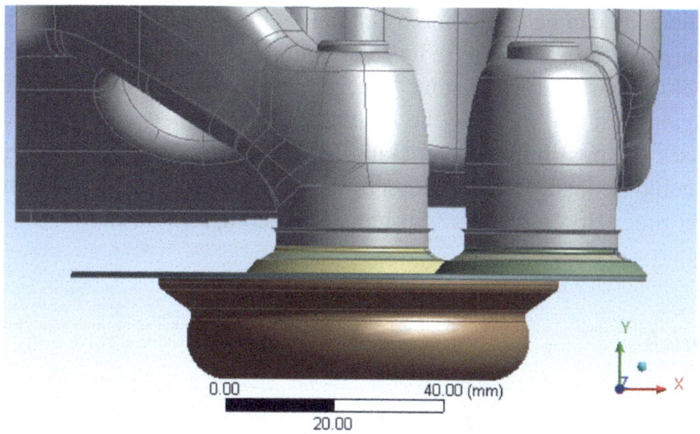

Fig. 1 Straight valve engine [1]

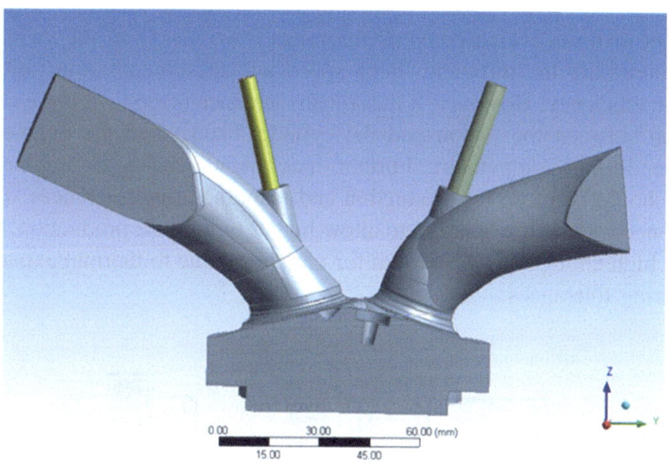

Fig. 2 Canted valve engine [1]

be as large as possible relative to the bore. Therefore, modern engines are designed with four valves (two intakes and two exhausts) instead of two. However, if the combustion chamber is flat, the surface area available gets limited for the valve layout to just the cross section. If the combustion chamber is hemispheric or pent roof, it opens up more surface area for the intake and exhaust valves, allowing them to be larger and more efficient. The deficiency in such a design is that the combustion chamber volume and surface area are large. This implies that the flame front for combustion has to travel a longer distance, increasing the chance of incomplete combustion. Further, the compression ratio will decrease since there is a larger volume at the top centre. A larger wall surface area increases the heat losses during combustion. This will affect combustion efficiency.

This deficiency can be overcome by changing the piston shape from the flat to a domed or any desirable shape to reduce the volume. This means that the flame front has to travel around the piston dome to reach all parts of the combustion chamber volume. Again, this will increase the time taken for complete combustion which in turn increases the possibility of knocking in SI engines. To overcome this defect, the piston could be incorporated with a bowl in the centre, which would reduce the flame travel time. However, this will reduce the compression ratio.

5.3 Squish

In addition, geometric design consideration is to have a "squish" region. Squish is nothing but the region around the perimeter of the combustion chamber with as minimum clearance volume as possible between the piston and the cylinder head at top centre (refer Fig. 3). As mentioned earlier, a small or low squish region causes the swirling air flow to get squeezed out into the combustion chamber. This causes a turbulent jet to form, which converts the mean flow energy in the swirl into turbulence, increasing the turbulent flame speed and the compression ratio and the combustion efficiency. However, if the squish clearance is too low, there is a chance of collision between the piston and the cylinder head when the engine material expands due to high temperature. Further, mass manufacturing of engines requires that there should be room for production and manufacturing tolerances. Therefore, the squish needs to be low enough to allow higher turbulence production, but at the same time high enough to allow room for variability due to thermal expansion and manufacturing tolerances.

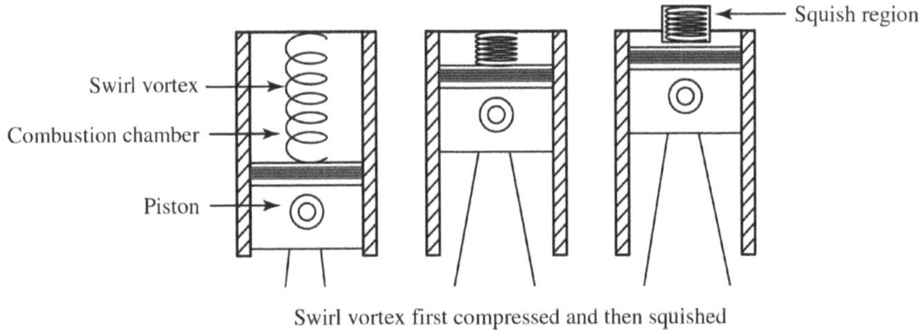

Fig. 3 Illustration of swirl and squish

5.4 Compression Ratio

The compression ratio is defined as the ratio of the total cylinder volume to the clearance volume (refer Fig. 4). It is a critical factor in combustion efficiency and pollutant formation. It may be noted that a high compression ratio enhances the combustion efficiency. At the same time, the higher temperatures due to high compression ratio cause more NO_x to form, thus increasing the emissions.

In the 1970s, automotive engines had much higher compression ratios, since the emissions norms were not stringent. Over the period with stricter environmental regulations on emissions, the compression ratios were reduced to meet the new standards. However, in the 1990s, technological advances in catalytic converters and increase in combustion efficiency allowed higher compression ratios again and improved fuel economy. An additional consideration, especially for diesel engines, one should take care of the materials used for the piston and combustion chamber to withstand the peak temperatures and pressures encountered with high compression ratios and high boost levels.

Thus, intake/exhaust ports designs, along with heads, valves and pistons, involve interplay and trade-offs between the volumetric and combustion efficiency, pollutant formation and packaging considerations, materials choices and manufacturing tolerances. In this connection, the ability to accurately analyse the engine fluid dynamics plays a key role in optimizing the engine to effectively deliver the peak power, while meeting emissions standards and further taking into consideration packaging and manufacturing constraints.

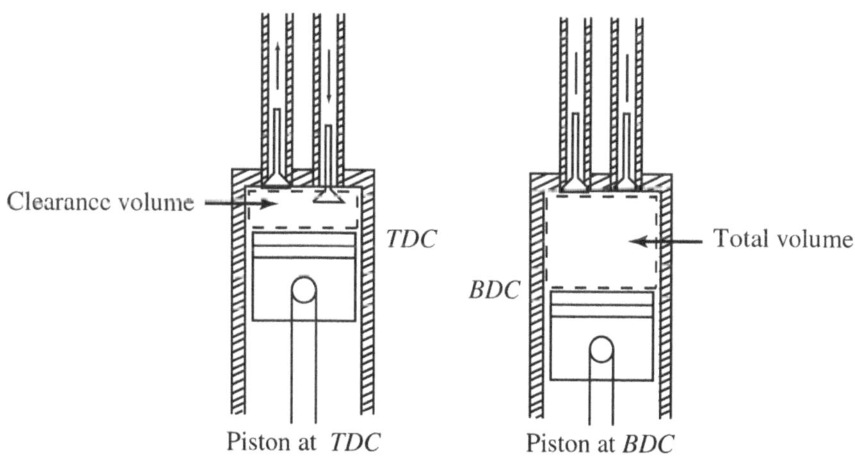

Fig. 4 Illustration of total volume and clearance volume

5.5 Design for Low Speed and Idle

In many cases, the performance of engines at idle or low speed is a critical design consideration. For automobile engines, the engine is designed for peak power at a specific speed. This speed is typically high. Further, the engine will still have to perform well at lower speeds and at idle speed also. Variable valve timing, which allows the valves to have different lift profiles and opening and closing events for different engine speeds, is being tried in many modern engines. Here, the goal is to maximize the volumetric and combustion efficiency. This is achieved by producing optimal levels of swirl, tumble, and turbulence at both low and high speeds. This will enable the combustion charge, i.e. the air/fuel mixture, to get well mixed, and the turbulent flame speed is high enough for complete combustion. There could be additional geometric design changes to the ports and valves for low speed or idle conditions. For example, valve masking and valve shrouding are used to impart additional swirl to the air flow and increase the jet velocity (refer Fig. 5).

5.6 Spark and Injection Timing

During the engine cycle, the spark timing in SI engine and start of injection in CI are to be optimized to provide the desired power or torque with minimum pollutant formation. Strategies, such as exhaust gas recirculation (EGR), are used to minimize the peak temperature in the engine by increasing the thermal mass of the intake air. This reduces the NO_x production, which is strongly a function of high temperatures. For diesel engines, there is a trade-off between soot production and

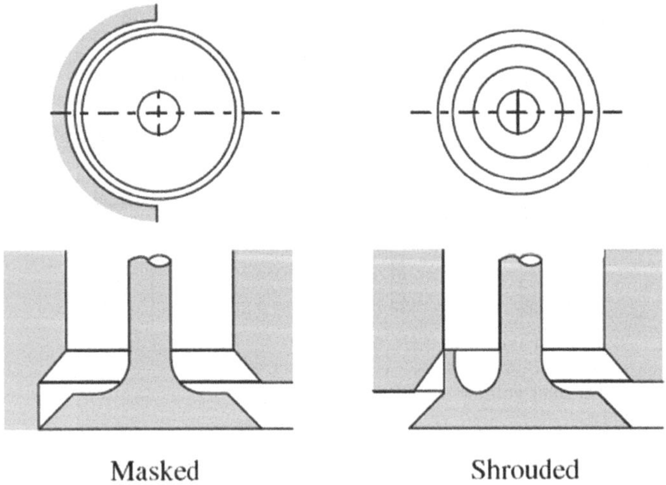

Fig. 5 Inlet valves used to induce high swirl at low engine speeds and low valve lift

NO_x. Note that soot production decreases with temperature, whereas NO_x production increases with temperature. Thus, designing high efficiency engines to meet performance and emissions standards requires trade-offs which should take engine fluid dynamics into account.

6 Application of CFD in IC Engine Design

This section emphasizes the role of CFD modelling in IC engine design and performance analysis. The information in this section is divided into the following subsections:

- the role of CFD in engine design;
- types of CFD in engine performance.

6.1 *The Role of CFD in Engine Design*

As already described, IC engines involve complex fluid dynamic interactions between air flow and fuel injection. Further, moving geometries and combustion play a vital role. Fluid dynamics phenomena such as fuel jet formation and wall impingement are to be considered. Further, swirl and tumble and turbulence are equally crucial for obtaining high efficiency and meeting emissions norms. The design problems that are encountered include port flow design, combustion chamber shape design, variable valve timing, injection and ignition timing, and design for low or idle speeds.

There are several tools available in practice during the design process. These include:

1. experimental investigation using test or flow bench set-ups;
2. one D codes;
3. analytical models;
4. empirical/historical data;
5. computational fluid dynamics (CFD).

Of these, CFD has very high potential for providing useful and detailed information. It can provide insights that can be fed back into the design process. This is because in CFD we solve the fundamental equations that describe fluid flow directly on a mesh which describes the three-dimensional geometry. Further, incorporation of submodels for fuel injection, chemistry, combustion and turbulence enhances the capability of predictions. A number of techniques are available for modelling motion of the moving geometry and its effect on fluid flow.

From the CFD results, it is possible to visualize the flow phenomena on 3D geometry and can be analysed numerically. This can provide excellent insight into the complex interactions that take place inside the engine. This will enable the designer to compare different designs and quantify the trade-offs. Various trade-offs, such as soot versus NO_x, swirl versus tumble and impact on turbulence production, combustion efficiency versus pollutant formation, provide a platform for optimal designs. Hence, nowadays, CFD analysis is used extensively as part of the design process in power generation, in transportation and especially in automotive engineering. With the rise of modern and inexpensive computing power due to the availability of superfast digital computers and 3D CAD systems, it has become much easier for analysts to perform CFD analysis. In increasing importance, the CFD analyses performed can be classified into:

- Port flow analysis: it provides quantification of swirl, tumble and flow rate, with static engine geometry by placing the piston at different locations during the suction stroke of the cycle.
- Cold flow analysis: here, moving geometry is brought into picture and the air flow can be evaluated. However, no fuel injection or reactions are considered.
- In-cylinder combustion simulation: in this expansion and exhaust strokes with fuel injection, mixing, reactions, ignition and pollutant formation with moving geometry are considered.
- Full cycle simulation: entire engine cycle with air flow, fuel injection, combustion, and reactions intake and exhaust are predicted.

However, the CFD analysis of various processes for engines is highly complex, time-consuming and error-prone. Typically, the researcher has to go through several steps to set up the problem. Any minor error can lead to failure of the simulation. Once the analysis has been set up, it takes many hours or even days of computation to get the solution depending upon the grid density and evaluation of the results. The results coming out of simulation are fairly complex, with large data sets. It requires time and effort to analyse and get useful and meaningful information, which can be fed back to the design stage.

Automation and process compression thus become a critical need. In the next section, we will further evaluate the practical issues facing engineers in conducting a CFD analysis on IC engines. Following that, we will explain the results obtained from an integrated environment such as STAR-CD workbench and analyse some typical results obtained for the IC engine analysis.

6.2 Port Flow Analysis

In port flow analysis, the geometry of the ports valves and cylinders is "frozen" at critical points during the engine cycle. The air flow through the ports is analysed using CFD for the various case studies undertaken. The flow rate through the engine volume, swirl and tumble in the cylinder and turbulence levels are evaluated. Fluid dynamics phenomena such as separation, jet formation, valve choking, wall impingement and reattachment, as well as the secondary motions, can be visualized and analysed.

The results provide snapshots of the fluid dynamics throughout the engine cycle and can be used to modify the port geometry to produce desired behaviour of the air flow. Simulation validation wherever possible has to be performed using the data available in the literature. The results do not capture dynamic phenomena such as expansion and compression of air due to piston movement and turbulence production from swirl and tumble.

In practice, conducting port flow analysis at a single point is fairly straightforward because the geometry is static, which fits well with the workflow and capabilities in CFD software. One can start with the port, valve and cylinder geometry at a particular position, create a mesh, specify the mass flow rate or pressure drop for the compressible flow and a turbulence model and compute the results. The RANS-based turbulence models are used to compute the effect of turbulence. Since the turbulent flow interactions with the walls are critical, mesh refinement in the near wall region is necessary using inflation or boundary layers. However, when the number of critical positions and hence the number of cases increase, the problem complexity increases significantly. Setting up large numbers of static cases with identical mesh and flow settings is time-consuming, with scope for error. In the following sections, we will present three case studies on port flow.

6.2.1 Case Study 1: Steady Flow CFD Analysis for Improving Volumetric Efficiency

An industrial project was undertaken to analyse the flow through the inlet and exhaust manifolds of a popular two wheeler engine in India and then, if necessary, arrive at a modified design based on the CFD analysis. The objective of this project was to estimate the volumetric efficiency of the base engine and if necessary modify the design to arrive at better volumetric efficiency. The results presented below establish how usefully CFD can be applied for improving the engine port design.

Figure 6 shows the geometry of the intake port. A steady flow analysis at three different valve lifts was carried out to estimate the volumetric efficiency and also to examine the flow through the ports of the existing design. As transient analysis takes time in making fixed and moving meshes, steady flow analysis was thought of to get quick results and make design modifications if necessary for improving the flow and volumetric efficiency. Initially, a CAD model of the inlet port was developed and was meshed with one lakh and seventy-five thousand grid points (refer Fig. 6).

The meshed geometry for different valve lifts for the inlet manifold is shown in Fig. 7. Prescribing appropriate boundary conditions, viz. pressure boundary condition at inlet and exit boundary being the cylinder bottom centre, the flow for the

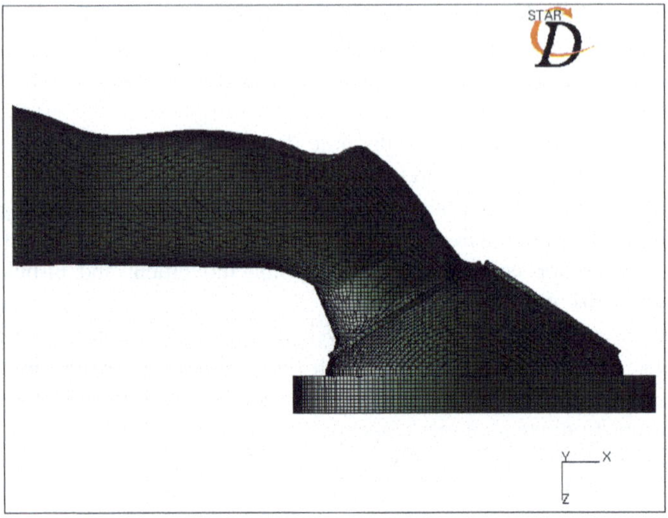

Fig. 6 CAD model of inlet port geometry with grid

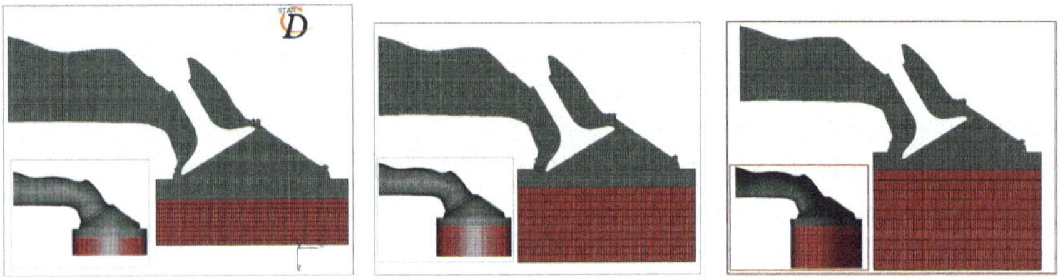

2mm Valve lift

4mm Valve lift

5.6mm Valve lift (Maximum)

Fig. 7 Meshed geometry of the inlet port at three-valve lift position

three-valve lifts has been studied. Figure 8 shows the velocity vector along with velocity contours for the three-valve lift positions.

As can be seen that the engine has pent-roof cylinder head and the piston being stationary as the flow pass through the inlet port initially even with 2-mm valve lift, there was a slight flow separation near the bend. As the vale lift increases from 2 to 4 mm, there is more separation and, at maximum lift of 5.6 mm, there is large flow separation at the bend region. Further, it causes recirculation behind the vale head. This can hamper volumetric efficiency of the engine. The pressure contours shown in Fig. 9 for the three valves were analysed and found that it is favouring separation in the bend region of the inlet manifold. There is a distinct pressure gradient at the bend region which helps to separate the flow (refer Fig. 9).

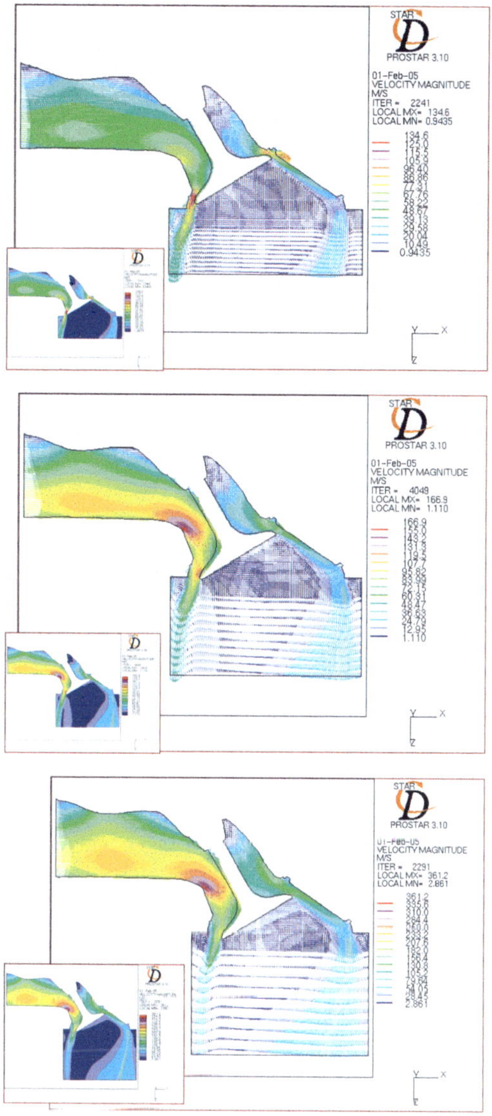

2 mm valve lift (IVO)

4 mm valve lift (IVO)

5.6 mm valve lift (Max. valve lift)

Fig. 8 Velocity vector and velocity contours for three-valve lifts during valve opening process

To reduce the effect of separation in the inlet manifold design, changes were made and the bend angle was changed from 60° to 45° and the flow analysis was carried out again. This helped considerably in reducing the recirculation in the inlet manifold, and the final design was arrived at. The volumetric efficiency improved from 82.5 to 86%. Thus, this case study has established that steady flow analysis using CFD can be successfully applied for the design changes of the inlet manifold

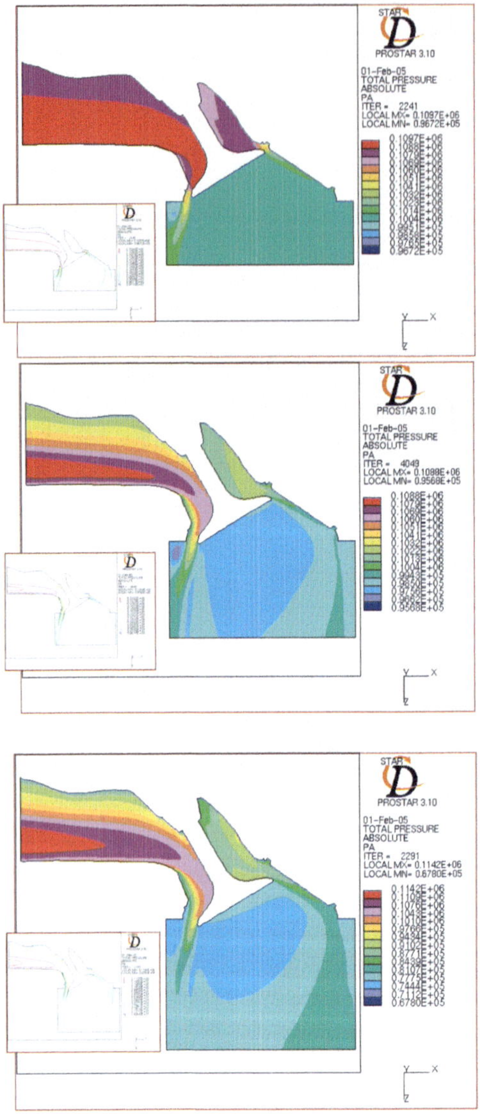

2 mm valve lift (IVO)

4 mm valve lift (IVO)

5.6 mm valve lift (Max. valve lift)

Fig. 9 Total pressure contours for the three-valve lifts

by which volumetric efficiency could be considerably improved and a final geometry was arrived at. Similar analysis was carried out for the exhaust manifold also, and the design changes helped to increase the scavenging efficiency. The meshed geometry of the exhaust manifold is shown in Fig. 10. The velocity vectors in the exhaust manifold are shown in Fig. 11.

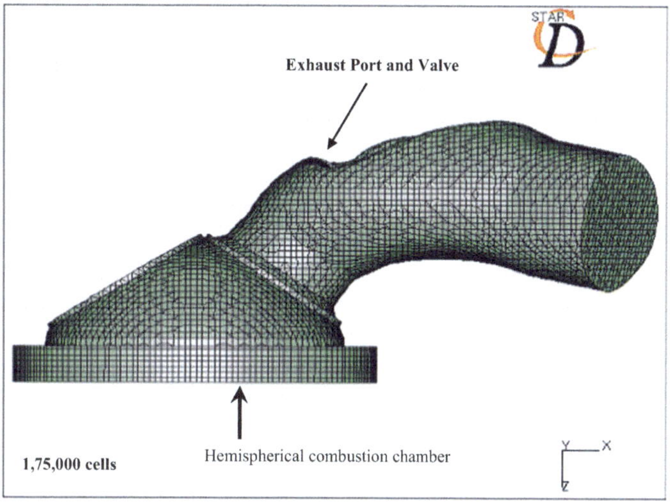

Fig. 10 Meshed geometry of the exhaust manifold

Based on the results obtained, similar modification was carried out in the exhaust manifold also and final design of the cylinder head with inlet and exhaust manifold is shown in Fig. 12. This case study has clearly established that CFD can be used as design tool along with flow analysis.

6.2.2 Case Study 2: Steady Flow Analysis on Multi-cylinder Inlet Manifold

Having got the confidence by means of the case study in the previous section in improving the design by CFD analysis in a single-cylinder engine, a multi-cylinder manifold analysis was carried out to improve the design and flow characteristics. This is also a typical industrial project. In this, a four-cylinder inlet manifold along with throttle body was taken for the analysis. The purpose of this study was to understand the flow behaviour of a multi-cylinder throttle body assembly where there is one inlet and multiple outlets. The main aim of this case study was to establish equal flow rate to all the cylinders. The main problem in a multi-cylinder SI engine is the maldistribution of fuel–air mixture in various cylinders. This can be easily analysed using CFD under steady flow conditions. Initially, a CAD model was developed and the corresponding meshed model is shown in Fig. 13. Fortunately for the engine under consideration, the valve timing was such that at any time only one cylinder valve was in the open position, whereas for other three cylinders it is in closed position. Flow studies have been carried out for different valve lift conditions (25, 50 and 100%). The flow results are presented for all the three-valve lift condition. Compared to the manifold in the case study 1, this manifold was more complex and meshing was a little more challenging.

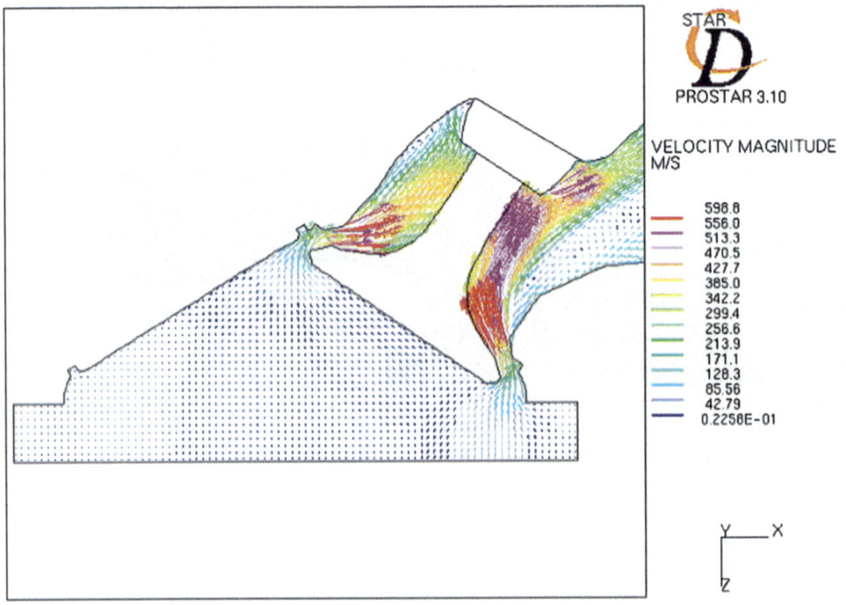

Exhaust Valve lift = 2 mm

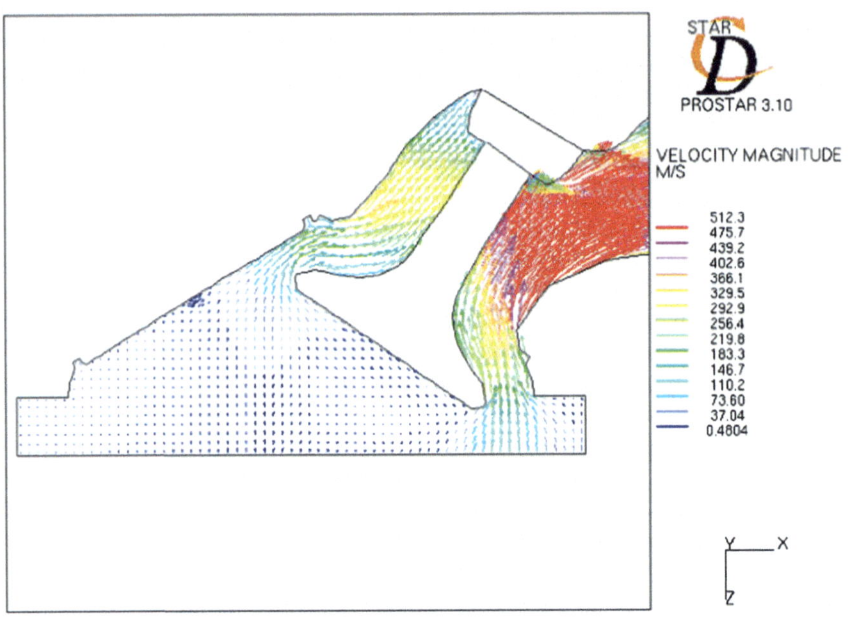

Exhaust Valve lift = 6 mm

Fig. 11 Velocity vectors in exhaust manifold for two-valve lifts

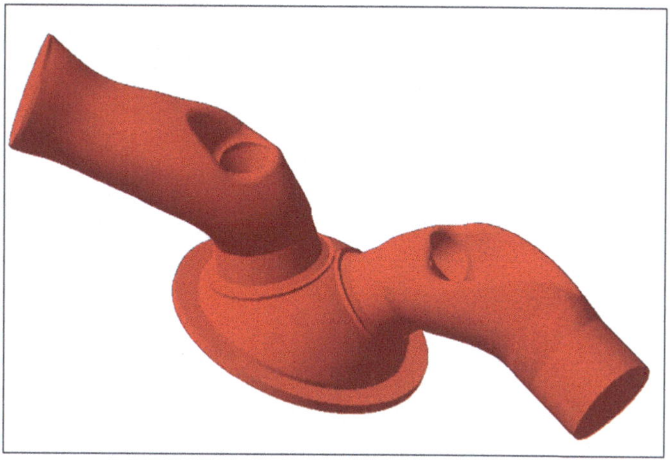

Fig. 12 Final design of intake and exhaust manifold

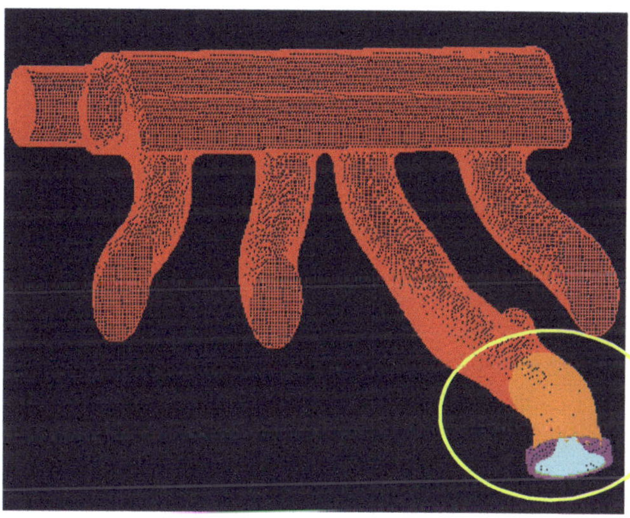

Fig. 13 Meshed geometry of the multi-cylinder inlet manifold

Because the manifold geometry is complex, it was decided to carry out the analysis dividing the flow domain in five regions as indicated in Fig. 14. The various regions were carefully chosen so that one can understand the complete flow characteristics. Region 1 is the one where the possibility of recirculation is there. Region 2 is the region where flow passes over the vale stem. Flow region 3 is one where the flow starts entering the cylinder. Region 4 is the one where flow separation can take place. Region 5 is the one where the bulk flow takes place. Thus, this configuration provides the possibility for all the flow complications and

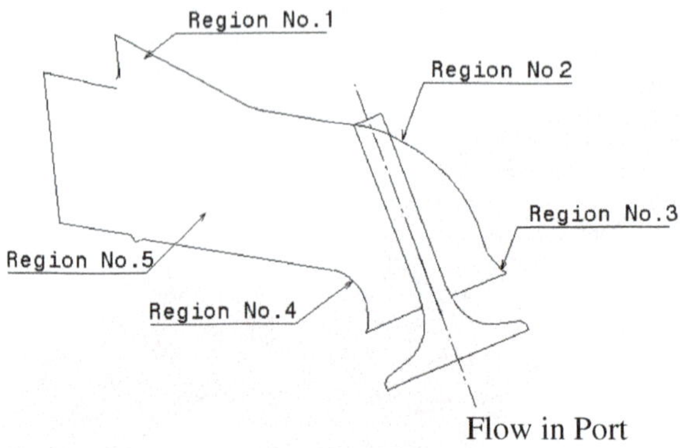

Fig. 14 Division of flow domain into number regions

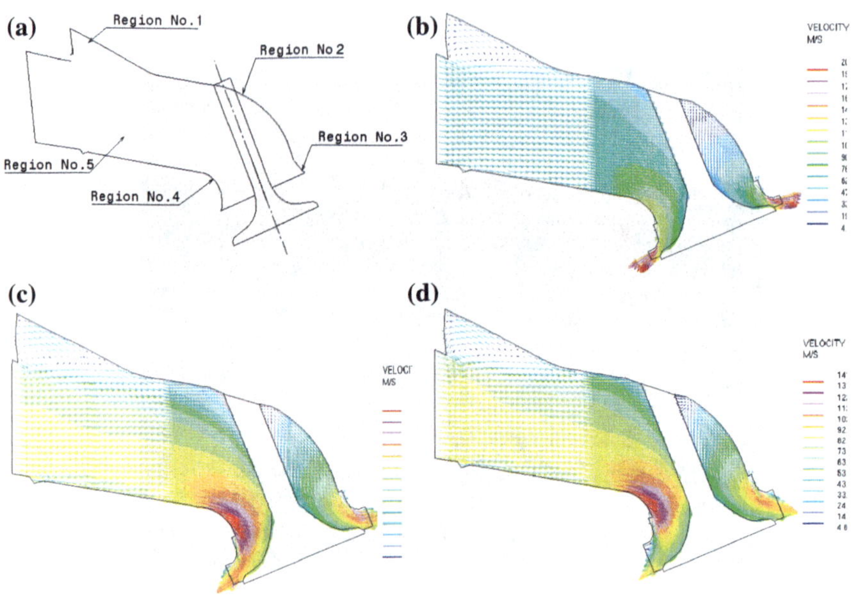

Fig. 15 Flow velocity vectors at various lift conditions in manifold 3

therefore it was a challenging job to undertake. The necessary boundary conditions were specified, appropriate equations were solved, and the flow velocity vectors were obtained for all three-valve lift conditions. The details are shown in Fig. 15. We will explain the results obtained for full-valve lift condition only.

Flow Analysis in Various Regions

As shown in Fig. 15, in region 1, there is a formation of strong recirculation zone. This is due to the projection in the inlet manifold. This will cause flow losses, and this is to be avoided. On analysing the region 2, there is a restriction imposed by valve stem. Because of the stem, the flow starts separating behind the stem. It may be noted that this cannot be avoided. Coming to region 3, as can be seen, there is possibility for flow separation because of the slight bend and due to the stepped configuration as shown in Fig. 15a. Even though the step increases the flow area if the separation takes place, it will nullify the positive effect. When region 4 is analysed, it is clearly seen that there is a strong separation. This is to be avoided. As shown in Fig. 15d up to 50% of the manifold length, the bulk flow is normal. Beyond that because of the manifold bend, valve stem as well as valve head effect, there is back pressure which decelerates the flow. The deceleration is minimum for full-valve lift conditions. Hence, this region should be looked into more carefully. Thus, the steady flow analysis could throw some useful flow structure and the design was modified taking into consideration all the details and drawbacks described, by which uniform flow distribution could achieved in all the four cylinders. Again, this study has established that CFD can be used as a tool for design modifications. As this was a steady flow analysis, the time required to obtain the solution was comparatively very much less.

From this study, we could find that the fourth manifold was getting less air since the air has to travel a longer distance. In order to get uniform flow, the plenum chamber design was slightly modified based on the CFD results and by which uniform distribution could be achieved. It should be emphasized here that because of the CAD model availability, the geometry could meshed accurately and flow vectors are shown in Fig. 15. This is the second case study by which we could establish that CFD can be applied for the design. The CAD model of the modified design is shown in Fig. 16.

6.2.3 Case Study 3: Steady Flow Analysis on Six-Cylinder V-Engine Inlet Manifold

Based on the confidence obtained through the above two case studies, we went on to evaluate a highly complex intake manifold with power valve of a six-cylinder V engine. Three manifolds are in the right bank, and three others are in the left bank. The manifold was very carefully designed, and the main purpose of this study was to evaluate the flow coefficients for all the six manifolds to see whether the flow is uniform. It is to be emphasized here that the manifold was designed after many iterations and was expected to give uniform flow distribution. Again, it was subjected to steady flow analysis and the flow coefficient values are evaluated.

Figure 17 shows the meshed geometry under consideration. Meshing was really a very challenging job, and it took considerable time to arrive at appropriate mesh size. Again, appropriate boundary conditions were prescribed and equations were

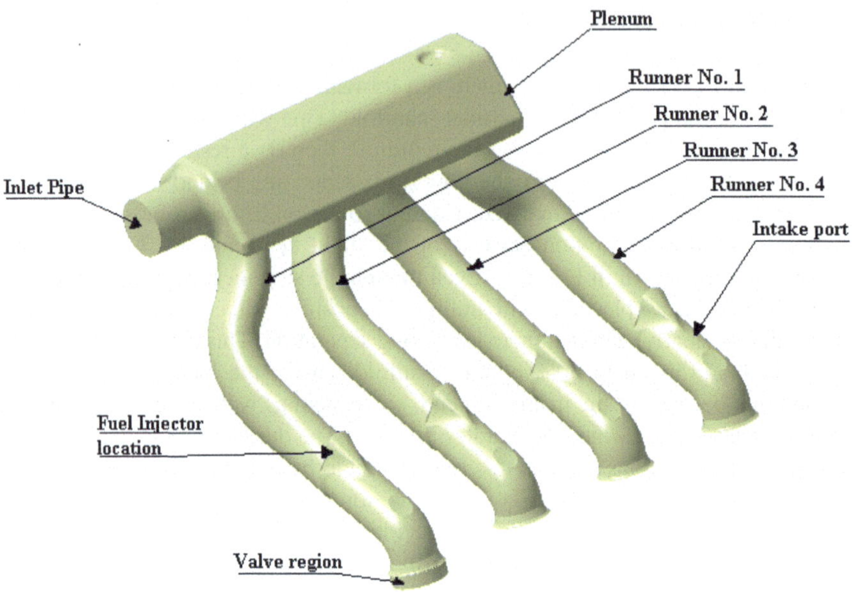

Fig. 16 CAD model of the modified design of the inlet manifold

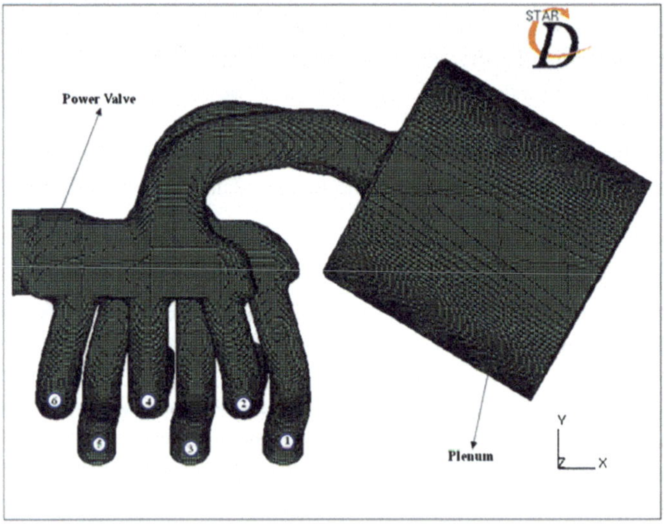

Fig. 17 Six-cylinder inlet manifold versus V type engine

solved. It may be noted there was a power valve which can be kept either in closed position or in open position. When it is kept in closed position, flow will pass through the left bank to all the six cylinders, whereas when it is in open position flow will take place through the right bank to all and the six cylinders will get the

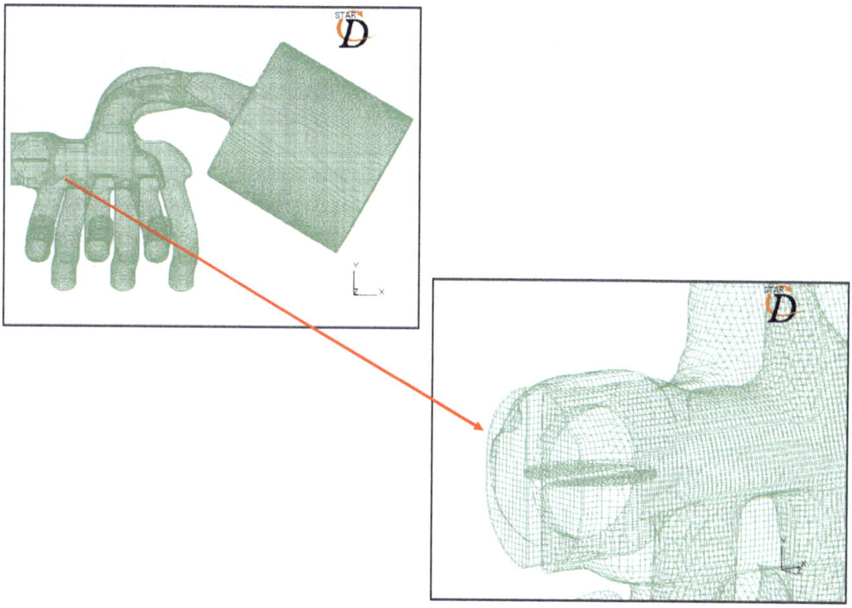

Fig. 18 Grid arrangement when the power valve is in open position

flow. As it is a complex design, it will be difficult to visualize the flow when one sees this for the first time.

Meshed geometry is shown in Fig. 18 when the power valve is open and flow velocity vectors for all the six cylinders are shown in Fig. 19. From the velocity vectors, flow coefficients were calculated and are given in Table 1. It is surprising to see that the flow coefficient K is same to first decimal accuracy. Establishing this by means of experiments would have been time-consuming and costly. Thus, by means of this CFD study, we could validate the design parameters. As this was a steady flow analysis, the results obtained were very fast and less costly.

Table 1 gives the flow coefficient based on the quantity of flow, and the pressure drop shows clearly that the flow coefficient is the same to the first decimal accuracy. Therefore, there was no necessity to make any changes in the design. This further establishes the use of CFD in confirming the design.

6.3 Cold Flow Analysis

Modelling of cold flow takes into account mainly the airflow and possibly the fuel injection without involving reactions. The main aim is to capture the mixture formation process. For this, accurate accounting of the interaction of moving piston with the fluid dynamics of the induction process. In this analysis, the changing

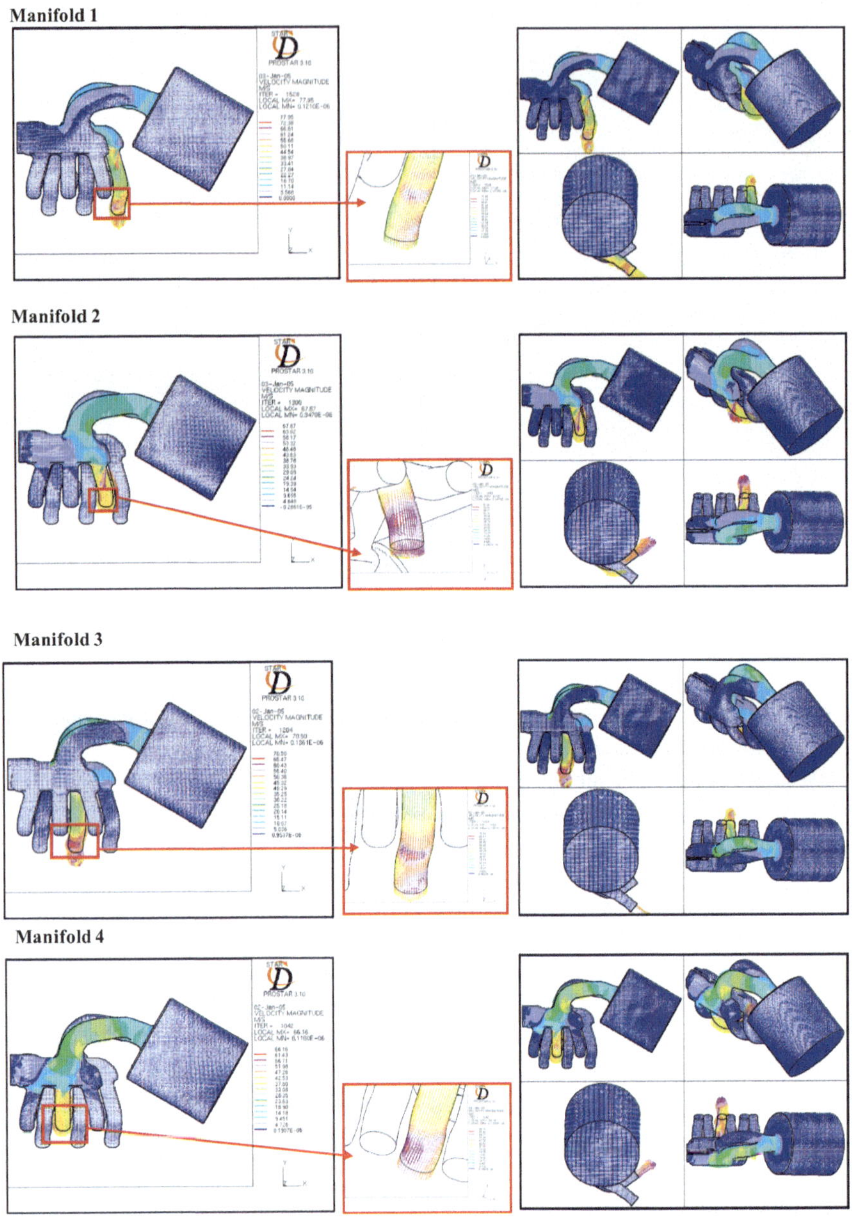

Fig. 19 Flow velocity vector for all the six manifolds

Manifold 5

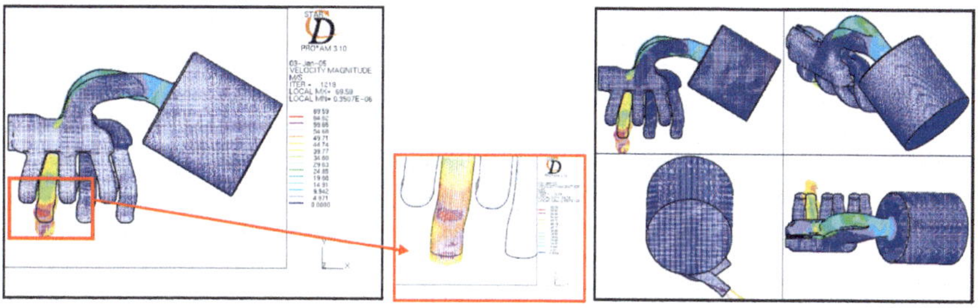

Manifold 6

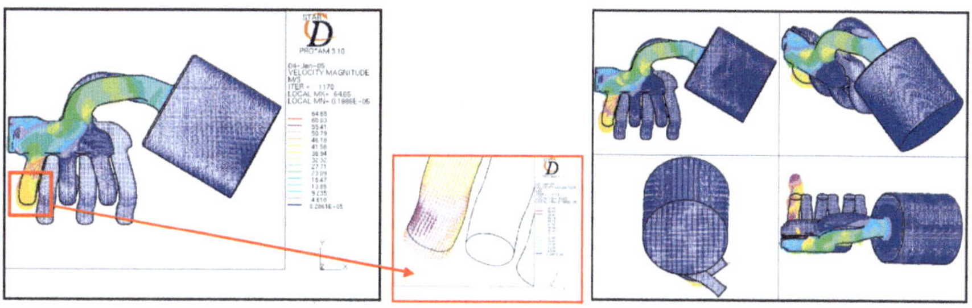

Fig. 19 (continued)

behaviour of the air flow which tumbles into the cylinder can be studied. Further, it can take into account swirl generated by intake valves and the exhaust jet coming out of the exhaust valves. Further, the turbulence generated from swirl and tumble due to compression and squish can be conveniently analysed.

This information will be very useful to decide whether the conditions in the cylinder at the end of the compression stroke are conducive for combustion and flame propagation. It is a known fact that turbulence enhances flame propagation. Further, it aids complete combustion during the power stroke. For the best power output, homogeneity of mixture with right air/fuel ratio throughout the combustion process is a must. CFD analysis can provide the degree of charge stratification.

However, cold flow simulations do not include the significant thermodynamic changes that accompany combustion. This is because the flow characteristics during the expansion and exhaust strokes do not reflect actual conditions. In terms of validation, experimental Particle Image Velocimetry (PIV) or LDV data in cycling engines is not easy to obtain as with port flow analysis. However, transparent pistons and cylinders can be used to gather velocity information for simple engine configurations.

CFD modelling of cold flow involves additional work. First of all, one should incorporate the valve and piston motion. Further, boundary conditions, turbulence models and other required submodels should be specified. Details of valve and piston geometry should be given, along with the lift curves. Engine geometric characteristics, such as bore, stroke, connecting rod length and compression ratio, should be given as input data in order to calculate piston position as a function of

Table 1 Flow coefficient details with power valve open and closed position

Power valve closed	Mass flux (kg/s)	Density (kg/m^3)	Q (m^3/min)	Inlet "P2" (Pa)	Outlet "P1" (Pa)	(P2 –P1) (Pa)	(P2 – P1) (mm of H_2O)	K
Manifold 1	0.0908265	1.16203	4.69040	99889	97058	2831	288.583	3.6218
Manifold 2	0.0901336	1.16209	4.65489	99855	97058	2797	285.117	3.6274
Manifold 3	0.0917929	1.16195	4.73994	99879	97,058	2821	287.564	3.5776
Manifold 4	0.0905110	1.16182	4.67427	99,843	97,058	2785	283.894	3.6046
Manifold 5	0.0935049	1.16208	4.82780	99,870	97,058	2812	286.646	3.5069
Manifold 6	0.0883281	1.16175	4.56181	99,850	97,058	2792	284.606	3.6981
Power valve open	Mass flux (kg/s)	Density (kg/m^3)	Q (m^3/min)	Inlet "P2" (Pa)	Outlet "P1" (Pa)	(P2 –P1) (Pa)	(P2 – P1) (mm of H_2O)	K
Manifold 1	0.0937697	1.16200	4.84181	99,911	97,058	2853	290.826	3.4221
Manifold 2	0.0961538	1.16214	4.96431	99,898	97,058	2840	289.501	3.4274
Manifold 3	0.0952559	1.16212	4.91804	99,914	97,058	2856	291.131	3.4693
Manifold 4	0.0970425	1.16222	5.00985	99,916	97,058	2858	291.335	3.4069
Manifold 5	0.0961706	1.16223	4.96479	99,915	97,058	2857	291.233	3.4373
Manifold 6	0.0958784	1.16217	4.94996	99,921	97,058	2863	291.845	3.4512

crank angle. Since the volume in the cylinder is changing throughout the engine cycle, the mesh change should take place accordingly. Methods to automatically modify the mesh during piston motion also need to be incorporated. The CFD calculation is an inherently transient computational problem. It involves moving deforming or dynamic mesh. All the geometric motion is a function of a single parameter, viz. the position of the crankshaft in its rotation, or crank angle.

The preprocessing from geometry to solver set-up is typically time-consuming and challenging. Here, we should separate or decompose the geometry into moving and stationary parts. Typically, the intake ports are split off from the cylinder and valves. The region between the valve margin and valve seat, which opens and closes during valve motion, should be separated. The combustion chamber and piston region may be also decomposed or separated into smaller parts. Then, each part can be meshed accordingly for the solver set-up. Any errors at this stage can lead to failures downstream during the solution process.

The run times for solver runs can be fairly long since the motion is typically resolved with small time steps (approximately 0.25 crank angle) to get accurate results, and the simulation is run for two or three cycles to remove the initial transients. Finally, the large volume of transient data that results from the CFD solution needs to be post-processed to obtain useful insight and information. Thus, cold flow analysis would also benefit from design automation and process compression.

6.3.1 Case Study 4: Cold Flow Transient Analysis on a Single-Cylinder Engine

As is evident that in the previous case studies, only steady flow analysis has been carried out. It will provide first-cut information for the design modification requirements. However, true picture will emerge only if a transient flow analysis with moving piston is carried out. In this case study, a CFD analysis has been carried out to investigate the effect of intake port configurations such as intake port cross-sectional area, bend angle, engine speed, and intake pressure on the flow field inside the cylinder of a DI diesel engine under cold flow condition. Further, to evaluate consistency of in-cylinder flow structure results, a grid independence study has been conducted to arrive at the optimum grid density that can be used for subsequent simulation studies.

In-cylinder fluid dynamics in direct injection CI (diesel) engines play a vital role during combustion process. In particular, in-cylinder fluid dynamics contribute to the fuel-air mixing, which is one of the most important factors for the control of the fuel burning rate. It significantly affects the ignition delay, magnitude of the pre-mixed burn, magnitude and timing of the diffusion burn, and the emissions of nitric oxide and soot. In order to have better mixing, swirl is generated during the intake stroke as a result of intake port shape and its orientation. In IC engines, the fuel evaporation and mixing processes are strongly influenced by the turbulent nature of the in-cylinder flows. The velocity gradient in the mean flow is one of the major reasons for the turbulent fluctuations. The air-jet created by flow during the intake process interacts with the cylinder wall and moving piston to generate large-scale rotating flow, both in the vertical and in the horizontal planes. The behaviour of the in-cylinder turbulent flow can be characterized by monitoring the kinetic energy and the integral length scale variation of turbulent eddies that contribute to turbulence production during intake and compression processes. In-cylinder flow characteristics at the time of fuel injection and subsequent interactions with fuel sprays and combustion are the fundamental considerations for the engine performance and exhaust emissions of a diesel engine.

7 Engine and Operating Conditions

A single-cylinder DI diesel engine having a toroidal combustion chamber with two direct intake ports whose outlet is tangential to the wall of the cylinder and two exhaust ports to simulate the flow field inside the cylinder has been chosen. The geometry with different intake port configurations such as constant cross-sectional area (CCS), varying cross-sectional area (VCS), combination of both (CVCS) and varying cross-sectional area intake ports with different bend angles (20°, 30° and 40°) have been investigated. However, the results pertaining to varying cross-sectional area (VCS) with 20° bend angle only will be presented and discussed. Geometrical details and operating conditions of the engine are given as

Table 2 Geometrical details of the engine

Bore	130.0 mm	Static bumping clearance	1.55 mm
Stroke	150.0 mm	Compression ratio	15.5:1
Connecting rod length	275.0 mm	Intake valve diameter	44.4 mm
Displacement	1991 cc	Intake valve opening	9° bTDC
Rated speed	2000 rpm	Intake valve closure	62° aBDC
Bowl diameter	76.0 mm	Exhaust valve opening	43° bBDC
Bowl depth	29.0 mm	Exhaust valve closure	10° aTDC
Bowl volume	113.5 cc		

input, and appropriate fuel injection strategy has been adopted. The geometrical details of the engine are given in Table 2.

8 Methodology

The CFD code, *STAR-CD,* has been used as solver. The code solves the discretized Navier–Stokes equations. For physical modelling, the *RNG k–ε* turbulence model with standard wall function has been employed. The solution algorithm is based on the pressure-correction method. It uses the *PISO* algorithm with the first-order upwind differencing scheme (UD). For the solution of the momentum, energy and turbulence equations, the implicit temporal discretization method is employed. Spatial discretization is done using the UD. For solving momentum, turbulent kinetic energy/dissipation and temperature, the appropriate equations with central differencing scheme are used.

The calculations begin at TDC of the intake stroke and complete at 30° after TDC (aTDC) of compression. This means a cold flow analysis has been carried out. The results of different variables plotted are presented in non-dimensional form. Normalization is done with respect to mean piston speed which facilitates the comparison. The various equations are represented through Eqs. (2–6).

$$\text{Swirl ratio, SR}(\theta) = \frac{v(\theta)}{\frac{2\pi N}{60} r} \quad (2)$$

$$\text{Radial velocity} = \frac{u(\theta)}{V_p} \quad (3)$$

where v = tangential velocity in m/s, θ = crank angle in degrees, N = engine speed in rpm, r = distance from cylinder axis in m, u = radial velocity in m/s, V_P = mean piston speed in m/s.

$$\text{Turbulence intensity} = \frac{u'(\theta)}{V_p} = \frac{\left[\frac{2}{3}k\right]^{1/2}}{V_p} \tag{4}$$

where u' = fluctuating velocity in m/s and k = turbulent kinetic energy in m^2/s^2.

Swirl is defined as the organized rotation of the charge in the cylinder with respect to the cylinder axis. With an initial angular momentum in the incoming charge, swirl is created. Swirl ratio is defined as the ratio of total angular momentum of the in-cylinder fluid about the cylinder axis to the total moment of inertia of the fluid about the cylinder axis:

$$\begin{aligned} \text{where } H_Z &= \sum_{i=1}^{N_{\text{Cells}}} \{v_i(x_i - x) - u_i(y_i - y)\}m_i = \sum_{i=1}^{N_{\text{Cells}}} \{v_i x_i - u_i y_i\}m_i \\ \text{and } M_Z &= \sum_{i=1}^{N_{\text{Cells}}} \{(x_i - x)^2 + (y_i - y)^2\}m_i = \sum_{i=1}^{N_{\text{Cells}}} \{x_i^2 + y_i^2\}m_i \end{aligned} \tag{5}$$

$$\text{Mass averaged turbulence intensity} = \frac{u'(\theta)}{V_P} = \frac{\sum_i^n \left[\frac{2}{3}k_i\right]^{1/2} V(\theta)_i \rho(\theta)_i}{V_P \sum_i^n V(\theta)_i \rho(\theta)_i}. \tag{6}$$

where H_z represents the total angular momentum of the in-cylinder fluid about the cylinder axis (z-axis), M_z is the total moment of inertia of the fluid about the cylinder axis, N is the angular speed of the crankshaft in rpm, u_i and v_i represent the local components of velocity in x and y directions, respectively, m_i is the mass accumulated in each cell, V is the cell volume, ρ is the density, θ is crank angle, *and* x_i and y_i represent axes in radial and tangential direction.

9 Modelling Procedure

A hexahedral block structure mesh is employed for the entire computational domain of the engine with 505,542 cells. The preprocessor *GAMBIT* is used to create the entire computational domain of the engine including intake and exhaust ports. *STAR-CD* is used for the solution of governing equations and post-processing the results. The meshed intake and exhaust port geometry is shown in Fig. 20.

10 Boundary and Initial Conditions

The boundary and initial conditions (Table 3) employed are as follows: by neglecting dynamic effects, constant pressure boundary conditions are prescribed to both intake and exhaust ports. Smooth wall option is used in order to take it into

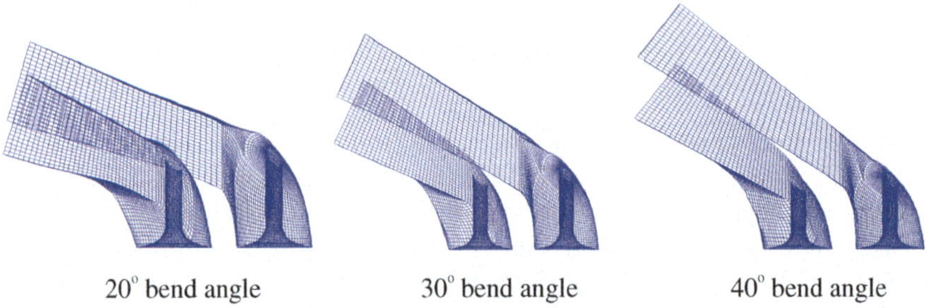

Fig. 20 Varying cross-sectional area with different bend angles

Table 3 Boundary and initial conditions

Boundary conditions		Initial conditions	
Intake and exhaust ports		Intake and exhaust ports	
Type Temperature Turbulent intensity Length scale	Constant pressure 1.01 bar 303 K 5%[a] 1 mm[a]	Pressure Temperature Turbulent intensity Length scale	1.01 bar 303 K 5%[a] 1 mm[a]
Cylinder–valve interface		Cylinder	
Type	Attach	Pressure Temperature Turbulent intensity Length scale	1.0 bar 303 K 5% 1 mm
Wall			
Type Wall heat	Smooth wall, no slip Adiabatic		

[a]*Source* www.adapco-online.com

account frictional effects at the walls. Walls are considered as adiabatic. The initial values of pressure and temperature prescribed are assumed to be in the whole domain. The residual swirl of the flow in the cylinder at the end of the exhaust stroke is considered as negligible. This means that the flow is supposed to be quiescent initially. The initial turbulent intensity is assumed as 5% of the mean flow. The integral length scale is calculated using Prandtl's mixing length model. The walls in the entire are considered adiabatic.

In-cylinder fluid dynamics in DI diesel engines plays a vital role during combustion process. In particular, in-cylinder fluid dynamics plays a vital role in controlling the fuel burning rate. Burning rate in turn affects the ignition delay, rate of the premixed as well as the diffusion burning. Further, it controls the emissions of NO_x and soot. For better mixing, during the intake, stroke swirl is generated. This is achieved by designing appropriate intake port shape and its orientation.

Fuel evaporation and mixing process in an in IC engines are strongly influenced by the in-cylinder turbulence. The turbulent fluctuations are created by the velocity gradient in the mean flow. The jet action by flow during the intake process interacts with the cylinder wall and moving piston. This generates large-scale rotating flow, both in horizontal and in the vertical planes. In-cylinder turbulent flow behaviour can be characterized by the kinetic energy and the integral length scale variation of turbulent eddies. These turbulent eddies contribute to turbulence production during both intake and compression processes. Engine performance and exhaust emissions of a diesel engine are controlled by in-cylinder air motion at the time of fuel injection and subsequent interactions with fuel sprays and combustion.

The objectives of this case study are to investigate the effect of inlet port configuration on the in-cylinder flow dynamics. The results obtained through simulations to meet the objective are presented in this section in the following order:

- validation of the code used and the grid independence study;
- study of the effect of variable cross-sectional area intake port with different bend angles under cold flow motoring condition.

11 Validation of the Code

For the validation of the CFD results, a 20° bend angle having varying cross-sectional area intake ports is considered. The predicted results are compared with the experimental results of Payri et al. [57, 58]. This configuration has been chosen since experimental results are available for such a configuration. The operating speed of the engine, intake pressure and temperature are 1000 rpm, 1.01 bar and 303 K, respectively. The results obtained for swirl ratio, turbulent intensity and radial velocity variations at different crank angles are compared with Laser Doppler Velocimetry (LDV) measurements of the same engine. The engine under consideration is equipped with toroidal combustion chamber in which measurements have been made at different locations at a radial distance of 34 mm from the axis of the cylinder and at a distance of 4, 8, 12 and 16 mm (Z1, Z2, Z3, Z4), respectively, from the cylinder head the details of which are shown in Fig. 21. One of the exhaust valves of the cylinder under consideration is made inoperative. A thick quartz window is inserted in the space for measurement purpose.

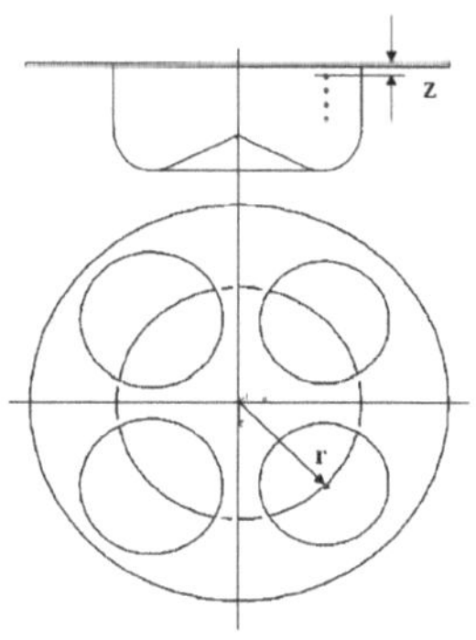

1 : 4 mm from the cylinder head
2 : 8 mm from the cylinder head
3 : 12 mm from the cylinder head
4 : 16 mm from the cylinder head
$r = 34$ mm

Fig. 21 Measuring locations

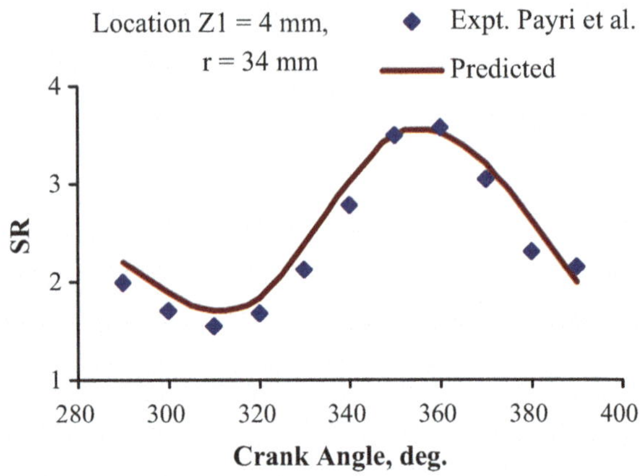

Fig. 22 Swirl ratio versus crank angle at various locations (Z1 = 4 mm and Z2 = 8 mm from the cylinder head) compared with experimental LDV data for varying cross-sectional area with 20° bend angle

Figures 22 shows the distribution of swirl ratio at various crank angles (70° during compression stroke and 30° during expansion stroke) around compression TDC for the two measuring locations. It is shown that swirl ratio decrease is observed during compression stroke. It may be due to swirl getting converted into squish slowly. The wall friction also may play a part. As can be seen from the figure as the piston approaches TDC, swirl is enhanced. It is due to the flow acceleration which causes the reversal of its angular momentum within the smaller diameter piston bowl. During the expansion stroke as the flow starts exiting, reverse squish action takes place. The sudden fall in the swirl ratio is caused by the piston bowl and wall friction. As shown in Fig. 22, the predicted results are in good agreement with experimental results at all locations. During the expansion stroke, there is a faster decay of swirl at locations especially near the cylinder head. This is due to reverse squish–swirl interaction. As the piston moves away from TDC, this interaction becomes less and less.

Figure 23 shows the distribution of non-dimensional turbulent intensity, which decays almost linearly as it approaches TDC at 4- and 8-mm location. This is due to squish and swirl interaction. As the bowl has an open geometry at the top, the radial motion gets weakened and the squish effect becomes small. It is true particularly at the measurement points. As can be seen from the figure, the measuring points are located relatively far away from the walls. It is evident that turbulence dissipation rate is higher compared to the turbulence generation rate. The predicted results are in good agreement with experiments.

12 Grid Independence Study

Results presented here has been carried out with varying cross-sectional area intake ports with a 20° bend angle which is used for validation, to study the effect of grid density on the simulated results. For this purpose, the computational domain

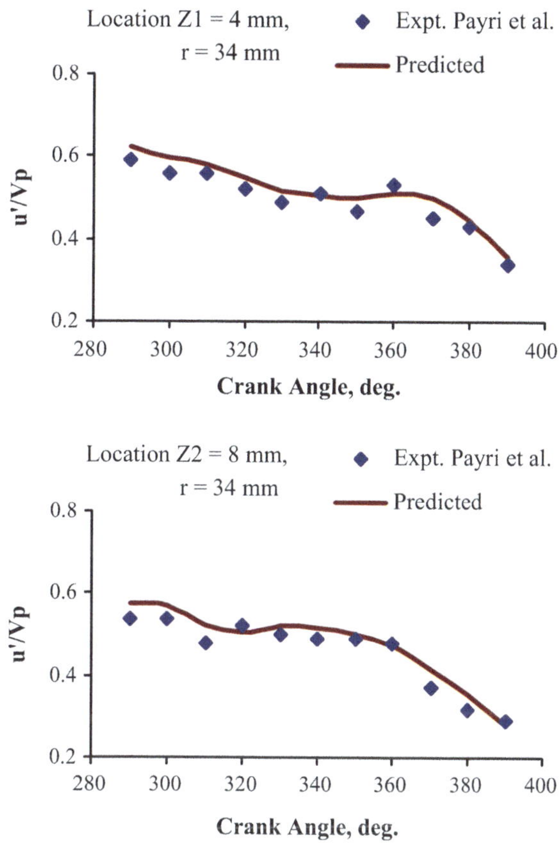

Fig. 23 Non-dimensional turbulent intensity versus crank angle at various locations (*Z*1 = 4 mm and *Z*2 = 8 mm from the cylinder head) compared with experimental LDV data for varying cross-sectional area with 20° bend angle

consisting of 337,978 (grid *A*), 505,542 (grid *B*), 691,354 (grid *C*) and 912,680 (grid *D*) cells has been used. Figure 24 shows mass averaged swirl ratio (SR) and non-dimensional turbulent intensity during suction and compression strokes for various grid densities. These mass averaged variables are considered to be the most appropriate to characterize the flow within the cylinder. The swirl in the cylinder is generated during the intake stroke, and the swirl increases with increasing piston speed. After reaching the maximum piston speed, the piston starts decelerating towards BDC, the mass averaged cylinder velocity decreases, and swirl starts decreasing slowly during the rest of the intake stroke (say up to 180°). This reducing trend continues in the first part of the compression stroke due to slow acceleration and compression effect. However, while approaching TDC, swirl is enhanced as the flow accelerates in preserving its angular momentum within the smaller diameter piston bowl. During the expansion stroke, there is reverse squish, as the flow exits from the piston bowl which is mainly responsible for the sudden fall in swirl velocity. A similar behaviour can be observed in case of the turbulent intensity also. There is a rapid increase at the beginning of the intake stroke caused

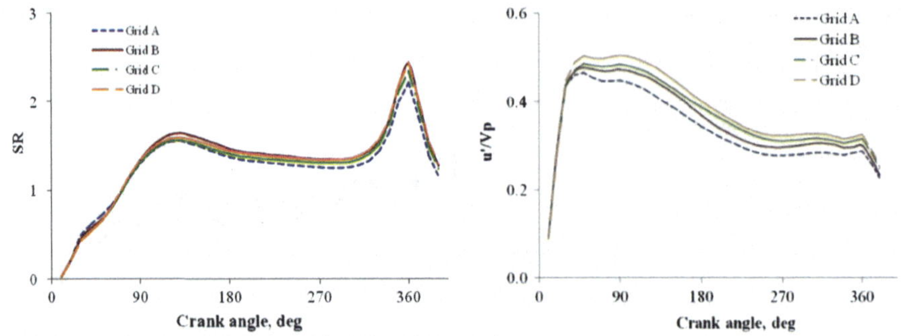

Fig. 24 Temporal distribution of mass averaged swirl ratio for different grids

by the shear stresses associated with the intake jet entering the cylinder and piston acceleration. This is followed by a reduction of turbulent intensity up to a point just before TDC and a slight rise up to TDC and drop during early expansion. With higher and higher cell density, there is slight increase in swirl ratio near compression TDC and slight decrease between maximum valve lift and earlier to TDC. However, a small increase in turbulent intensity is observed throughout the stroke. The comparison of flow quantities such as the mass averaged swirl ratio and non-dimensional turbulence intensity obtained at 115 CAD corresponds to maximum valve lift, 345 CAD near the point of fuel injection and at compression TDC. Grids B and D show almost same maximum swirl ratio at all the three crank angles considered, and grid B shows very small reduction in turbulent intensity compared to higher grid densities. Considering the computational expense and time, the grid *B* with 505,542 cells is used for further computations.

13 Parametric Studies

Based on the confidence gained by the validation, simulation has been carried out to study the effects of intake port cross-sectional area, bend angle, engine speed and intake pressure on in-cylinder flow structure. Three intake port cross sections CCS, VCS and CVCS are considered. To start with a 20° bend angle has been taken for constant cross-sectional area intake ports. Three engine speeds (1000, 1500 and 2000 rpm) and different intake pressures (1.01, 1.1, 1.2, 1.4, 1.6 and 1.71 bar) have been studied. The main aim is to present the results of a comprehensive CFD study on the flow characteristics inside the cylinder of a heavy-duty DI diesel engine and provide an insight to the influence of above-said parameters on flow characteristics near TDC just before the start of combustion. *However, only VCS configuration with 20° bend angle will be presented.*

13.1 *Effect of Intake Port Cross-Sectional Area*

Simulation is carried out to study the effect of intake port cross-sectional area on flow structure with initial values of pressure and temperature of 1.01 bar and 303 K, respectively, at an engine speed of 1000 rpm. The flow in the cylinder during the intake and compression stroke is analysed. The volumetric efficiency estimated from the flow is 88.79%. Figures 25 and 26 show the temporal variation of mass averaged swirl ratio and non-dimensional turbulent intensity during suction and compression strokes. The mass averaged swirl ratio and non-dimensional turbulence intensity are analysed for three crank angles, viz. 115 CAD corresponds to maximum valve lift, 345 CAD near the point of fuel injection and at compression TDC. Consider the swirl ratio at 115 CAD where the intake valves are fully opened. The swirl ratio at this crank angle is higher due to maximum valve lift condition. The reason for the increase in swirl ratio is pressure drop due to convergent cross-sectional area, which results in increased flow velocity. Hence, the flow accelerates and because of the 20° bend angle there is flow deflection. Due to acceleration, the momentum increases, which manifests itself as swirl. Further, it is interesting to note that with respect to 115 CAD (full-valve lift position), the swirl ratio correspondingly at 345 CAD and 360 CAD is 23.13 and 51.88% more for varying cross sections. This is due to enhancement of swirl by squish effect and also due to high initial velocity. Turbulence intensity will fall with commencement of intake valve closure, and turbulence fluctuations again increase beyond the intake valve closure due to increased fluid movement. Both swirl ratio and turbulent intensity fall after compression TDC due to reverse squish–swirl interaction.

Figure 27 shows the distribution of non-dimensional velocity and turbulent intensity, respectively, at a cross section passing through the axis of intake valves at 35 CAD (35° aTDC). It is the initial stage of intake stroke at which intake valves have just started opening. As can be seen, two toroidal vortices appear near the wall of the cylinder as well as at the centre of the bowl because strong interaction of

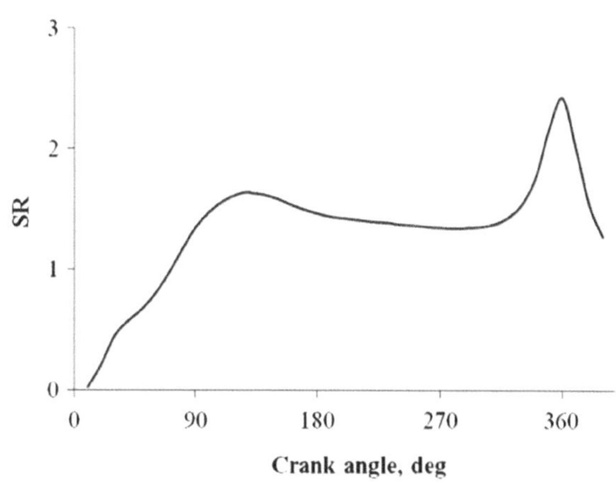

Fig. 25 Temporal distribution of mass averaged swirl ratio in variable area intake port cross sections with 20° bend angle

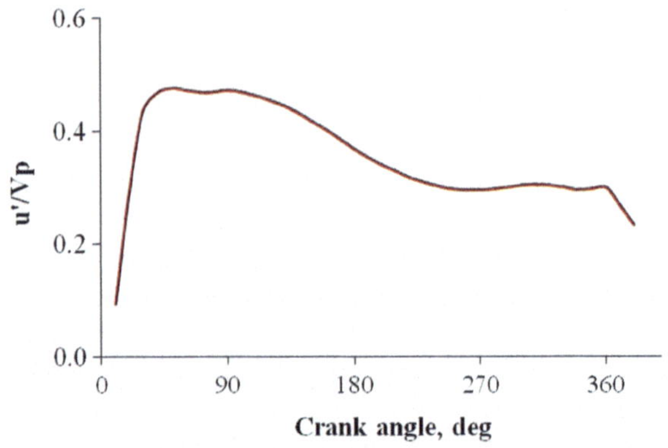

Fig. 26 Temporal distribution of mass averaged non-dimensional turbulent intensity in different intake port cross sections with 20° bend angle

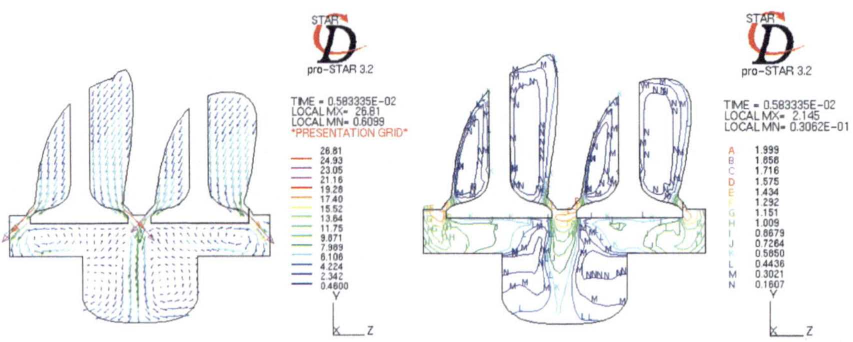

Fig. 27 Non-dimensional velocity field and turbulent intensity in axial midsection at 35 CAD (35° aTDC) in a computational domain with varying cross-sectional area intake ports with 20° bend angle

intake jets impinges on the walls of the cylinder and the piston during the induction. In the initial stage of intake stroke, the space in the small volume contained between the cylinder head and the piston bowl is not sufficient for a vertical flow to develop clearly in any direction. During the induction, the intake jets impinge on the walls of the cylinder and the piston. Due to this strong interaction of intake jets, toroidal vortices appear near the wall of the cylinder. The two induction jets collide and impinge on the bottom surface of the combustion chamber creating two more toroidal vortices (recirculation zones) inside the combustion chamber bowl. In all the cases, maximum value of turbulent intensity observed near the cylinder wall and at the centre of the combustion chamber where the jet impinges. The turbulent intensities are highly non-uniform with maximum values in the strong recirculation zone where the jets collide.

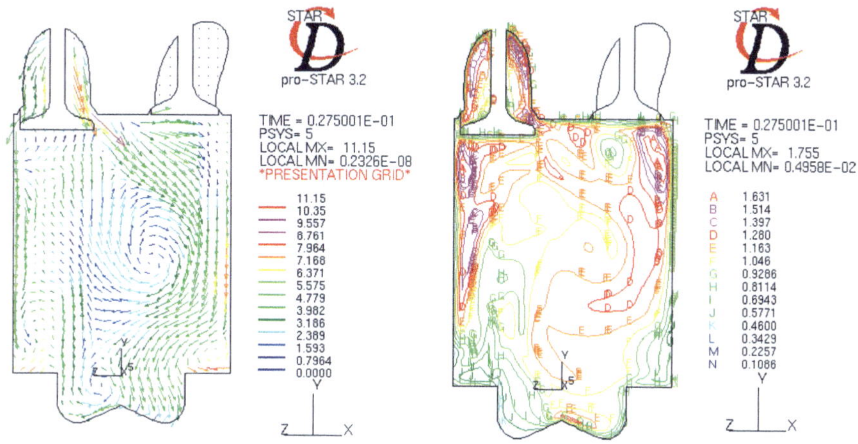

Fig. 28 Non-dimensional velocity and turbulent intensity field at 165 CAD (165° aTDC) in a computational domain with varying cross-sectional area intake ports with 20° bend angle

Figure 28 shows the distribution of non-dimensional velocity and turbulent intensity at 165 CAD (165° aTDC). It is the final stage of intake stroke at which intake valves are starting to close. This crank angle has been chosen to study the flow structure development near the intake valve closing. The vortex at the centre of the cylinder seems to have stabilized and elongated at the same location, and there is no other vortex observed below the intake and exhaust valves have been disappeared. The vortex, nearer to left-side cylinder wall and above the piston top surface, is closer to cylinder wall, and the other one is closer to the corner adjacent to cylinder wall and piston top surface. One more additional vortex is observed in the centre of the combustion chamber. When the piston is approaching BDC, the piston speed is low and the inlet velocity of the flow entering the cylinder is also low. This makes the vortex observed earlier gets stabilized at the same location.

The turbulent intensity distribution is almost uniform. The maximum value of turbulent intensity still appears around the valves. A small zone of strong turbulence observed near the intake and exhaust valves. After the closing of intake valves, the turbulence becomes more uniform during remaining part of the compression stroke because the discharge velocities and the shear stresses induced in the top portion of the cylinder gradually decay.

Figure 29a, b shows the distribution of non-dimensional velocity and turbulent intensity, respectively, at 345 CAD (near the point of fuel injection). There are two vortices with opposite rotational directions observed near the centre of the combustion chamber. The maximum value of turbulent intensity is observed in the intake side, centre and exhaust side of the combustion chamber. Maximum values of velocity and turbulent intensity are observed in this configuration.

Figure 30a, b shows the distribution of non-dimensional velocity and turbulent intensity at compression TDC (360 CAD). For cases *a* and *c*, two toroidal vortices with opposite rotational directions near the combustion chamber wall and one almost at the

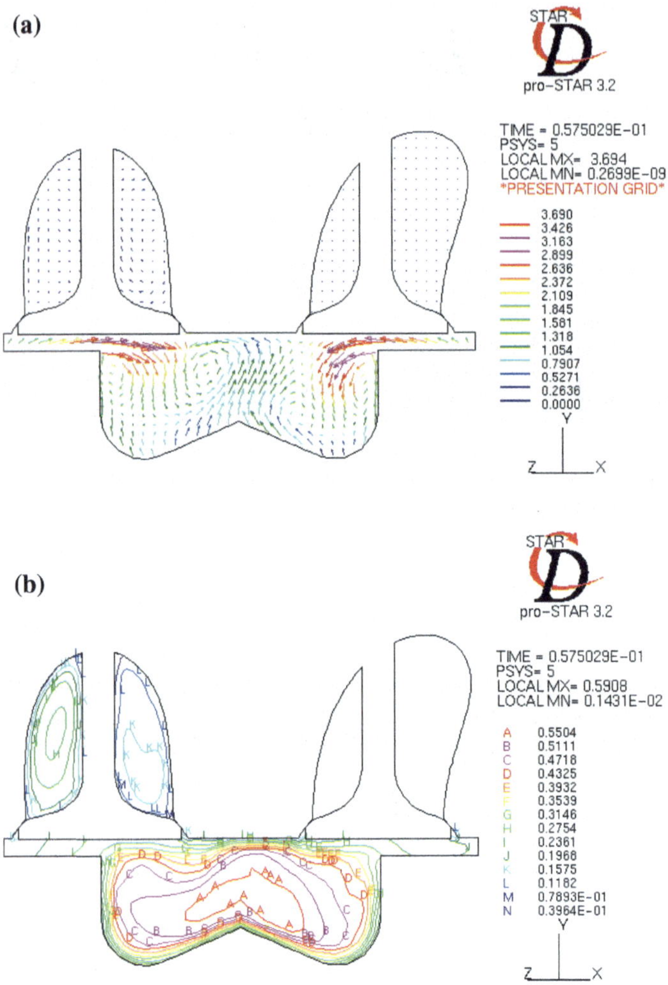

Fig. 29 **a** Non-dimensional velocity field at 345 CAD in a computational domain with varying cross-sectional area intake ports with 20° bend angle. **b** Non-dimensional turbulent intensity field at 345 CAD in a computational domain with varying cross-sectional area intake ports with 20° bend angle

centre of the combustion chamber are observed. For the configuration under consideration, there are four vortices with opposite rotational directions observed in the combustion chamber. The appearance of toroidal vortices with opposite rotational directions is due to strong interaction between squish and swirl near TDC and also due to central projection in the bowl. The centrifugal force caused by the tangential vortex impedes the flow from entering radially towards the central zone of the cylinder. The maximum value of turbulent intensity is observed in the centre of the combustion chamber. From the results so far presented, it may be concluded that the varying cross-sectional area intake port configuration provides plausible results.

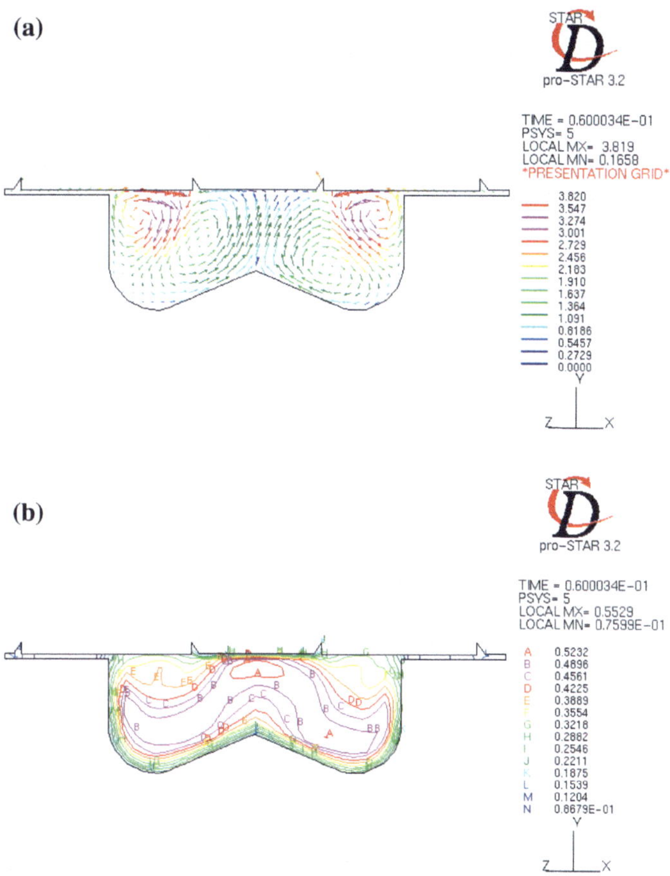

Fig. 30 **a** Non-dimensional velocity field at TDC (360 CAD) in a computational domain with varying cross-sectional area intake ports with 20° bend angle. **b** Non-dimensional turbulent intensity field at TDC (360 CAD) in a computational domain with varying cross-sectional area intake ports with 20° bend angle

13.2 In-Cylinder Combustion Simulation

Combustion process involves simulation of the power stroke during the engine cycle, starting from closing of valves to the end of the expansion stroke. Since the valves are closed, the combustion chamber is the main flow domain and the piston is the sole moving part. These simulations are also known as "in-cylinder combustion". Combustion simulation in multidimensional modelling is less complicated geometrically than a port flow simulation. If the geometry is rotationally symmetric, the entire domain can be modelled as a sector to speed up the calculation.

Typically, the initial flow field at this stage can be obtained from

- a cold flow simulation if 3D geometry is used;
- patching-in can be done based on a cold flow analysis;
- running the piston putting off the combustion to obtain charge compression.

As with cold flow, a moving deforming mesh model can be used for the piston motion. Since only the piston motion is included, geometric decomposition is not required in the simulation. The modelling of the fluid dynamics in the valve port region and their effect on combustion need not be included in the in-cylinder combustion simulations.

Models pertaining to the fuel spray, combustion and pollutant formation should be included. In direct injection engines, the fuel spray from the tip of the injector nozzle should be included at the specific crank angle and duration using appropriate spray model. For port fuelled engines, it may be assumed that the combustion charge is well mixed. A chemical mechanism describing the reaction of fuel vapour with air is used to describe combustion. Models for turbulence–chemistry interaction should be specified. Submodels for NO_x and soot formation must be incorporated to calculate pollutant formation, which can be coupled with the combustion calculation or calculated as a post-processing step.

The main challenge in modelling in-cylinder combustion process lies in the physics for spray modelling and combustion. The spray can be considered as a column of liquid entering the domain at high speed which subsequently breaks into droplets due to aerodynamic forces. These droplets can coalesce into larger droplets or break into even smaller droplets. During this process, droplets exchange mass with the surrounding gases. Submodels for coalescence and break-up, as well as heat and mass transfer, are used to capture spray dynamics. The most important point is that the CFD mesh should be sufficiently fine to capture the coupling between the liquid droplets and the gases in the cylinder accurately. If liquid spray impinges upon the cylinder walls, it is possible to form a thin liquid film. This will undergo its own processes of movement and vaporization and will require a separate treatment.

To evaluate combustion, detailed chemical mechanisms for pure fuels that constitute the components of diesel fuel and gasoline involve hundreds of species and thousands of reactions. These reactions are coupled with the fluid dynamics due to the similar timescales of fluid mechanical motions and chemical reactions. The energy release during the combustion process increases pressures and temperatures for the working medium. This in turn affects the fluid motions inside the cylinder. A direct computation of this coupled interaction, while including detailed chemistry is staggeringly expensive in terms of computation time and is impractical for complex geometries. Therefore, submodels should be included to take care of this.

Reduced order mechanisms can capture most of the essential chemistry in a narrower range of temperature and equivalence ratio. They should be used along with a submodel for turbulence–chemistry interaction. One such model is the probability density function (PDF) approach which allows an efficient computation of turbulence–chemistry interaction. Flame propagation is modelled using a progress variable-based approach which calculates the transient flame front speed and location. These approaches enable computation of the combustion process on large meshes in complex geometries with a reasonable computational power.

Simplified mechanisms can be used to compute the NO_x formation due to

- high temperatures (thermal NO_x);
- nitrogen in the fuel (fuel NO_x);
- fuel reactions in the flame front (prompt NO_x);
- sulphur oxides in the fuel (SO_x);
- soot formation.

These pollutants are generally a very small percentage of the total mass in the cylinder. Therefore, these calculations can be decoupled from the calculation of the main energy release. In some cases, this is done during the post-processing operation at the end of the simulation. However, it is more accurate to include the pollutant formation stage in the simulation. This is true especially for the pollutants arising from incomplete combustion that oxidize later in the cycle.

13.2.1 Case Study 5: Combustion Analysis in a DI Diesel Engine

In the previous case study, we investigated the effect of port bend angle in cylinder flow field. I found that a variable area port with 20° bend angle was giving better in cylinder air motion. In this case study, we will move one step forward to simulate in cylinder combustion process on a single-cylinder DI diesel engine. First, we will explain the combustion model used and also give models for NO_x and soot formation, and then, the predicted results will be presented and explained.

13.3 Modelling of Combustion Process

The process of combustion is modelled using extended flame coherent model. It is named as ECFM-3Z model which is a general-purpose combustion model capable of simulating the complex mechanisms of turbulent mixing, flame propagation, diffusion combustion and pollutant formation. The flame propagation phase is modelled by the flame surface density (FSD) transport equation which is given by

$$\frac{\partial \Sigma}{\partial t} + \nabla.(u\Sigma) - \nabla.\left[\left(D + \frac{\mu_t}{Sc_t}\right)\nabla\left(\frac{\Sigma}{\rho}\right)\right] =$$
$$\sum\left[C_{\mathrm{div}u}\frac{2}{3}\nabla.u + C\alpha\Gamma\frac{\varepsilon}{k} + \frac{2}{3}\frac{\rho_u}{\rho_b}U_l\sum\frac{l-c}{c} - \beta U_l\sum\frac{1}{1-c} - \frac{2}{3}\frac{1}{\gamma p}\frac{\partial p}{\partial t}\right] + S_{\mathrm{conv}}$$

where D is the molecular diffusivity, $C_{\mathrm{div}u}$ is an empirical parameter with default value of 1.0, C is a correction factor that takes into account the flame chemical timescale and the flame interaction with the walls, Γ is the ITNFS (Net Flame

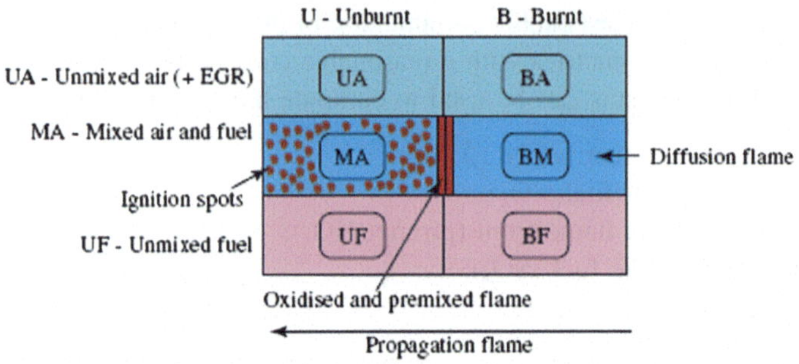

Fig. 31 ECFM 3Z combustion model

Stretch) function, ρ_u and ρ_b the density of the unburnt and burnt gases, U_l the effective laminar flame speed, γ is the isentropic coefficient, S_{conv} is an additional contribution to the FSD from convection at the spark plug, α and β are empirical coefficients for the production and destruction terms, μ_t is the turbulent viscosity, p is the thermodynamic pressure, and c is the Reynolds-averaged progress variable.

It can also be used for in-cylinder analysis in a multi-injection environment and for multi-cycle simulations. ECFM-3Z model is recommended for general combustion simulation of both gasoline and diesel engines with homogeneous or non-homogeneous fuel–air mixtures. "3Z" stands for three zones of mixing, namely the unmixed fuel zone, mixed gases zone and the unmixed air plus EGR zone which is shown in Fig. 31. The three zones are too small to be resolved by the mesh and are therefore modelled as subgrid quantities. The mixed zone is the result of turbulent and molecular mixing between gases in the other two zones and is where combustion takes place.

13.4 NO_x Model

The principal reactions for the formation of thermal NO_x is recognized and proposed by Zeldovich mechanism. The rate of formation of NO_x is significant only at higher temperature since the thermal fixation of nitrogen requires breaking of strong N_2 bonds. This effect is represented by the high activation energy of reaction, which makes the reaction as the rate-limiting step of Zeldovich mechanism.

13.5 Soot Model

The formation and emission of carbonaceous particles is a process that is often observed during the combustion of hydrocarbons. In general, the soot formations in IC engines are due to incomplete combustion of fuel. Apart from the loss of fuel and combustion efficiency, it affects the health of human life and environment. Hence, there are strict legislative demands to produce cleaner engines. There are many models proposed for soot formation. In Chapter "Characterization of Ringing Operation in Ethanol Fueled HCCI Engine Using", the details of all models, viz. RNG k–ε, droplet break-up, ECFM-3Z combustion model for diesel combustion, NO_x model, soot model, are explained.

Initially, a cold flow analysis was carried out and, from that analysis, it was found that out of eight-piston top configurations studied, flat with central bowl and pent roof with offset bowl pistons exhibits higher velocity, TKE, TR, turbulence intensity and length scale as compared to those of other piston top configurations. Hence, CFD analysis is extended for combustion studies with these piston top profiles (called as optimized piston top profiles). However, only results pertaining to flat piston-with-centre-bowl will be presented and discussed.

13.6 Spray Structure

The most significant feature of the diesel engine is the use of liquid hydrocarbon as a fuel. The liquid fuel injected through the nozzle breaks up, atomizes and evaporates in a high temperature surrounding air and burns as being mixed with air. The behaviour of fuel spray during the break-up, atomization and vaporization mainly depends on the air entrainment with a higher velocity and the higher temperature inside the engine cylinder [16].

Figure 32 depicts the plot of fuel injection and spray evaporation for the flat piston-with-centre-bowl at the end of fuel injection (717 CAD). Figure 32a shows the interaction of flow field with diesel fuel injection. Figure 32b shows the exploded view of diesel fuel spray evaporation having larger droplets in head region and a very less number of evaporated droplets near the tail region inside the engine cylinder.

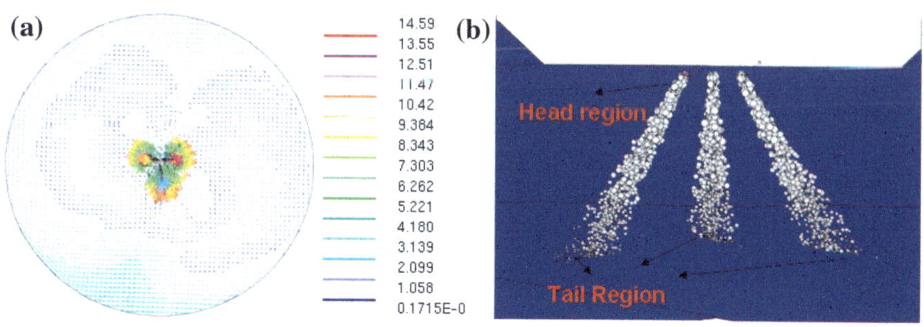

Fig. 32 Plot of fuel injection on flow field and spray evaporation for flat piston-with-centre-bowl

13.7 Diameter and Droplet Velocity

Figure 33 shows the variation of fuel droplet diameter with crank angles for the flat piston top profiles with central bowl during the period of injection. From Fig. 33, it is clear that fuel droplet diameter is around 1.8 μm during the start of injection. As more fuel gets accumulated between 705° and 715° crank angle and because of adhesion, the diameter of droplets gets increased. Near the end of injection as the fuel nozzle getting closed, the fuel break-up is not proper and therefore the diameter of the droplets shoots up.

Figure 34 depicts the variation of fuel droplet velocity with crank angles for the flat piston top profiles with central bowl during the period of injection. From Fig. 34, it is noted that the flat piston-with-centre-bowl gives a maximum fuel droplet velocity of about 75 m/s.

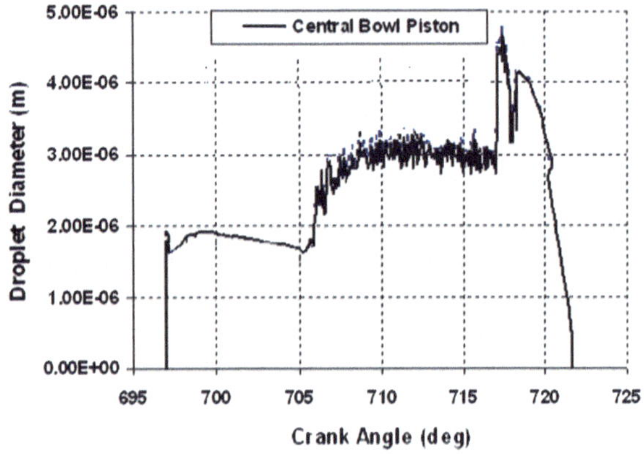

Fig. 33 Variation of droplet diameter with crank angle with flat piston-with-centre-bowl

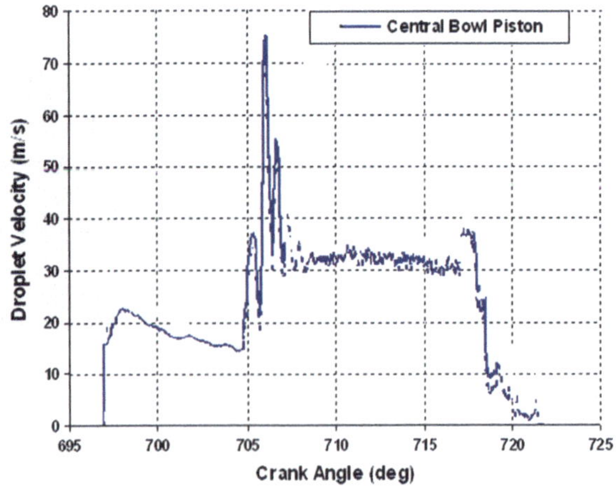

Fig. 34 Variation of droplet velocity with crank angle for flat piston-with-centre-bowl

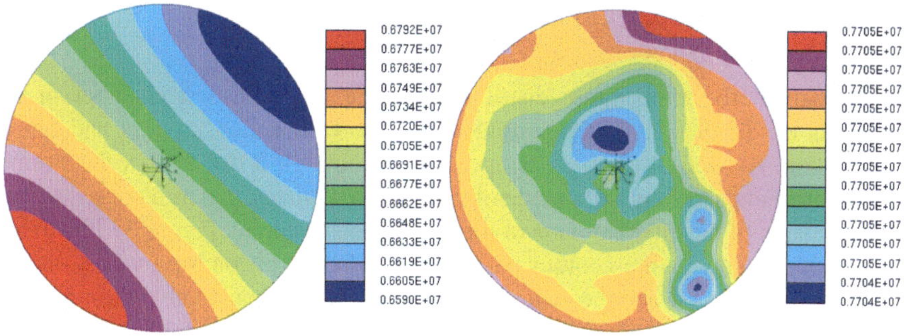

(a) At 710 CAD (b) 720 CAD

Fig. 35 Pressure contours for flat piston-with-centre-bowl

13.8 Peak Pressure

Figure 35 depicts the pressure contours for flat piston-with-centre-bowl under firing condition at an engine speed of 1500 rpm with a conventional start of fuel injection timing at 23° bTDC. It is observed that flat piston-with-centre-bowl develops peak pressure of 67.9 and 77 bar at 710 and 720 CAD under full load condition which is reasonably a good value for such a configuration.

13.9 Heat Release Rate

Figure 36 shows the variation of heat release rate with crank angles for flat piston-with-centre-bowl at an engine speed of 1500 rpm. From Fig. 36, it shows the early start of heat release during premixed combustion period and slower rate of combustion during diffusion combustion period. The maximum heat release rate of 78 J/°C A is obtained with the flat piston.

Fig. 36 Heat release rate versus crank angle

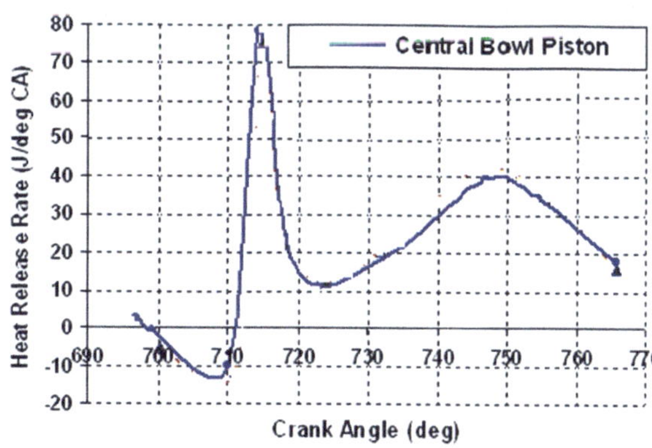

13.10 Peak Temperature

Knowledge of the temperature distribution inside a combustion chamber of an engine is the fundamental step to understand the combustion process and flame propagation. As well, if the peak temperature is too high it can disrupt the engine functioning, since the engine is made of material that dilates, and as a consequence, the dimension of the different parts of the engine gets distorted. Figure 37 illustrates the computed temperature contour plot for two different piston top profiles on the midplane passing between the centrelines of intake and exhaust valve through axis for different crank angle degrees (say 720–1080 CAD in steps of 30 CAD). From these figures, it is observed that flat piston-with-centre-bowl gives peak an about 2369 K. Further, it beautifully depicts the process of combustion during the expansion stroke. As expressed earlier, even though such a process to visualize

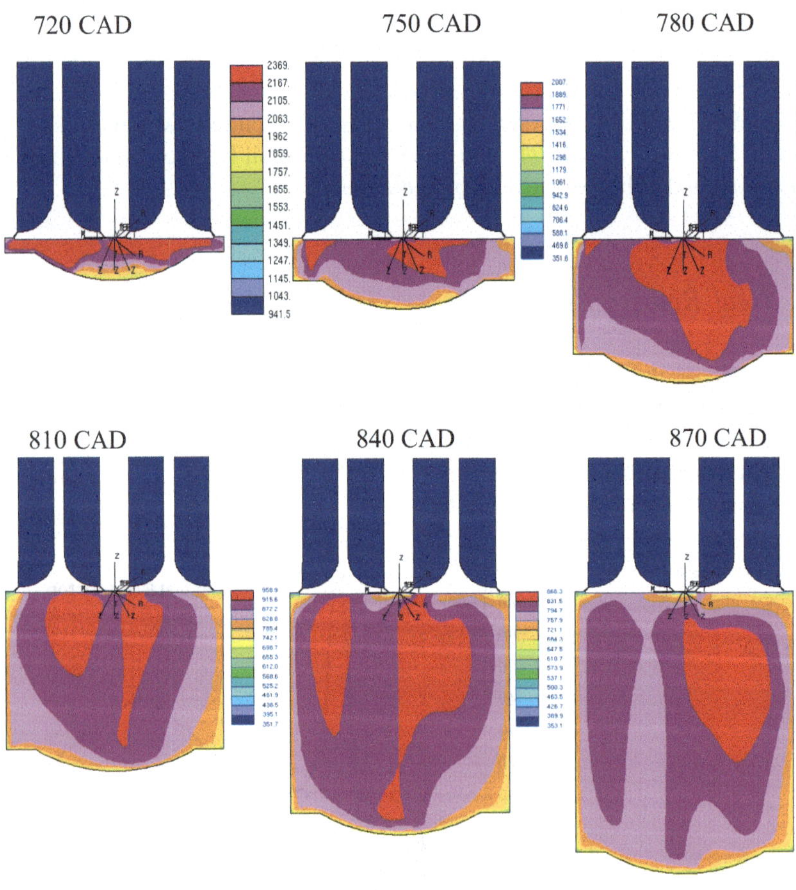

Fig. 37 Temperature contours for flat piston-with-centre-bowl

experimentally is extremely difficult, CFD analysis throws a good light on theoretical visualization in the process of combustion.

14 Full Cycle Simulations

As the name indicates, full cycle simulations essentially involve all the five processes, viz. suction, compression, combustion, expansion and exhaust. This means the prediction of all details right from cold flow analysis, in-cylinder combustion and complete simulations of the entire engine cycle. This is nothing but a transient computation of turbulent airflow, spray and combustion, and exhaust with moving valves and pistons. The initial flow field is obtained from a cold flow simulation or by running the engine without combustion for a cycle before turning on spray and combustion.

The advantage of full cycle simulations is that they provide the full picture of engine performance, taking into account intake and exhaust valve fluid dynamics, mixing, turbulence production, spray, combustion and flame propagation, and pollutant formation. It is to be emphasized here that they are extremely complex to set up and expensive to run.

The geometry preparation is a complex exercise. It includes geometry from the throttle body, ports, valves, combustion chamber, cylinder and the piston, making it difficult to perform clean-up, decomposition and meshing. The solver set-up has to include moving mesh, airflow, turbulence, spray, turbulence–chemistry and flame propagation, and pollutant formation.

From the above, the need is for process compression and automation all the way from geometry to post-processing. If properly done, it will considerably reduce the time needed for problem set-up and post-processing. Further, accurate and efficient models for chemistry, spray and combustion, as well as efficient solver techniques, are required to get the solution in the shortest time possible.

14.1 *Case Study 7: Expansion Process with NO_x and Soot Prediction*

So far, we have illustrated up to expansion process in the previous case studies. In this section, we will present the results for the exhaust process so that we will have the satisfaction of presenting all the processes of an IC engine by different case studies, which is the aim of this chapter. Figure 38 shows the complete exhaust process starting from 900 to 1080 CAD in terms of 30 CAD. As the piston moves from bottom centre to top centre, it pushes the product of combustion. During the process, the temperature of the working fluid in the cylinder drops which is well presented in this figure.

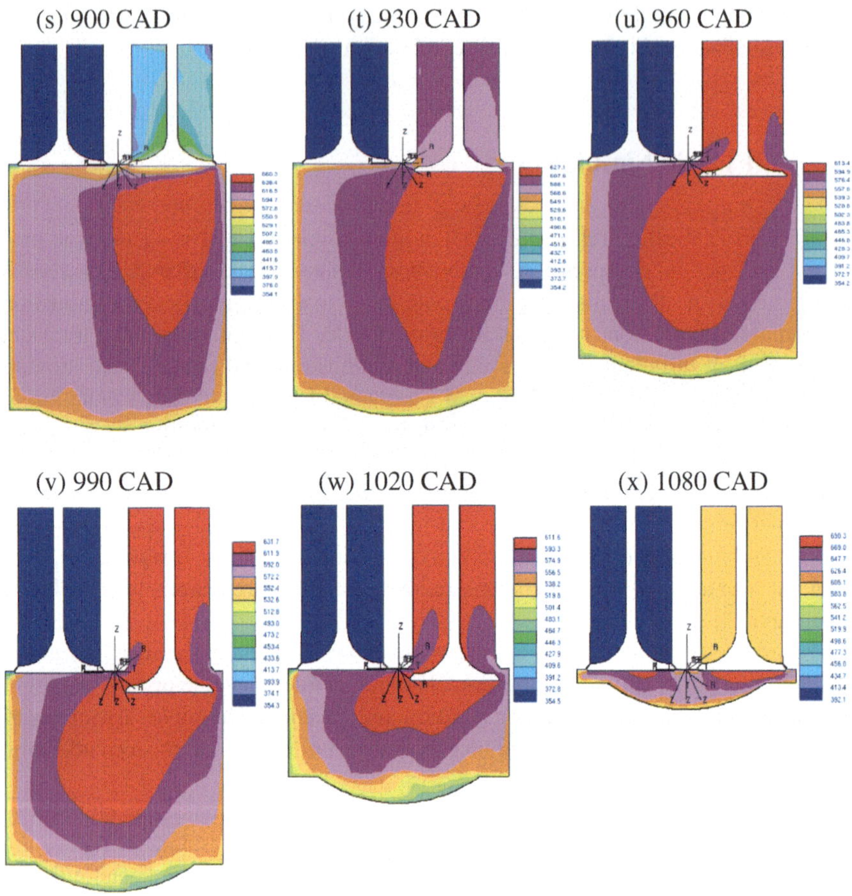

Fig. 38 Temperature contours for flat piston-with-centre-bowl during exhaust

14.2 NO_x and Soot Emissions

Comparison of Figs. 39 and 40 shows a trade-off between NO_x and soot emissions for the flat piston-with-centre-bowl. NO_x formation at the beginning of injection is higher. This is due to the rate of temperature rise is higher during the early part of combustion.

In order to lower pollutant emissions of diesel engines, enormous efforts have been taken to overcome the trade-off between NO_x and soot emissions. As it is well known, due to the NO_x and soot trade-off relation, it is difficult to reduce these two pollutants simultaneously. For example, changing the combustion chamber shape to achieve a better in-cylinder air motion is an effective way to reduce soot formation by increasing the peak cylinder temperature and pressure. However, this usually results in an increase in NO_x production.

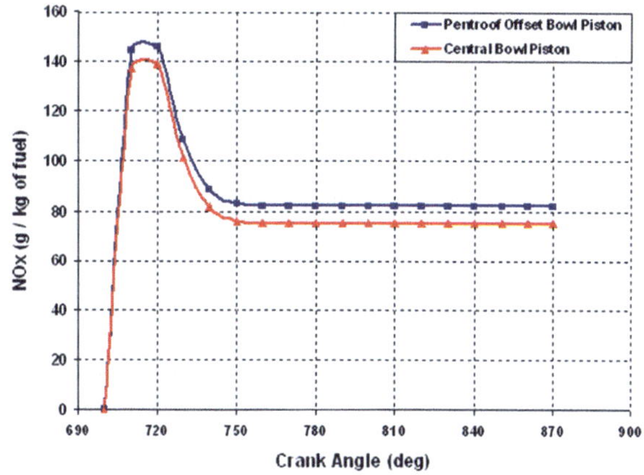

Fig. 39 NO_x emission versus crank angle

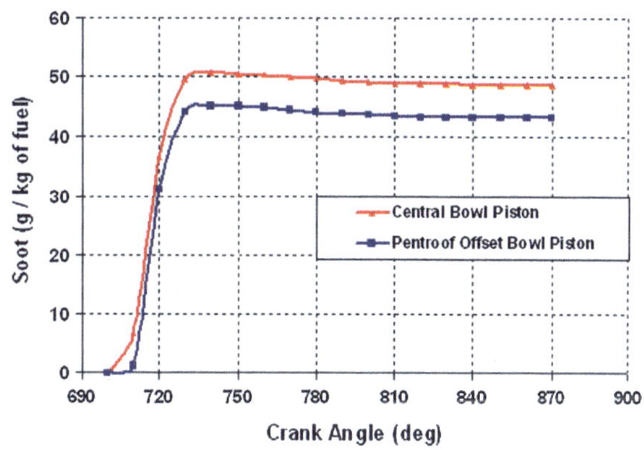

Fig. 40 Soot emission versus crank angle

15 Conclusion

As mentioned in the early part of this chapter, the main aim of this chapter is to introduce to the readers the application of CFD in IC engine analysis and design has been achieved by means of different case studies. Consciously, equations have been avoided. Different case studies have been presented purposely to establish that it can be applied to different geometries with different input conditions. From the case studies presented, it can be concluded that mentioned aim has been achieved.

Acknowledgements The results presented here are from the theses of our **P**unctilious, **h**ard working and **D**edicated (PhD) scholars. We gratefully acknowledge their efforts in producing the results. Authors are thankful to CD-ADAPCO for providing licence free of cost for performing CFD studies.

Bibliography

1. Amin M, Saray RK, Shafee S, Ghafouri J (2013) Numerical study of combustion and emission characteristics of dual-fuel engines using 3D-CFD models coupled with chemical kinetics. Fuel 106:98–105
2. Andreas WB, Josefsson G, Ekdahl R, Ogink R, Grandin B (2011) The effect of tumble flow on efficiency for a direct injected turbocharged downsized gasoline engine, SAE Paper No. 2011-24-0054
3. Andreassi L, Cordiner S, Rocco V (2003) Modelling the early stage of spark ignition engine combustion using the KIVA-3V code incorporating an ignition model. Int J Engine Res 4(3): 179–192
4. Assanis DN, Hong SJ, Nishimura A, Papageorgakis G, Vanzieleghem B (2000) Stududies of spray break-up and mixture stratification in a gasoline direct injection engine using KIVA-3V. J Gas Turbine Power 122:485–492
5. Auriemma M, Corcione FE, Diana S, Police G, Valentino G (2001) Investigation of the intake tumble flow in a prototype GDI engine using a steady state test rig, SAE Paper No. 2001-01-022
6. Auriemma M, Corcione FE, Martino UD, Valentino G (2001) Analysis of the intake flow in a diesel engine head using dynamic steady flow conditions, SAE Paper No. 2001-01-1307
7. Auriemma M, Caputo G, Corcione FE, Valentino G, Riganti G (2003) Fluid-dynamics analysis of the intake system for a HDDI diesel engine by STAR-CD code and LDA technique, SAE Paper No. 2003-01-0002
8. Barths H, Hasse C, Peters N (2000) Computational fluid dynamics modeling of non-premixed combustion in direct injection diesel engines. Proc IMechE Vol 1 J Engine Res
9. Beard OC, Miche M (2003) Improved modeling of DI diesel engines using sub grid description of spray and combustion, SAE Paper No. 2003-01-0008
10. Bedford Frederick C (2001) Modeling production and abatement of nitrogen oxide pollutants in direct injection diesel engines. University of Wisconsin-Madison
11. Benny P (2008) Effect of inlet manifold configuration on in-cylinder air motion and combustion in a DI diesel engine, Ph.D. thesis, IIT Madras, India
12. Bertrand HD (2002) Practical diesel engine combustion analysis. SAE international, Warrendale, PA, USA
13. Bianchi GM, Cantore G, Fontanesi S (2001) Turbulence modelling in CFD simulation of ICE intake flows, SAE Paper No. 2001-01-049
14. Cipolat D (2007) Analysis of energy release and NO_x emissions of a CI engine fuelled on diesel and DME. J Appl Therm Eng
15. Corgard DD, Reitz RD (2001) Effects of alternative fuels and intake port geometry on HSDI diesel engine performance and emissions, SAE Paper No. 2001-01-0647
16. Cyril C (2002) Combustion process in diesel engine. Ph.D. thesis, University of Brighton
17. Franz XT, Srinivasan S (2009) CFD-based optimization of fuel injection strategies in a diesel engine using an adaptive gradient method. Appl Math Model 33:1366–1385
18. Ganesan V (2000) Computer simulation of compression—ignition engine processes. Universities Press, Hydrabad
19. Ghandhi JB, Herold RE, Shakal JS, Strand TE (2005) Time resolved particle image velocimetry measurements in an internal combustion engine, SAE Paper No. 2005-01-3868
20. Gnana Sagaya Raj A (2014) In-cylinder air motion and combustion in diesel engines, Ph.D. thesis, IIT Madras, India
21. Gnana Sagaya Raj A, Mallikarjuna JM, Ganesan V (2012) Flow investigation on different combustion chamber configuration in a DI diesel engine—a CFD approach. Int J Adv Mater Res 505:420–423
22. Gnana Sagaya Raj A, Mallikarjuna JM, Ganesan V (2013) Analysis of in-cylinder air motion in a DI diesel engine with four different piston bowl configuration—a CFD and PIV comparison, SAE Paper No. 2013-01-2786

23. Gnana Sagaya Raj A, Mallikarjuna JM, Ganesan V (2013) CFD prediction of combustion on direct injection diesel engine with two different combustion chamber configurations, SAE Paper No. 2013-01-2804
24. Gnana Sagaya Raj A, Mallikarjuna JM, Ganesan V (2013) Energy efficient piston configuration for effective air motion—a CFD study. Appl Energy 102:347–354
25. Goldsworthy L (2006) Computational fluid dynamics modeling of residual fuel oil combustion in the context of marine diesel. IMechE J Engine Res 7:181–199
26. Heywood JB (1988) Engines: an introduction. McGraw-Hill Inc, NY
27. Hossainpour S, Binesh AR (2009) Investigation of fuel spray atomization in a DI heavy-duty diesel engine and comparison of various spray breakup models. Fuel 88:799–805
28. Hountalas DT, Papagiannakis RG (2000) Development of a simulation model for direct injection dual fuel diesel-natural gas engines, SAE Paper No. 2000-01-0286. https://www.sharcnet.ca/Software/Ansys/17.0/en-us/help/wb_icom/icom_intro_fd.html
29. Huan YC, Zhou L, Li B, Jiang DM (2001) Experimental study on fuel spray development history near the spark plug gap in the direct injection circularly stratified charge combustion system. Trans Chin Soc Int Combust Engines 19(6):497–501
30. Huang RF, Huanga CW, Changa SB, Yanga HS, Lina TW, Hsub WY (2004) Topological flow evolutions in cylinder of a motored engine during intake and compression strokes. J Fluids Struct 20:105–127
31. Jaichandar S, Annamalai K (2012) Influences of re-entrant combustion chamber geometry on the performance of Pongamia biodiesel in a DI diesel engine. Energy 44:633–640
32. James Gunasekaran E (2008) Modeling of flow and combustion in a four stroke DISI engine, Ph.D. thesis, IIT Madras, India
33. Jayashankara B (2009) Multi-dimensional modeling of CI engine processes, Ph.D. thesis, IIT Madras, India
34. Jayashankara B, Ganesan V (2010) Effect of fuel injection timing and intake pressure on the performance of a DI diesel engine—a parametric study using CFD. Energy Convers Manag 51:1835–1848
35. Kar MP, Kiat H, Gan S (2012) In-cylinder diesel spray combustion simulations using parallel computation: a performance benchmarking study. Appl Energy 93:466–478
36. Kawahara N, Tomital F, Kasaharal D, Sumida M (2004) Liquid sheet break-up of high-pressure swirl injector for DISI engine. In: Comodia 2004
37. Kulkarni Y, Aghav Y, Sohi NS, Dani AD (2005) Optimization of inlet port performance on emission compliance of naturally aspirated DI diesel engine, SAE 2005-26-010
38. Kyungnam K, Arai M (2002) Diesel spray and adhering fuel on an impingement wall, SAE Paper No. 2002-01-1628
39. Lappas P, Evans RL (2006) A numerical and experimental study of the squish-jet combustion chamber. ImechE Int J Engine Res 7:471–487
40. Lee BH, Song YJ, Chang JH, Jeon CH (2010) Effect of the number of fuel injector holes on characteristics of combustion and emissions in a diesel engine. Int J Autom Technol 11(6): 783–791
41. Lee D, Rutland CJ (2001) Multidimensional modeling of a six mode diesel test cycle using a PDF combustion model, SAE Paper No. 2001-01-0585
42. Li Y, Zaho H, Leach B, Ma T (2004) Development of a fuel stratification spark ignition engine. Proc Inst Mech Eng J Autom Eng 219(7):923–934
43. Li Y, Zaho H, Leach B, Ma T, Ladommatos N (2004) Characterization of an in-cylinder flow structure in a high-tumble spark ignition engine. Int J Engine Res 5(5):375–400
44. Li Y, Zhao H, Peng Z, Ladommatos N (2002) Particle image velocimetry measurement of in-cylinder flow in internal combustion engines—experiment and flow structure analysis. Proc Inst Mech Eng 216(Part D):65–81
45. Lin L, Shulin D, Jin X, Jinxiang W, Xiaohong G (2000) Effects of combustion chamber geometry on in-cylinder air motion and performance in DI diesel engine, SAE Paper No. 2000-01-0510

46. Liu Y, Lu F, Reitz RD (2005) The use of non-parametric regression to investigate the sensitivities of high-speed direct injection diesel emissions and fuel consumption to engine parameters. IMechE Vol 7 J Engine Res
47. Lu L, Shulin D, Jin X, Jinxiang W, Xiaohong G (2000) Effects of combustion chamber geometry on in-cylinder air motion and performance in DI diesel engine, SAE Paper No. 2000-01-0510
48. Lumley JL (1999) Engines: an introduction. Cambridge University Press
49. Micklow GJ, Gong W (2001) Combustion modeling for direct injection diesel engines. Proc Inst Mech Eng 215(Part D):191–205
50. Micklow GJ, Gong WD (2007) Intake and in-cylinder flowfield modelling of a four-valve diesel engine. Proc IMechE J Autom Eng 221(Part D):1425–1440
51. Mikalsen R, Roskilly AP (2009) A computational study of free-piston diesel engine combustion. Appl Energy 86:1136–1143
52. Miles PC (2000) The influence of swirl on HSDI diesel combustion at moderate speed and load, SAE paper No. 2000-01-1829
53. Mohamed AJ, Kantchev G, Abid MS (2011) Influence of intake manifold design on in-cylinder flow and engine performances in a bus diesel engine converted to LPG gas fuelled, using CFD analyses and experimental investigations. Energy 36:2701–2715
54. Muralikrishna B, Mallikarjuna JM (2008) Effect of engine speed on intake valve flow characteristics of a diesel engine—an analysis using particle image velocimetry. In: First international conference on emerging trends in engineering and technology, IEEE, New York. https://doi.org/10.1109/ICETET.2008.97
55. Muralikrishna B, Mallikarjuna JM (2009) Tumble flow analysis in an unfired engine using particle image velocimetry. World Acad Sci Eng Technol 54:430–435
56. Paul CM, Collin R, Hildingsson L, Hultgvist A, Andersson OI (2007) Combined measurements of flow structure, partially oxidized fuel, and soot in a high-speed, direct-injection diesel engine. In: Proceedings of the combustion institute, vol 31, Issue 2
57. Payri F, Benajes J, Margot X, Gil A (2003) CFD modeling of the in-cylinder flow in direct-injection diesel engines. Comput Fluids 33:995–1021
58. Payri F, Desantes JM, Pastor VJ (1996) LDV measurements of the flow inside the combustion chamber of a 4-valve D.I. diesel engine with axisymmetric piston-bowls. Exp Fluids
59. Prasad BVVSU, Sharma CS, Anand TNC, Ravikrishna RV (2011) High swirl-inducing piston bowls in small diesel engines for emission reduction. Appl Energy 88:2355–2367
60. Ra Y, Hruby EJ, Reitz RD (2005) Parametric study of combustion characteristics in a direct-injection diesel homogeneous charge compression ignition engine with a low-pressure. Proc IMechE J Engine Res 6:215–230
61. Rakopoulos CD, Hountalas DT, Rakopoulos DC, Giakoumis EG (2005) Experimental heat release rate analysis in both chambers of an indirect injection turbocharged diesel engine at various load and speed conditions, SAE Paper No. 2005-01-0926
62. Rakopoulos CD, Kosmadakis GM, Pariotis EG (2010) Investigation of piston bowl geometry and speed effects in a motored HSDI diesel engine using a CFD against a quasi-dimensional model. Energy Convers Manag 51:470–484
63. Ramdasi SS, Marathe NV, Moorthi S, Balasubramanian AK (2005) An integrated design approach for robust engine development: a case study, SAE Paper No. 2005-26-002
64. Sagaya Raj A, Mallikarjuna J, Venkitachalam G (2013) CFD prediction of combustion on direct injection diesel engine with two different combustion chamber configurations, SAE Technical Paper 2013-01-2804. https://doi.org/10.4271/2013-01-2804
65. Saijo K, Kojima T, Nishiwaki K (2005) Computational fluid dynamics analysis of the effect of mixture heterogeneity on combustion process in a premixed charge compression ignition Engine. IMechE J Engine Res 6(5)
66. Sangjin H, Wooldridge MS, Hong GI, Assanis DN, Pitsch H (2005) Development and application of a comprehensive soot model for 3D CFD reacting flow studies in a diesel engine. Combust Flame 143:11–26

67. Shabbir S, Gokhale N, Thatte V, Gandhi N, Deshmukh B, Aghav Y, Kumar M, Nandgaonkar M, Babu MK (2011) Efficient approach for optimization of piston bowl shape, compression ratio and EGR for DI diesel engine, SAE Paper No. 2011-24-0013
68. Shi Y, Reitz RD (2008) Optimization study of the effects of bowl geometry, spray targeting, and swirl ratio for a heavy-duty diesel engine operated at low and high load. ImechE Int J Engine Res 9:325–346
69. STAR-CD Version 324 user guide, Computational Dynamics ltd. London
70. Star Methodology 3.2, CD-ADAPCO Ltd
71. STAR-CD Version 324 Methodology guide, Computational Dynamics ltd. London
72. Steeper RR, Stevens EJ (2000) Characterisation of combustion, piston temperatures, fuel sprays and fuel air mixing in a DISI optical engine, SAE paper 2000-01-2900
73. Thornhill D, Li H, Fleck R, Cuningham G (2006) Modeling of unsteady gas-dynamic flow in a pipe and at its exit using CFD. Proc IMechE J Autom Eng 220(Part D)
74. Tonini S, Gavaises M, Theodorakakos A (2008) The role of droplet fragmentation in high-pressure evaporating diesel sprays. Int J Therm Sci 48:554–572
75. Towers DP, Towers CE (2004) Cyclic Variability measurements of in-cylinder engine flows using high speed particle image velocimetry. Meas Sci Technol, 1917–1925
76. Treea DR, Svenssonb KI (2006) Soot processes in compression ignition engines. Prog Energy Combust Sci
77. Vanderwedge VA, Lounsberry TH, Hochgreb S (2000) Numerical modelling of fuel sprays in DISI engines under early operating conditions, SAE paper 2000-01-0273
78. Versteeg H, Malalasekara W (2008) An introduction to computational fluid dynamics—the finite volume method, 2nd edn. Pearson Publication
79. Vincent M, Barton I, Angelberger C, Poinsot T (2004) CFD evaluation of effects of split injection on combustion and emissions in a DI diesel engine, SAE No. 11PFL-0029
80. Vincent M, Barton I, Angelberger C, Poinsot T (2004) Towards large eddy simulation in internal-combustion engines: simulation of a compressed tumble flow, SAE Paper No. 2004-01-1995
81. Vitek O, Polasek M (2002) Tuned manifold systems-application of 1-D pipe model, SAE Paper No. 2002-01-0004
82. Wagner GJ, Liu WK (2000) Turbulence simulation and multiple scale subgrid models. Comput Mech 25:117–136
83. Willy D, Dupont A, Baby X, Charnay G, Boree J (2003) PIV measurements of internal aerodynamic of diesel combustion chamber, SAE Paper No. 2003-01-3083
84. Willy D, Dumont P, Dupont A, Baby X, Charnay G, Boree J (2004) Airflow cyclic variations analysis in diesel combustion chamber by PIV measurements, SAE 2004-01-1410
85. Wu HW, Peng SW (2004) Numerical analysis of thermal turbulent flow in the bowl in piston combustion chamber of a motored engine. Int J Therm Sci 43:1011–1023
86. Xu H (2001) Some critical technical issues on the steady flow testing of cylinder heads, SAE Paper No. 2001-01-1308
87. Young J, Munoz RH, Anderson RW, Lavoie G (2000) Study of stratified charge DISI engine with an air force fuel injection system, SAE paper 2000-01-2901
88. Yun H, Reitz RD (2005) Combustion optimization in the low-temperature diesel combustion regime. Proc IMechE Int J Engine Res 6(Part D)
89. Zhang Y, Tomoaki I, Nishida K (2001) Characterisation of mixture formation in split-injection diesel sprays via laser absorption-scattering (LAS) technique, SAE Paper No. 2001-01-3498
90. Zhao H, Li Y, Peng Z, Ladommatos N (2002) Tumbling flow analysis in a four valve spark ignition engine using particle image velocimetry. Int J Engine Res 3(3):139–155
91. Zhen L, Wang T, Li X, Li L, Zhang G (2012) Parametric design of the tangential intake port in diesel engines. Proc IMechE J Autom Eng 221(Part D):1–13
92. Zheng OP, Zhang HM, Zhang DF (2005) A computational study of combustion in compression ignition natural gas engine with separated chamber. Fuels 84

93. Zhu Y, Zhao H, Ladomatos N (2005) Computational fluid dynamics study of the effects of the RE_Entrant lip shape and toroidal radii of piston bowl on a high-speed direct injection diesel engine's performance and emissions. Proc IMechE J Autom Eng 219(Part D)
94. Zhu Y, Zhao H, Melas DA, Ladommatos N (2004) Computational study of the effects of the re-entrant lip shape and toroidal radii of piston bowl on a HSDI diesel engine's performance and emissions, SAE Paper No. 2004-01-0118
95. Zhu Y, Zhao M, Ladommatos N (2003) Computational study of the effects of injection timing, EGR and swirl ratio on a HSDI multi-injection diesel engine emission and performance, SAE Paper No. 2003-01-0346

Part V
Next Step for Indian Automotive Industry

Future Mobility Solutions of Indian Automotive Industry: BS-VI, Hybrid, and Electric Vehicles

Tadveer Singh Hora, Akhilendra Pratap Singh and Avinash Kumar Agarwal

Abstract Worldwide scientists and researchers are concerned about climate change and global warming. Automotive vehicles are a major source for emission of greenhouse gases (GHG) and particulate matter (PM). Complying with strict BS-VI emission norms require improvised engine calibration, complex after-treatment (DOC, SCR, and DPF) system calibration, infrastructure development and engine validation. BS-VI will significantly reduce GHG and atmospheric PM, but with long-term perspective, an alternate solution is required to develop zero-emission vehicles. Recently, government planned to debar gasoline and diesel vehicles by 2030. In this scenario, Indian automotive industry has to be future-ready. The future disruptions in Indian automotive would include implementation of hybrid, electric, and fuel-cell vehicles. Government has started working in infrastructure development of hybrid and electric vehicles such as charging units, battery development, charging infrastructure development. However, currently hybrid and electric vehicles are significantly costlier and are required to be economically feasible. It can be assumed that conventional gasoline engines will be used in hybrid vehicles. Diesel engines would also be difficult to be phased out since implementation of hybrid/electric vehicles in long-haul vehicles and high-tonnage vehicles would remain challenging, where diesel engines are currently used virtually unchallenged by any other technology options.

Keywords Emission norms · BS-VI · Advanced combustion technologies
After-treatment devices · Hybrid vehicles

T. S. Hora · A. P. Singh · A. K. Agarwal (✉)
Indian Institute of Technology Kanpur, Kanpur 208016, India
e-mail: akag@iitk.ac.in

T. S. Hora
e-mail: tadveers.h@gmail.com

A. P. Singh
e-mail: akhips@iitk.ac.in

D. K. Srivastava et al. (eds.), *Advances in Internal Combustion Engine Research*,
Energy, Environment, and Sustainability,
https://doi.org/10.1007/978-981-10-7575-9_14

1 Introduction

Energy is one of the main requirements for the economic development of any country. Rapidly increasing energy consumption rate in developing countries is resulting in more petroleum import and increasing dependence on fossil fuels such as coal, oil, and natural gas. Therefore, India needs some immediate action for developing a sustainable path for energy supply. Biofuels seem to have the potential to contribute significantly to India's energy security. However, a clear choice needs to be made on priorities. The fuel costs are rising and are dependent on crude oil pricing. Further, availability of uniform quality gasoline and diesel throughout the country is a major issue. Distribution infrastructure of fuels needs to be strengthened. Sustained availability of fossil fuels has to be ensured. Fuel quality is also affected by issues such as adulteration. Development of high-performance additives is essential. Availability of gaseous fuels like CNG needs to be improved. Infrastructure for dispensing alternative fuels like CNG and LPG and charging electric and hybrid vehicles needs to be established. Use of new synthetic fuels such as DME, methanol, GTL need to be encouraged.

Fossil fuels supply more than 80% energy for global consumption and more than 95% energy for transport sector globally. Vehicular pollution from combustion of these fossil fuels in IC engines cannot be avoided because the pollutants are emitted at the ground level, close to human breathing level. Severity of vehicular pollution is reflected in increased human mortality and morbidity and increased symptoms like cough, headache, nausea, irritation of eyes, various bronchial problems, and visibility in affected population. Vehicular pollution affects human health adversely due to the presence of CO, unburned HC, oxides of nitrogen (NO_x), soot, suspended particulate matter (SPM), and aldehydes, among others in the engine exhaust. Apart from these harmful pollutants, CO_2 leads to various long-term global problems including greenhouse effect. Almost all countries are working on the technology options for CO_2 emission reduction from engine tailpipe to combat this menace. This motivates the fuel researchers to develop techniques to utilize alternative energy resources.

2 Challenges for Transport Sector

Diesel engines emit NO_x and particulate in significant amounts and hydrocarbon (HC) and carbon monoxide (CO) in smaller amounts in comparison with its gasoline counterpart. The main problem involved with the diesel engines is NO_x and PM emissions. Diesel engines are one of the most important sources of NO_x, which leads to the formation of secondary organic aerosols (SOA). NO_x formation takes place at very high-temperature region during combustion in diesel engines. In the diesel engine, fuel is injected into the combustion chamber filled with hot and compressed air, when piston is about to finish the compression stroke. This type of mixing results in a heterogeneous distribution of fuel in the combustion chamber. The heterogeneity of mixture is different at various locations in the combustion chamber, and it changes at a certain point with time. Mixing of fuel and air is dependent on fuel injection parameters, fuel characteristics, and combustion chamber design. Diesel engine power output is mainly governed by the injected fuel quantity. Hence, formation of pollutants is highly dependent on the injected fuel quantity. Formation of HC, CO, and PM takes place in different regions of combustion chamber during the two main combustion phases, namely "premixed" and "mixing controlled" combustion phases, in the combustion chamber. NO_x formation takes place in the high-temperature region of the spray flame. HC formation takes place either by flame quenching near the chamber wall or by late fuel injection toward the completion of compression stroke, which results in incomplete or partially burnt fuel. Heterogeneous mixing of fuel and air produces a fuel-rich region in the fuel spray with high temperature and pressure, which results in formation of soot. In the later stage of combustion, soot gets oxidized partially, when it comes in contact with oxygen and other oxidizing agents.

HC emissions produce serious consequences, when diesel engine starts. During engine cold start, HC emissions are significant and produce white smoke sometimes. Diesel odor comes because of some HCs present in the exhaust gas. Diesel engines are not considered as a major source of CO. During combustion, the concentration of pollutants is actually different with the measured concentration in the exhaust, where chemical equilibrium is considered. For proper investigations of these emissions, it is important to analyze detailed kinetics of the processes and chemical mechanisms of the formation of these pollutants. In this process, a vast number of chemical species are formed and destroyed in fraction of milliseconds, some of which survive and come out into the atmosphere in detectable concentrations. Some of these compounds are very harmful, carcinogenic, teratogenic, and mutagenic. Unregulated pollutants emitted from IC engines are also very harmful for human health. Some of the important unregulated species emitted from IC engines and their harmful health effects are given in Table 1.

In response to the challenging fuel-economy standards, higher fuel prices, and tighter emissions legislations, automotive companies are motivated to develop advanced IC engine technologies, which can deliver higher thermal efficiencies and make road transport more environment friendly. Following subsection describes the main issues related to the transport sector and possible solutions to resolve these issues.

2.1 Emerging Engine Technologies

Two major streams of technology in SI engines are two-stroke and four-stroke technologies. Two-stroke engine technology has been popular in the 1990s due to its simplicity and low cost. However, two-stroke technology was phased out in favor of the more fuel-efficient four-stroke engine technology. Also the

Table 1 Health and environmental effects of unregulated emission species [59]

Unregulated emission species	Possible health and environmental effect
Methane (CH_4)	• GHG with greenhouse index 21 times that of CO_2 • Simple asphyxiant, when inhaled • Leads to headache, dizziness, weakness, nausea, vomiting, and loss of consciousness
Normal octane (n-C_8H_{18})	• Giddiness, vertigo, skin redness and rashes, brain irritation, or apneic anoxia • Causes throat and lungs problems and headache
Ethylene (C_2H_4)	• Causes headache, drowsiness, dizziness, weakness, and unconsciousness • Causes irritation to respiratory system and alters carbohydrate metabolism • Acts as ozone formation agent
Acetylene (C_2H_2)	• Causes suffocation, dizziness, headache, unconsciousness, and nausea • Inhalation results in high blood pressure, fits, and abnormal heart rhythm
Benzene (C_6H_6)	• Drowsiness, dizziness, rapid or irregular heartbeats, headache, tremors, confusion, unconsciousness, and carcinogenic to humans • Chromosomal aberrations in human peripheral lymphocytes
Formaldehyde (HCHO)	• Irritation in eyes, nose, and throat, coughing, and skin irritation • Considered as human carcinogen, cause asthma-like respiratory problems • Affects pregnancy and reproductive system
Acetaldehyde (CH_3CHO)	• Irritation of skin, eyes, mucous membrane, throat, respiratory tract, nausea, vomiting, and headache • Probable carcinogen
Formic acid (HCOOH)	• Causes teary eyes, running nose, coughing, sore throat, bronchitis, shortness of breath, pulmonary edema, liver and kidney damage • Burns tissues and membrane of the skin, and respiratory tract
Sulfur dioxide (SO_2)	• Higher concentration (>100 ppm) causes danger to life and health • Burning sensation in nose and throat, breathing difficulties, and severe airway obstructions

two-wheelers are moving from carburetion to fuel injection technology, due to more efficient control of fueling, thereby offering superior fuel economy, higher power output and torque, and lower emissions. Both, the passenger car and SUV segment use SI and CI engine technologies. It can be noted that the passenger car engine technologies have completely shifted from carburetion to fuel injection. Single-point and multi-point fuel injection (MPFI) technologies are used currently. The future trend is to move toward gasoline direct injection (GDI) technology. Further incorporation of technologies such as exhaust gas recirculation (EGR), CNG-LPG, multi-layer catalytic converters will improve fuel economy and reduce emissions.

For heavy-duty vehicles, diesel engine technology shifted completely from indirect injection (IDI) to direct injection (DI). Further common rail direct injection (CRDI) technology was developed and introduced. The future trend is to move toward higher-pressure CRDI engine technology. Further incorporation of advanced technologies such as EGR, turbocharging, multi-valve, and after-treatment systems like DOC and DPF is inevitable. CI engines for passenger cars will move toward CRDI and use other technologies such as thermal management, downsizing, advanced after-treatment systems, and hybridization in future. Modern HCV and LCV engines require higher specific power and torque output with faster response, lower fuel consumption, and reduced emissions. Various powertrain technologies being adopted in vehicles are shown in Table 2.

2.2 *Stricter Emission Norms*

For reducing CO_2 emissions globally from the transport sector, different policies have been developed, which targets reduction of 30% in CO_2 emissions by 2020 and 50% by 2035. To satisfy these goals, overall emission reduction in all spheres of human activities is targeted, including the road transport sector. Reduction in emissions from commercial vehicles can be achieved by using cleaner diesel having low sulfur. Since emission levels are shrinking, engineering solutions to attain such low levels of emissions are becoming more complex and are closely linked with the engine and vehicle parameters. To achieve the target, India has moved from BS-I regulations to BS-IV regulations in last fifteen years. Past and current emission norms applicable in India are given in Table 3.

Enforcement of uniform emission norms throughout India will be possible only after April 1, 2017. Currently metro cities have emission norms, which are at a level higher than the rest of the country. On-road vehicle emission control should be affected by strengthening the PUC checking system, which is quite rudimentary and non-functional as of now. Fuel quality also affects the emissions; therefore, effective fuel quality measures need to be taken up. Enforcement of emission control devices such as TWC, DPF, SCR, and DOC on vehicles is required to curb vehicular pollution. Research should be undertaken to develop superior catalysts, which are tolerant to high-sulfur fuels. Fuel-economy standards need to be introduced along with measures for reducing CO_2 emissions. Efforts for improving fuel economy of fleets should be encouraged. Inspection and maintenance regime should be introduced for all vehicles to ensure good emission performance of all vehicles.

2.3 *Depletion of Petroleum Reserves*

Presently, the road transport sector is mainly dependent on mineral diesel and gasoline. Two-wheelers and light-duty vehicles are mainly operate on gasoline and

Table 2 Powertrain technologies

SN	Name of technology	Description	Remarks	Investment required
1	MPFI	Multi-point fuel injection	Technology used for three-wheelers, passenger cars, LCVs, and HCVs	Investment required for indigenous development
2	GDI	Gasoline direct injection	Technology used for passenger cars, LCVs, and HCVs	Investment required for indigenous development
3	CRDI	Common rail direct injection	Technology used for three-wheelers, passenger cars, LCVs, and HCVs	Technology can be imported
4	Engine downsizing and derating	Reducing size of the engine for achieving same power or reducing power from same size	Technology used for SUVs and LCVs	Nil
5	Hybridization	Incorporation of electric battery and motor to supplement the IC Engine	Used for some luxury cars and being introduced for small cars, SUVs	Yes, investment required to produce low-cost electronics and batteries
6	Advanced Combustion	Advanced combustion concepts like HCCI, PCCI.	Technology used for SUVs and LCVs	Investment required for indigenous development
7	Variable compression, variable valve lift	Variable concepts in engines for higher fuel efficiency and performance	Used for some luxury cars and being introduced for small cars, SUVs	Technology can be imported
8	High-performance Engine Materials	Advanced materials like plastics, ceramics, high strength alloys.	Used for some luxury cars and being introduced for SUVs	Investment required for indigenous development
9	Combustion modeling	CFD and chemical kinetic modeling	Used for all types of engine development	Investment required for development of software and workstation facilities
10	Powertrain electronics and sensors	Used for Advanced engines like CRDI, GDI.	Used for all advanced engines	Technology can be imported

Table 3 Emission norms for heavy commercial vehicles in India [60]

Emission standard	Year	CO (g/kWh)	NO_x (g/kWh)	HC (g/kWh)	PM (g/kWh)
BS-I	1998	4.9	9.0	1.23	0.40
BS-II	2000	4.0	7.0	1.1	0.15
BS-III	2005	2.1	5.0	0.66	0.1
BS-IV	2010	1.5	3.5	0.46	0.02
BS-VI	2020	1.5	0.4	0.13	0.01

commercial transport like buses, trucks, and heavy-duty vehicles mainly operate on diesel. However, combustion of these fossil fuels in the IC engines produces different types of pollutants, which are harmful to human health as well as environment. In order to comply with these increasingly stringent emission norms, researchers are exploring use of alternative fuels, which can be adopted in current generation engines with minimal hardware modifications. It is expected that the emissions from such alternate fuels should be lower than the emissions from conventional fuels. In this quest, several alternative fuels have been searched and developed. Researchers are primarily looking at low-carbon fuels in order to reduce greenhouse gas emissions.

2.3.1 Gaseous Fuels CNG, LPG, and LNG

These gaseous fuels contain more hydrogen and less carbon; therefore, the emission of greenhouse gases (GHG) and fine particulate is lower from these fuels. Nylon and Lawson (2000) found that diesel combustion emitted 84 mg/km of fine PM compared to only 11 mg/km from CNG. The absence of PAHs, airborne toxins, and SO_2 in CNG/LPG/LNG makes these fuels cleaner and more environment friendly. CNG-/LPG-/LNG-fueled vehicles produce less vibrations and odor compared to conventional diesel engines. However, issues such as higher initial cost, shorter driving trip, expensive distribution, and storage network related to CNG/LPG/LNG vehicles make them less popular for commercial applications.

2.3.2 Biofuels

Biofuel production from renewable sources is widely considered to be one of the most sustainable alternatives to petroleum fuels and a viable means for environmental and economic sustainability. Biofuels are also an alternative for powering vehicles, and replacing traditional petroleum-based fuels. The complete landscape of different generations of biofuels is shown in Fig. 1.

Main source of first-generation biofuels are agricultural products such as beetroots, rapeseed. However, they compete with food production, which is the major

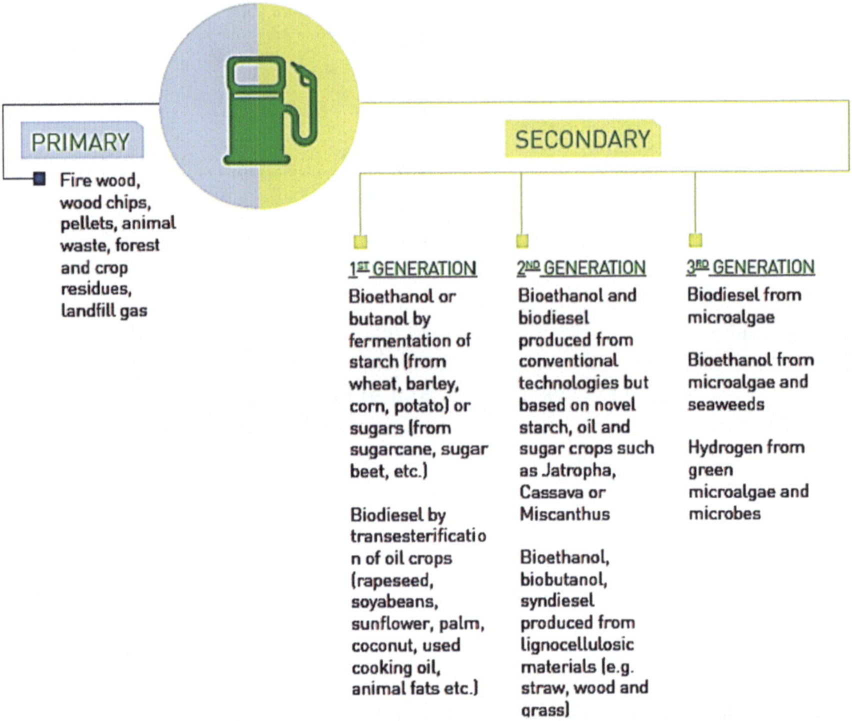

Fig. 1 Landscape of different generations of biofuels [38]

obstacle for the commercial production of first-generation biofuels. Second-generation biofuels are produced using the non-edible parts of plants (straw, wood, plant waste). This is the main advantage of second-generation biofuels. First- and second-generation biofuels can be used directly in conventional vehicles and have significant potential to reduce CO_2 emissions. First- and second-generation biofuels like ethanol and biodiesel have a number of inherent limitations, which make them less suitable for a long-term replacement of fossil fuels. Therefore, there is a need to develop third-generation biofuels in order to avoid this "food vs. fuels" dichotomy.

Microalgae are currently being promoted as an ideal third-generation biofuel feedstock. Microalgae can be used to produce liquid transportation fuels, such as biodiesel and bioethanol. Several advantages related to microalgae such as rapid growth rate, CO_2 fixation ability, and high production capacity of lipids make it suitable for long-term replacement of mineral diesel and gasoline. Algae do not compete with food-crops for resources and can be easily produced on non-arable land.

3 Possible Solutions for Future Road Transport Sector

Future Indian road transport will depend on gasoline/diesel for foreseeable future years and may slowly move to include hybrid and electric vehicles. Diesel engines emit harmful emission, and stricter norms are implemented to reduce toxic gases. Gasoline engines enjoy benefit of clean engines since PM emissions are minimal. Before the gradual spread of hybrid/electric vehicles, dual-fuel technology and usage of alternative gaseous fuels like CNG and hydrogen in IC engines can be widespread. Implementation of innovative combustion technologies like low-temperature combustion can also replace existing heterogeneous combustion diesel engines.

3.1 Innovative Engine Technology

Stricter emission norms of BS-VI/Euro-VI require improvement of in-cylinder combustion along with after-treatment. In-cylinder combustion improvement needs innovations in peripheral component technology of EGR, boost pressure control, throttling, fuel injection pressure, multiple injections, etc. Some of the innovations in engine technology include following combustion improvement technologies to meet BS-VI norms.

3.1.1 High-Pressure and Low-Pressure EGR

EGR is required to control the NO_x emissions from an engine. In an EGR system, certain amount of exhaust gas is mixed with fresh intake air before entering the intake manifold. EGR absorbs some of the heat released during combustion and restricts the peak temperature to enter the NO_x formation zone. Though EGR controls NO_x formation, it increases PM formation in the diesel engines. EGR rate should be properly calibrated in an engine based on combustion and emission optimization. EGR rate is determined based on intake and exhaust CO_2 concentrations. EGR rate is basically the amount of EGR in total intake charge (fresh air + EGR) to engine. Figure 2 shows a typical EGR valve and an EGR cooler. Size and specifications of EGR components varies from light-duty to heavy-duty engines.

$$\text{EGR rate} = \frac{\text{Intake } CO_2 - \text{Ambinent } CO_2}{\text{Exhaust } CO_2 - \text{Ambient } CO_2}$$

EGR can be inducted into engine in two forms, either as high-pressure EGR or as low-pressure EGR. In high-pressure EGR, exhaust gas is inducted from the exhaust gas pipe before the turbocharger. This exhaust gas is cooled in the EGR

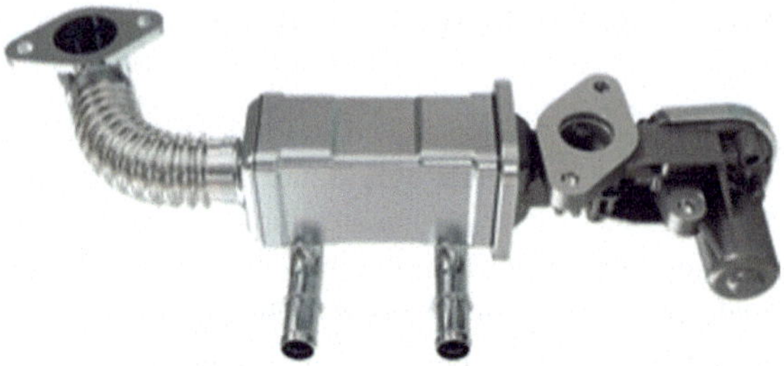

Fig. 2 Typical EGR valve and cooler [39]

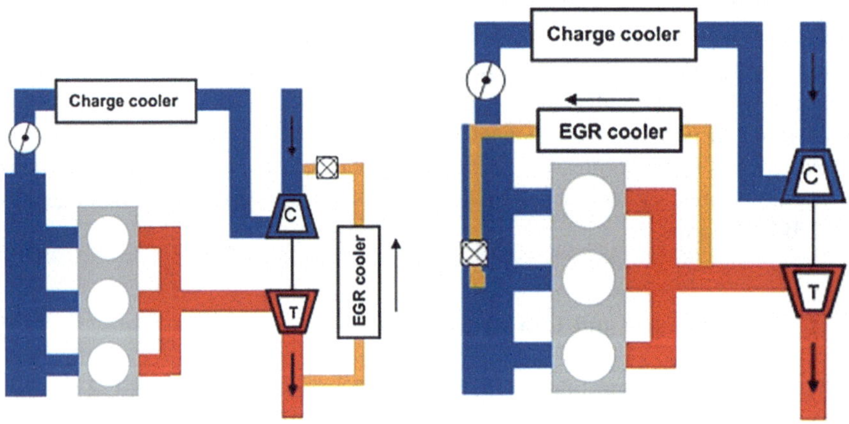

Fig. 3 Low-pressure EGR system (Left) and high-pressure EGR system (Right) [40]

cooler, and its amount is controlled by EGR valve before mixing it with fresh intake air. EGR cooler is a heat exchanger with corrugated tubes. High-temperature EGR is cooled using the heat exchange by the engine coolant. The pressure difference exists between intake air and EGR due to high pressure of gases in the exhaust manifold. Figure 3 shows a typical high-pressure EGR system and low-pressure EGR system layouts.

In a low-pressure EGR system, exhaust gas is inducted from the exhaust flow downstream of the after-treatment system. Downstream of the after-treatment system, gas flow pressure is low; thus, EGR is mixed with the air at intake of the turbocharger compressor inlet.

3.1.2 Turbocharger Boost Control

Lower emission levels require sufficient amount of air for proper and complete combustion. Also, lean-burn combustion reduces the NO_x formation. Application of turbocharger also increases the engine power output. Commercial naturally aspirated engines are rare, and most engines are equipped with turbochargers. Complying with BS-VI norms is difficult without a turbocharger. Figure 4 shows simple working principle of a turbocharger, which converts exhaust gas residual energy to rotate a shaft and the compressor wheel to generate boost pressure. Innovative turbocharger (TC) control technologies are also available. These are waste gate, variable geometry turbines (VGT), E-waste gate. Turbocharging in an engine can be single stage or double stage.

Figure 5 shows a typical waste-gate turbocharger and inner geometry of a VTG turbocharger. Waste-gate turbocharger is fixed geometry, and boost pressure is controlled by bypassing certain amount of exhaust gas. Waste gate is controlled either pneumatically or electronically. VTG is variable geometry turbocharger, which includes movable vanes in turbocharger wheels. The vane movement is controlled to limit the boost pressure and turbine wheel speed.

Major parameters for a turbocharger operation in an engine are its maximum working temperature and turbine speed. Most turbochargers are limited to temperature within 800 °C. Engine calibration is adjusted such that turbocharger does not exceed its maximum speed. Compressor and turbine maps are generated for each turbocharger, which defines surging, choking, compressor efficiency, etc. Appropriate pressure ratios should be maintained across the compressor such that turbocharger does not operate in surging and choking conditions.

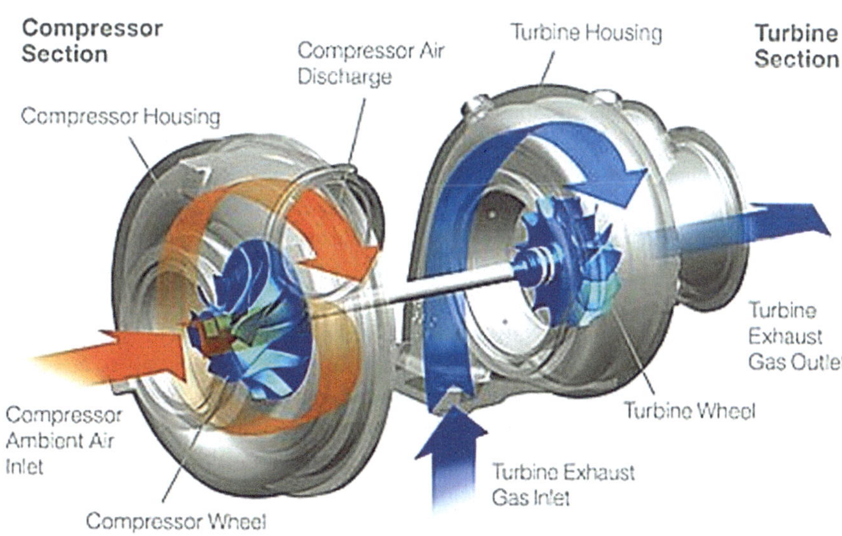

Fig. 4 Working principle of a turbocharger [41]

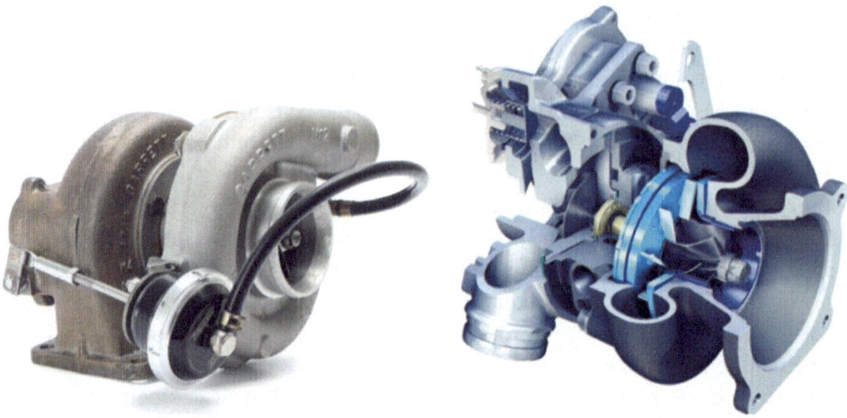

Fig. 5 Typical waste-gate turbocharger [42]

3.1.3 Intake and Exhaust Throttle

Intake throttle is used in gasoline and CNG engines to control the amount of air taken in. In gasoline engines, intake throttle is used to maintain stoichiometric air–fuel ratio. Future BS-VI/EU-VI engines, especially diesel engines, require either intake throttle valve or exhaust throttle valve. Throttle valve is required to control engine combustion thermodynamics, which helps in efficient working of exhaust gas after-treatment. Exhaust throttle valve is mounted in exhaust layout after the turbocharger, and intake throttle valve is mounted after the air-charge intercooler in an engine. For efficient functioning of after-treatment system such as diesel particulate filter (DPF) and selective catalytic reduction (SCR), certain minimum temperatures need to be maintained across the after-treatment system. For passive regeneration of DPF, soot burning temperature is achieved through the engine exhaust gas; thus, exhaust temperatures are maintained at higher levels by controlling the air–fuel ratio and post-injection. Air–fuel ratio is controlled through the throttle valve. Intake throttle valve directly controls the airflow to the engine, while exhaust throttle valve regulates the engine back-pressure to control the air–fuel ratio. Depending on engine application and thermodynamics, either intake or exhaust throttle valve is chosen. Figure 6 shows a typical electronic throttle valve body.

Intake and exhaust throttle valves should be fast and accurate in response, in order to meet the transient operation of engine. Both the throttle valves are electronically controlled through engine ECU. Proper calibration of engine decides the percentage opening of throttle valves. These valves can communicate to the ECU either through CAN signals or PWM signals. Exhaust throttle valve is manufactured to withstand high exhaust gas temperature and corrosion.

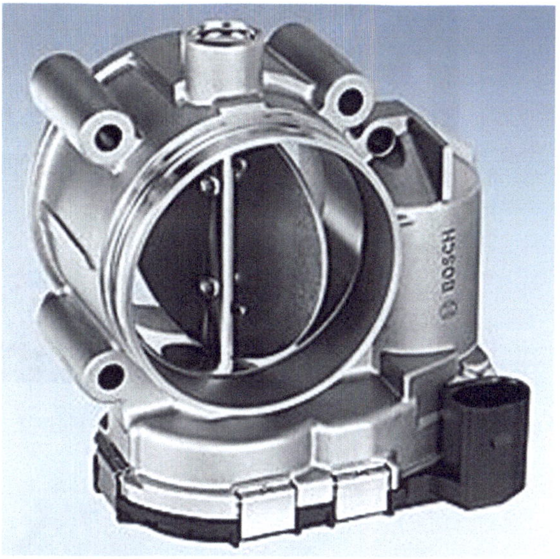

Fig. 6 Intake throttle valve [43]

3.1.4 Fuel Injection System

Most passenger and commercial vehicles in Indian market BS-IV and above norms are equipped with CRDI system. CRDI system generates high fuel injection pressure, which assists in improving in-cylinder combustion. Figure 7 shows a typical CRDI system. CRDI system includes high-pressure pump (FIP), high-pressure common rail, high-pressure fuel lines, injectors, and injector quills. CRDI systems are electronically operated to control the rail pressure.

Future cleaner emission norms of BS-VI require improvements in combustion to produce lower soot quantity. BS-VI engines require high fuel injection pressure of 2500 bar. High fuel injection pressure assists in superior atomization, thus produce low smoke and soot. Major common rail suppliers are Delphi, Bosch, Denso, and Continental, which are capable of manufacturing high-pressure common rail system for BS-VI applications.

3.1.5 Cylinder Head: Swirl Ratio and Exhaust and Intake Valves

To get the lowest possible smoke from the engine, cylinder head should be designed to allow sufficient airflow through the intake ports and minimum oil leakage through the intake valves. Intake ports are designed for optimum swirl ratios to allow proper mixing of air and fuel. High swirl ratio allows sufficient air available through liner surface and crevice volume for combustion of fuel. Figure 8 shows the effect of swirl during flow in the combustion chamber. Swirl creates a vortex circulation in the combustion chamber. Proper mixing of fuel with air

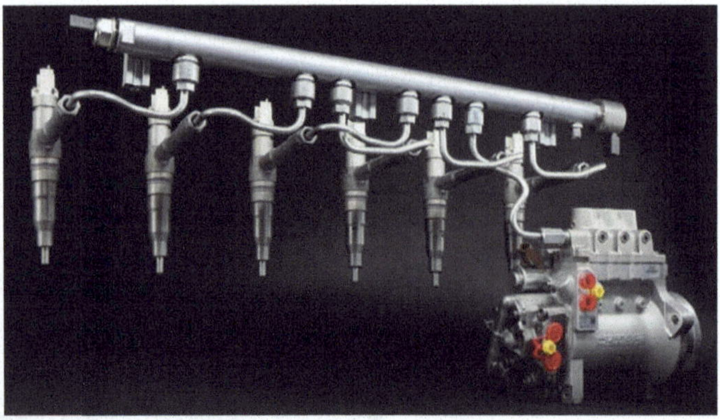

Fig. 7 Typical common rail direct injection system [44]

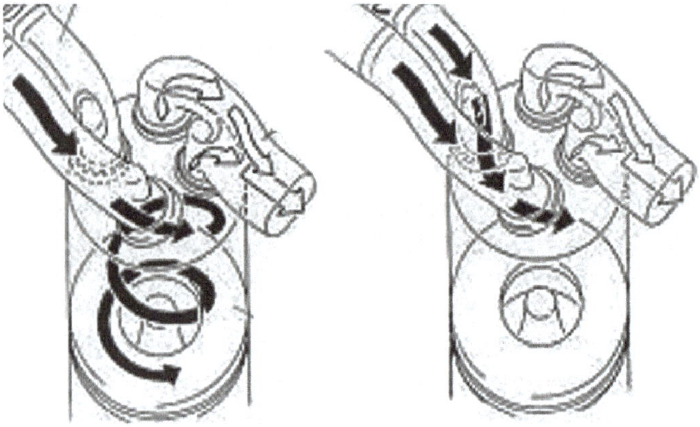

Fig. 8 Effect of high swirl (Left) and low swirl (Right) on vortex [45]

restricts formation of fuel-rich zones, especially in diesel engine by reducing formation of black soot.

Intake valves and exhaust valves should be designed to allow minimum possible oil leakage into the combustion chamber. Oil in combustion chamber produces unburnt hydrocarbons and releases white smoke in the engine exhaust. Valves must be durable to withstand fluctuating forces. Valve-stem seals are used in intake and exhaust valves to allow controlled lubrication for valve movement but restrict penetration of oil into the combustion chamber. Figure 9 shows the application of deflector and positive-type valve-stem seals.

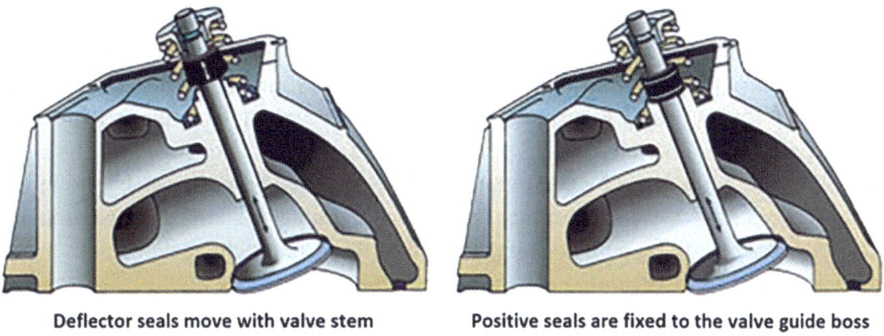

Fig. 9 Valve-stem seals [46]

3.1.6 Complex After-treatment Devices

Immediate solution of road transport sector is implementation of BS-VI for reducing emissions. Existing diesel and petrol engines can be modified with respect to combustion and exhaust after treatment to satisfy strict BS-VI emission norms. In diesel engines, diesel particulate filter (DPF) is used for the reduction of PM and particulate numbers, while selective catalytic reduction (SCR) system is used to reduce NO_x emissions. For a gasoline or diesel engine of lower cubic capacity, lean NO_x trap (LNT) is sufficient to reduce NO_x levels. Complex after-treatment calibration is required for efficient functioning of DPF and SCR. Figure 10 shows a typical layout of exhaust gas after-treatment system of BS-VI/EU-VI engine.

Basic function of after-treatment system is to reduce engine-out emissions of THC, CO, NO_x, and PM. Efficiency of after-treatment system is a critical parameter for sufficient reduction of emissions. Engine-out HC and CO are reduced by DOC, which also converts NO component of NO_x into NO_2. NO_2 reacts with soluble

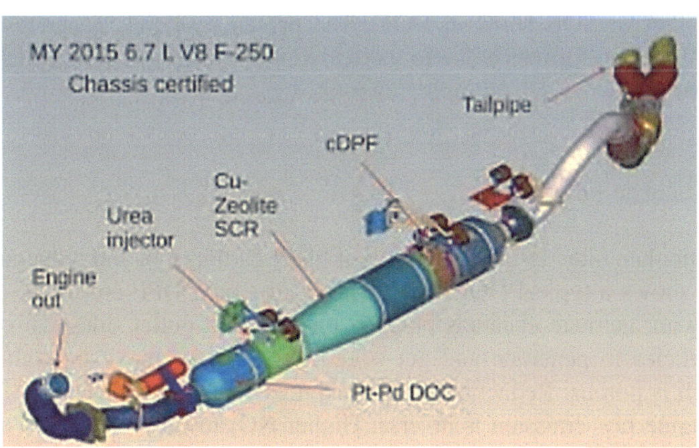

Fig. 10 Typical exhaust system layout [47]

organic fraction (SOF) and soot in DPF to reduce them to CO_2. Engine-out PM is thus burnt inside DPF. Only regulatory emission out of DPF is NO_x. NO_x is converted into nitrogen through the use of SCR. Some part of NH_3 remains unreacted with NO_x and is emitted out of SCR. Traces of NH_3 emitted through engine tailpipe is restricted to 15 ppm in EU-VI norms. This remaining NH_3 is controlled and neutralized by the ammonia slip catalyst (ASC). Figure 11 shows the complex layout of a typical exhaust gas after-treatment system with its subcomponents and sensors. Basic working principle of after-treatment system is described in the following subsections.

Diesel Oxidation Catalyst (DOC)

As the name signifies, DOC assists in oxidation of HC, CO, and NO_x. In a two-way catalytic converter, HC and CO get oxidized with the excess oxygen present in the exhaust gas to form water vapor (H_2O) and CO_2, while three-way catalytic converter also promotes oxidation of NO_x into N_2. To promote this oxidation, DOC consists of precious metals as catalyst. This catalyst usually contains platinum (Pt) and palladium (Pd). The catalyst is covered in a wash-coat material of alumina Al_2O_3 or silica Si_2O_3. The layer of wash-coat and catalyst is spread on DOC substrate. This substrate can be either ceramic or metallic in honeycomb structure. DOC efficiency is dependent on catalyst quantity, substrate size, cells per square inch (CPSI), and wall thickness of substrates. Depending on engine raw emissions, catalyst loading can typically vary from 5 to 50 g/ft^3. Higher the precious metal loading, higher will be the HC and CO conversion rates. Oxidation of HC and CO is also dependent on exhaust gas temperature. Figure 12 shows typical conversion rate with respect to DOC inlet temperature. For a typical DOC, reaction starts after a minimum temperature called light-off temperature. Light-off temperature varies according to precious metal loading and increases according to DOC poisoning and aging. Along with HC and CO oxidation, DOC assists in oxidation of NO to NO_2 and some part of SOF burns in DOC. The presence of SO_x generates negative impact on DOC performance due to its oxidation into sulfuric acid, which poisons the catalyst.

Diesel Particulate Filter (DPF)

Diesel particulate filter assists in burning of black carbon soot and reduction of PM. Figure 13 shows a typical DOC and its inner structure. DPF consist of a ceramic substrate with alternate channels plugged on inlet and outlet sides. This structure allows particles to penetrate and get stored, while gases move through the DPF wall, which is porous. Reduction of soot requires NO_2, but production of soot and NO_2 in engine raw emission is inverse. Higher NO_x means lower soot, and lower soot means higher NO_x. Thus, addition of NO_2 for reduction of soot is generated in the DOC. Therefore, a DOC is always used upstream of DPF in order to ensure

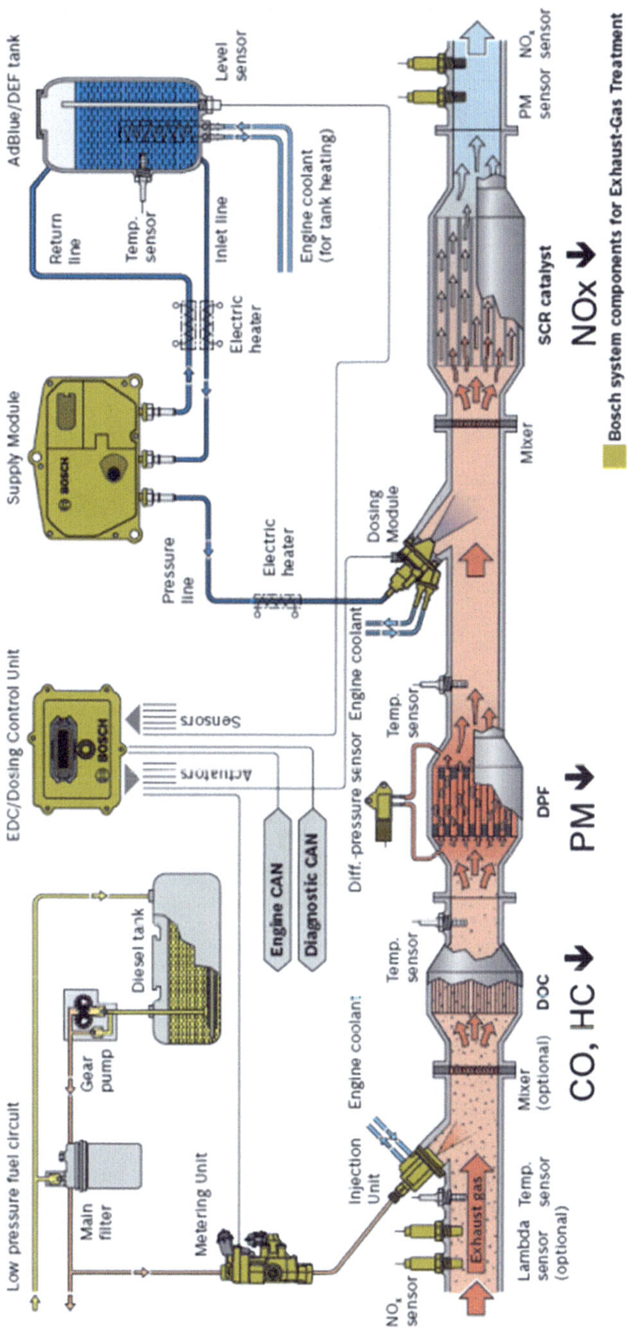

Fig. 11 Detailed exhaust gas after-treatment system layout of a BS-VI engine [48]

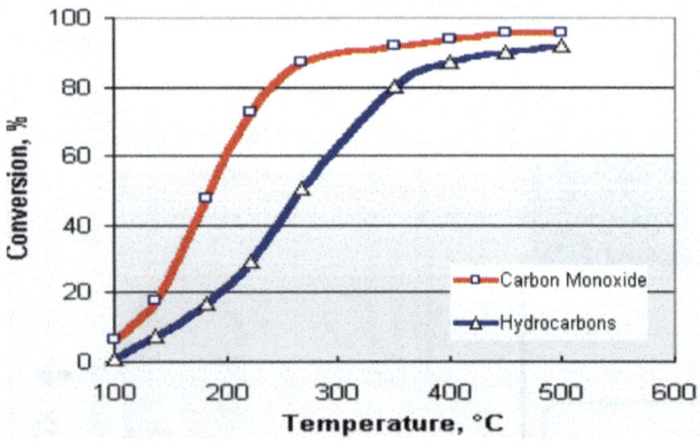

Fig. 12 Conversion efficiency w.r.t. exhaust gas inlet temperature [49]

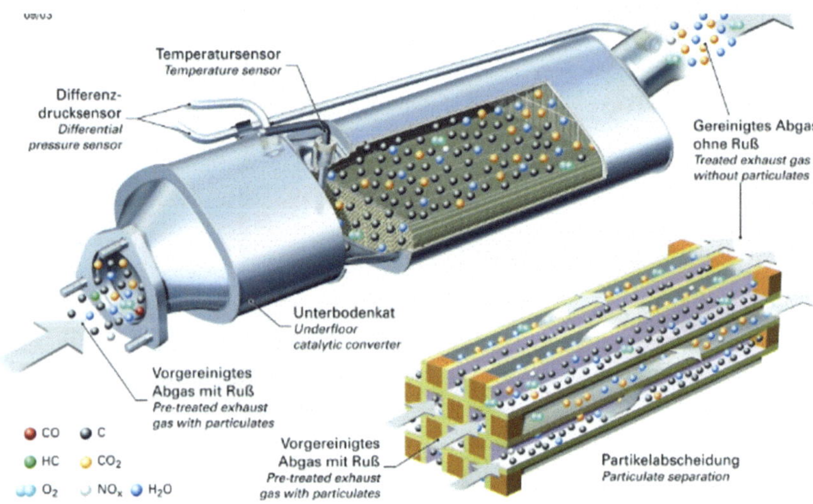

Fig. 13 Typical diesel particulate filter [50]

continuous regeneration of soot stored in DPF. Some DPF contains catalyst called catalyzed DPF (cDPF), which itself ensures sufficient availability of NO_2. DPF substrate is available in ceramic materials such as cordierite and SiC.

Deposited soot in DPF is to be periodically burnt. The instant of regeneration is decided through DPF regeneration strategy. This decision is based on increase in engine back-pressure experienced due to blockage created by soot loading of DPF, inputs from differential pressure sensor (ΔP sensor) feedback, and soot model estimation through ECU strategy. Engine management system and DPF strategy ensure the requirement of minimum combustion temperature since soot burning requires at least 500 °C. ΔP sensor is installed across the DPF, which measures the increase in pressure drop due to soot storage in DPF. Excess pressure drop requires

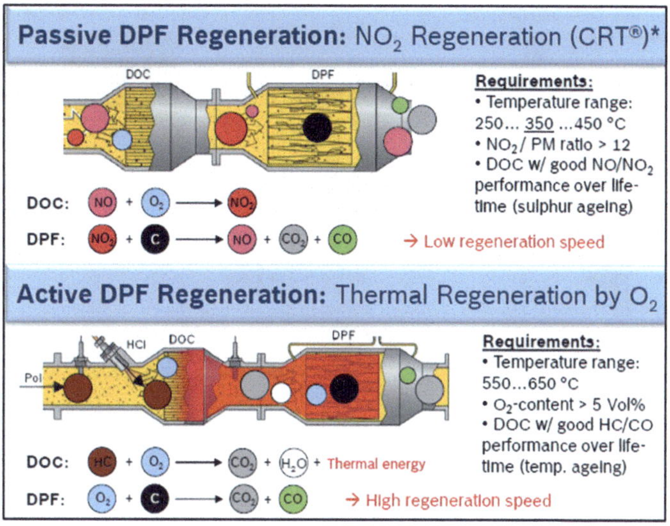

Fig. 14 DPF regeneration [48]

immediate soot regeneration. Also, soot model estimation in engine software calculates the soot loading. Increased soot loading from predefined soot loading capacity of DPF triggers regeneration. The regeneration of DPF can be either active regeneration or passive regeneration, as shown in Fig. 14.

In a passive regeneration strategy, soot is burnt continuously in the DPF, depending on the exhaust gas temperature. Regeneration starts at normal exhaust gas temperature and requires higher NO_2. Passive regeneration needs efficient DOC upstream of DPF and/or a catalyzed DPF (cDPF) to produce enough NO_2. Passive regeneration is simpler and does not cause inconvenience to driver or passer-by vehicles.

Figure 15 shows a layout of active regeneration. In active regeneration, exhaust gas temperature is increased to soot burning temperature either through engine thermal management or by external fuel injection upstream of DOC. Commonly, a HC dozer is used before the DOC, which increases the exhaust temperature. Engine calibration accounts for the amount of fuel injection during regeneration. Exhaust gas temperature can also be increased through late post-injection, rich A/F, retarded injection timing, etc., through the engine ECU. Thus, some of the EU-VI engines need intake throttling or exhaust throttling, in order to increase the exhaust gas temperature.

Selective Catalytic Reduction (SCR) System

Selective catalytic reduction (SCR) system operates by chemically reducing the NO_x (NO and NO_2) to nitrogen (N_2). This reaction is initiated through the injection

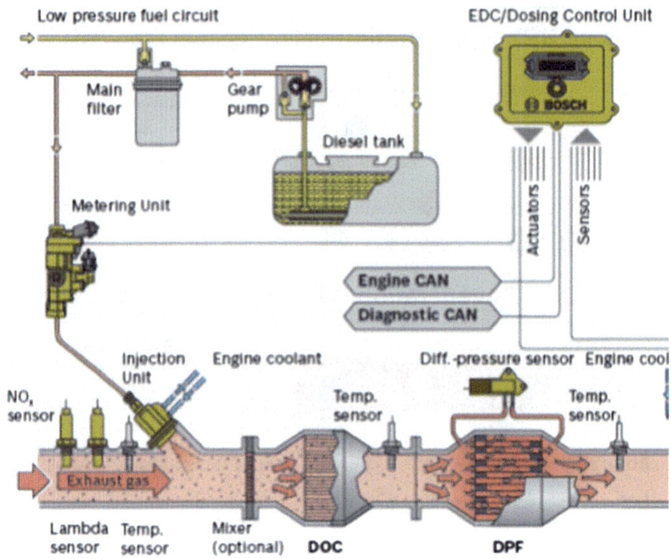

Fig. 15 Active regeneration system [48]

of reductant in the exhaust gas stream. This reductant is ammonia NH_3, which is generated in the exhaust stream through reaction of aqueous urea called AdBlue and also known as diesel exhaust fluid (DEF). Sufficient working temperature is required for SCR system to be functional, otherwise ammonia reacts to form un-intendent chemical components such as iso-cyanic acids, ammonium nitrate, or ammonium sulfate, which poison the SCR catalysts.

SCR substrate is coated with Cu-zeolite or combination of base metals such as vanadium, tungsten, and titanium oxides. These catalysts assist in hydrolysis, thermolysis, selective reduction, and oxidation reactions. NH_3 is formed from hydrolysis and thermolysis of urea. If sufficient temperature is not reached, urea residues are deposited in SCR causing reduction in efficiency. Base metal oxides-based SCR can function in low-temperature range, but Cu-zeolite-based SCR requires sufficient exhaust gas temperature. In some cases, SCR coating is integrated into DPF called as SDPF. SDPF lowers the exhaust system volume and temperature loss till SCR is avoided. Temperature benefits in SDPF results in higher conversion efficiency than SCR alone.

Figure 16 shows a closed-loop SCR system. A NO_x sensor is installed downstream and upstream of SCR and a NH_3 sensor after the SCR. NO_x emissions are continuously monitored, and the signal is transmitted to the SCR dosing control unit called DCU. Based on NO_x, NH_3, and exhaust gas temperature sensors, DCU decides the urea injection quantity. DCU has embedded NH_3 storage strategies and continuously predicts the remaining ammonia stored in SCR. Urea is stored in a tank and heated to avoid decomposition. Based on command from DCU, injector functions to release aqueous urea in the exhaust stream. Coolant is supplied to

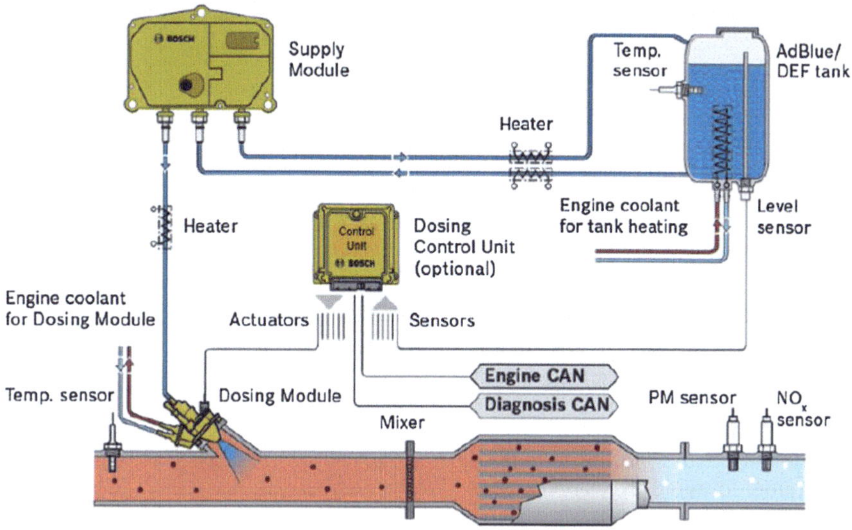

Fig. 16 Typical SCR system layout [48]

injector for releasing heat during the operation. A mixer is present before the SCR to ensure homogeneous mixing of urea in the exhaust stream. With a closed-loop system, efficiency of SCR exceeds 90%.

To enhance the efficiency of SCR at low temperature, NO_2 levels are increased in the exhaust stream. Following are the complex reactions occurring in a SCR system. Explosive ammonium nitrate and ammonium sulfate can form at low temperature.

Reduction reactions are as follows:

$$6NO + 4NH_3 = 5N_2 + 6H_2O$$
$$4NO + 4NH_3 + O_2 = 5N_2 + 6H_2O$$
$$6NO_2 + 8NH_3 = 7N_2 + 12H_2O$$
$$2NO_2 + 4NH_3 + O_2 = 3N_2 + 6H_2O$$
$$NO + NO_2 + 2NH_3 = 2N_2 + 3H_2O$$

If NO_2 is excess in the exhaust stream, it reacts to form N_2O.

$$8NO_2 + 6NH_3 = 7N_2O + 9H_2O$$
$$4NO_2 + 4NH_3 + O_2 = 4N_2O + 6H_2O$$

If quantity of ammonia is excess, it can lead to undesirable reactions such as:

$$2NH_3 + O_2 = N_2O + 3H_2O$$
$$4NH_3 + 3O_2 = 2N_2 + 6H_2O$$
$$4NH_3 + 5O_2 = 4NO + 6H_2O$$

Low temperatures can lead to undesirable SCR poisoning because of reactions with sulfur oxides:

$$2NH_3 + NO_2 + H_2O = NH_4NO_3 + NH_4NO_2$$
$$NH_3 + SO_3 + H_2O = NH_4HSO_4$$
$$2NH_3 + SO_3 + H_2O = (NH_4)_2HSO_4$$

Generally SCR calibration is performed by maintaining NH_3/NO_x ratio of 1 at SCR outlet to avoid excess NH_3 slip.

3.2 *Innovation in Combustion Technologies*

Automotive sector needs to include new engine technologies on gasoline and diesel engines to meet the future challenges. CI engines can include low-temperature combustion like premixed charge compression ignition (PCCI), homogeneous charge compression ignition (HCCI), and reactivity-controlled compression ignition (RCCI). HCCI combustion is an ideal LTC concept, in which homogeneous fuel–air mixture is supplied to the combustion chamber; however, in other derivatives of LTC such as PCCI or partially premixed combustion (PPC), fuel–air mixture is not homogeneous. In PPC, fuel is injected closer to the TDC and combustion is controlled by EGR. In PCCI combustion, fuel injection takes place early in the compression stroke, which leads to premixed fuel–air mixture formation. In RCCI combustion, a combination of low-reactivity and high-reactivity fuels is used to control the fuel–air chemical kinetics. Spark-ignited engines can include GDI and spark-ignited HCCI. Future technologies can also include dual-fuel combustion and opposed piston engines.

3.2.1 Homogeneous Charge Compression Ignition (HCCI)

HCCI is one of the early diesel combustion concepts, which combines the advantages of SI (homogeneous charge spark ignition) and CI (stratified charge compression ignition) combustion modes because it uses premixed charge like SI mode but relies on auto-ignition like CI mode. It has received substantial focus in past couple of decades due to its potential of simultaneously reducing NO_x and PM emissions together with attaining high thermal efficiency, even at part loads.

In HCCI combustion, fuel is injected well before the start of combustion, during the intake stroke such that sufficient time is available for the formation of homogeneous fuel–air mixture. This homogeneous mixture further triggers combustion simultaneously at multiple sites inside the combustion chamber unlike SI (flame propagation) or CI (locally rich flame front) combustion modes. Here, combustion phasing is separate from injection timing and combustion is mainly controlled by kinetics of chemical reactions. HCCI combustion concept has an advantage of fuel flexibility, due to which it can be applied in engines of any size.

This novel concept was initially proposed by Onishi et al. [1], who applied it in a gasoline-fueled two-stroke engine with an intention of increasing combustion stability at part load. They termed it as "active thermo-atmospheric combustion" (ATAC). They further reported a notable reduction in engine emissions together with lower engine noise and vibrations due to lean mixture combustion. Najt and Foster [2] further extended the study to four-stroke gasoline engines on the basis of results obtained from two-stroke engines. They reported that HCCI combustion basically suffered from the lack of control of the ignition process and could be applicable only for a small timing range. Thring et al. [3] further extended previous work in a four-stroke gasoline engine with an objective of conducting the performance analysis of a HCCI engine. Their results reported that the operating regime of this novel technique was limited to part-load operation and proper control of auto-ignition timing was considered as a critical issue. Several other researchers such as Stanglmaier et al. [4] and Maurya et al. [5] also demonstrated this novel combustion concept in SI engines successfully.

After successfully attaining and realizing its potential in gasoline engines, researchers examined the concept of diesel HCCI. Initially, early and late fuel injection strategies were tried by Gan et al. [6] and Dec et al. [7] for achieving diesel HCCI, but these strategies resulted in poor fuel–air mixture quality and menial combustion. Low volatility of diesel was found to be the main hurdle in achieving diesel HCCI. To alleviate this issue, Ryan et al. [8] applied external mixture preparation technique and supplied diesel into the intake airstream. This technique was further developed by Gray et al. [9]. Singh et al. [10] studied the combustion, emission, and performance characteristics of diesel HCCI combustion under varying EGR conditions using external mixture formation technique in a two-cylinder engine and reported superior emission characteristics of HCCI combustion compared to conventional combustion with slightly higher HC and CO emissions. Singh et al. [11] also presented very encouraging results in another similar investigation using biodiesel in a HCCI engine.

Despite benefits of HCCI combustion, this technique suffers from issues such as combustion phasing control, abnormal pressure rise with noise, limited window of operation, high levels of HC and CO emissions, cold-start issues, and challenges in homogeneous charge preparation. These barriers limit its commercialization in mass production engines and further encouraged researchers to develop a new LTC strategy named "PCCI" combustion in order to overcome these drawbacks of HCCI combustion and move closer to commercialization.

3.2.2 Premixed Charge Compression Ignition (PCCI)

Premixed charge compression ignition (PCCI) combustion is an advanced combustion strategy characterized by near-homogeneous mixing between diesel and air in a cylinder before compression ignition. In PCCI combustion, early fuel injection and higher EGR rates are used, which results in spontaneous combustion of nearly homogeneous fuel–air mixture. This leads to significantly lower NO_x emissions due to ignition at multiple points with lean premixed mixture, resulting in low combustion temperatures and lower particulate emissions due to the absence of fuel-rich zones inside the combustion chamber.

In PCCI combustion, unlike SI or CI combustion, ignition occurs simultaneously at multiple sites throughout the combustion chamber volume with quicker combustion duration and is controlled mainly by the chemical kinetics. In this combustion process, fuel can be injected into the combustion chamber through port fuel injection, advanced direct injection, multiple injections, and late direct injection. Port fuel injection and advanced direct injection result in incomplete fuel vaporization and fuel spray impingement on the cylinder walls, which acts as a source for high levels of HC and CO emissions together with fuel dilution. This can be tackled by adopting use of narrow spray cone-angle injectors and un-cooled EGR. Late direct injection circumvents fuel-wall impingement and provides good control over combustion phasing. Use of multiple injections has gained significant interest in recent years in conventional diesel combustion as a means to increase engine efficiency and to reduce engine combustion noise. Pilot injection reduces the ignition delay of the main injection and limits the premixed combustion duration.

Simescu et al. [12] and Weiskirch et al. [13] considered PCCI combustion as a middle path between conventional and clean-diesel combustion modes, since it is a compromise to avoid difficulties encountered in HCCI or LTC combustion modes. To overcome the challenges of combustion phasing control and homogenous mixture preparation in PCCI combustion, only a part of the fuel undergoes HCCI type of clean combustion, while the remaining fuel undergoes conventional combustion. Since major part of fuel experiences conventional combustion, combustion phasing control is dominated by injection timings. Since only a part of the fuel is used to prepare in-cylinder charge having enhanced homogeneity, mixture preparation for PCCI combustion is less challenging in contrast to HCCI combustion.

3.2.3 Reactivity-Controlled Compression Ignition (RCCI)

After massive research activities on LTC, researchers have demonstrated an advanced LTC technique, namely reactivity-controlled compression ignition (RCCI) combustion, which seems to be more promising compared to HCCI and PCCI. In RCCI combustion, high-reactivity fuels such as diesel, dimethyl ether (DME), are used to control the ignition and combustion of low-reactivity fuels such as gasoline, ethanol, methanol, butanol. RCCI combustion can be considered as a PCCI strategy based on dual-fuel operation. To achieve RCCI combustion, high-reactivity fuel is directly injected into the cylinder, and low-reactivity fuel is supplied to the intake manifold [14, 15]. RCCI combustion results in relatively higher brake thermal efficiency (BTE) compared to HCCI and other derivatives of LTC, while maintaining relatively lower NO_x and PM emissions [16].

For improving RCCI combustion, researchers investigated different parameters including EGR levels, swirl intensity levels of intake charge, FIP, SoI timings, and number of injections for both DI injectors and PFI injectors. Several researchers have proposed that RCCI combustion can be easily achieved by using alternate fuels. Various low-reactivity fuels, including natural gas [17–21], methanol/ethanol [22–26], and butanol [27–32] have been studied to explore their effects on load extension in RCCI combustion. The common conclusion is that stable and extended RCCI operation can be achieved with these low-reactivity alternate fuels.

Apart from these advantages, RCCI combustion has several issues such as high load extension, which is mainly restricted by either an excessive RoPR or by NO_x/soot emissions. Increasing r_p can resolve this issue partly; however, too high r_p resulted in gasoline-like HCCI combustion but was restricted by higher RoPR [33–37]. Several researchers proposed that RCCI operation can be extended by enhancing the in-cylinder mixture stratification, which can be achieved by injection strategy and system optimization.

3.2.4 Gasoline Direct Injection

GDI engine delivers improved efficiency and specific power in comparison with PFI engine. This difference arises from the fuel–air mixture preparation process. In a conventional SI engine, fuel–air mixture is prepared in the manifold outside the combustion chamber. This mixture is made stoichiometric, and there is sufficient time available to make this fuel–air mixture homogeneous. In a GDI engine, fuel–air mixture is prepared inside the combustion chamber. GDI engine works on two different modes of combustion. Figure 17 shows (a) homogeneous mode of combustion, and (b) stratified mode of combustion. In homogeneous mode of combustion, mixture is stoichiometric, and in stratified mode of combustion, mixture is lean.

There are three types of GDI engines (spray-, wall-, and air-guided) commercially available in the market. In spray-guided GDI engines, the gap between the spark plug and injector is comparatively less. The combustion chamber geometry, spray dynamics, FIP, fuel droplet distribution are the factors responsible for charge stratification. In the wall-guided GDI engines, the injector is mounted at one side and curved-shape piston guides the charge near the spark plug. In the air-guided GDI engines, interaction between the fuel spray and motion of the air charge inducted in the cylinder is responsible for charge stratification. In a GDI engine, one of the combustion system and both the modes are present.

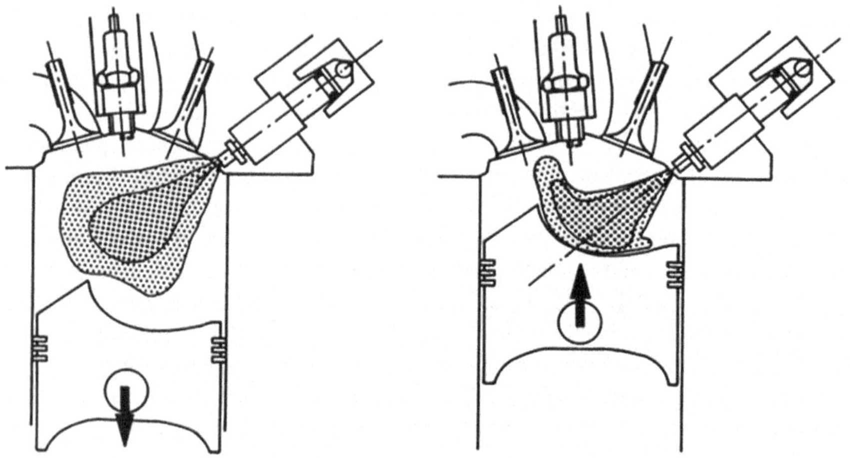

Fig. 17 Different DISI combustion modes [51]

4 Hybrid and Electric Vehicles

Fossil fuel depletion and resulting emissions are a major concern for climate change; hence, a shift to alternative fuel, hybrid vehicle, or electric vehicle is demanded. Future automotive sector will include hybrid, electric, and fuel-cell vehicle by 2030. Electric and fuel-cell vehicles will be completely dependent on battery and hydrogen, respectively. Currently, hydrogen infrastructure and battery technology in India are rather weak. However, hybrid engine technology requires smaller gasoline/diesel engines to charge onboard batteries. Diesel engines can be continued to be installed in hybrid applications because of their higher thermal efficiency and fuel economy. For Indian applications, electric and fuel-cell vehicles need immense infrastructure development related to charging stations, battery technology, hydrogen availability, etc. Figure 18 shows the global market share projected for light-duty vehicles based on different engine technologies like GDI and implementation of hybrid vehicles.

It is projected that by 2020, diesel vehicle growth will be constant while market share of gasoline PFI and GDI would increase. Beyond 2020, growth of hybrid and electric vehicles is projected. GHG emission reduction is a major concern to reduce effect of emissions on climate change. Government across the globe is implementing strict policies and targets aimed at automotive sector. Figure 19 shows the well-to-wheel efficiency comparison of 2010 GHG CO_2 emissions and huge projected drop by 2050. This figure also includes GHG reductions by hybrid and electric vehicles. Currently, plug-in electric vehicles show incremental trend in GHG reduction because CO_2 is emitted in the power plants to generate electricity used for charging the batteries.

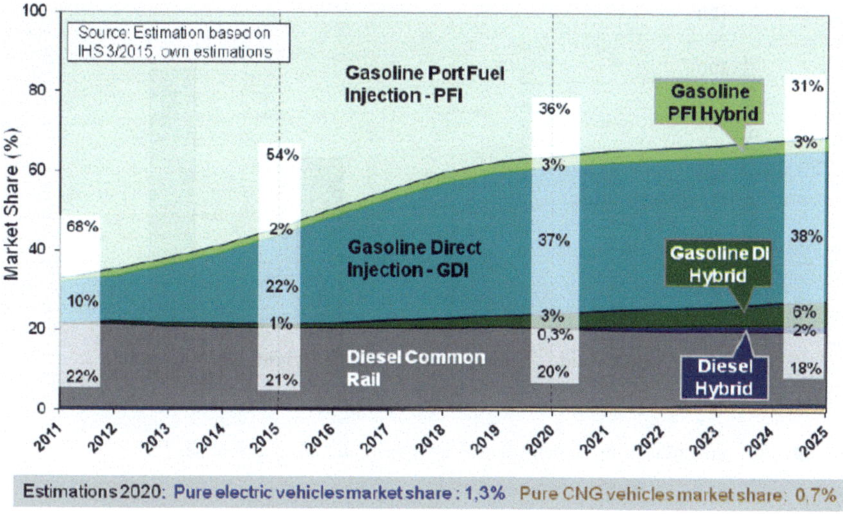

Fig. 18 Market share projection of fossil fueled and hybrid vehicles [52]

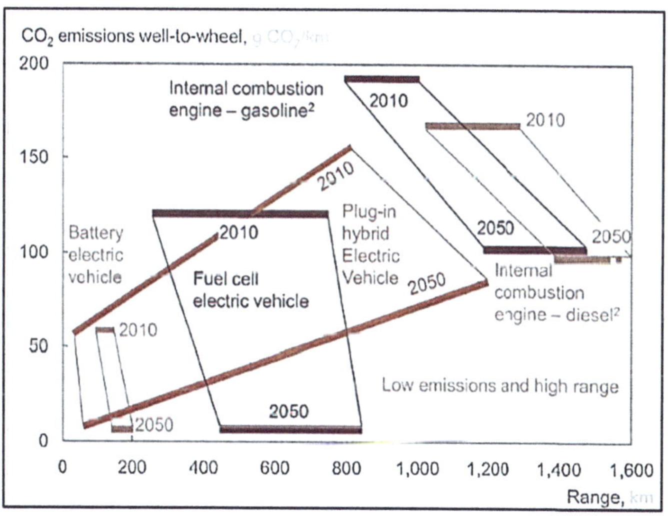

Fig. 19 Well-to-wheel analysis of future automotive sector [52]

The implementation of hybrid and electric vehicles will reduce the CO_2 levels immensely as depicted in Fig. 20. Figure 20 shows the CO_2 reduction potential through hybrid and electric vehicles compared to EU-VI gasoline vehicles. By the implementation of hybrid vehicles, 67% reduction is expected, while 100% reduction by complete implementation of electric vehicles.

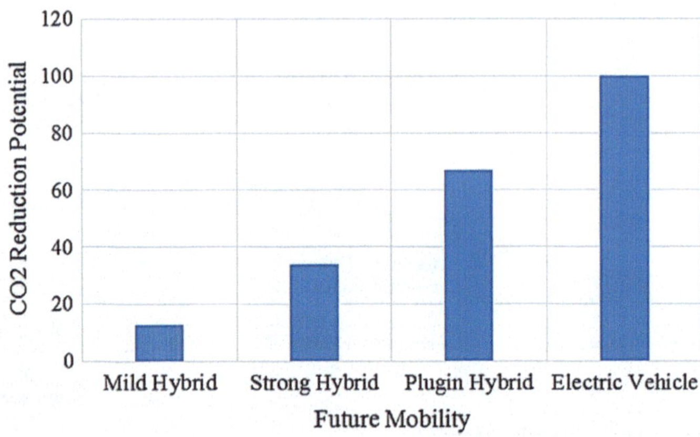

Fig. 20 CO_2 reduction from implementation of hybrid and electric vehicles [52]

Benefits of hybrid and electric vehicles are known to scientist and researchers, but its implementation requires huge infrastructure development, robust usage, and training to the users. Following sections describe basic technicalities in implementation of hybrid and electric vehicles.

4.1 Hybrid Vehicles

Figure 21 shows the fuel economy benefits with different hybrid configurations of diesel and gasoline engines relative to conventional gasoline engines. The fuel efficiency from hybrid configurations defers from city application to highway driving. Mild hybrid configurations do not provide improvement in all driving conditions while full hybrid vehicles with diesel engines provide better fuel economy compared to the conventional diesel engines also. Poor fuel economy of diesel engines in traffic conditions is improvised through hybrid diesel applications, but this configuration is rather expensive. For city applications, petrol engine hybridization provides significant fuel economy improvement.

Major components of hybrid vehicles are high-voltage battery to store electrical energy, generator to covert mechanical energy of IC Engine into electrical energy, inverter, and a transmission system, depending on hybrid configuration. Two important hybrid configurations are series and parallel hybrids. Figures 22 and 23 show the schematic of series and parallel hybrid configurations.

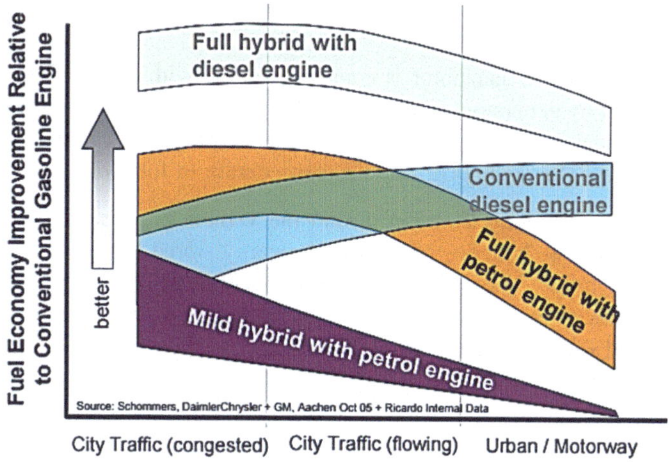

Fig. 21 Fuel economy improvement with hybridization [53]

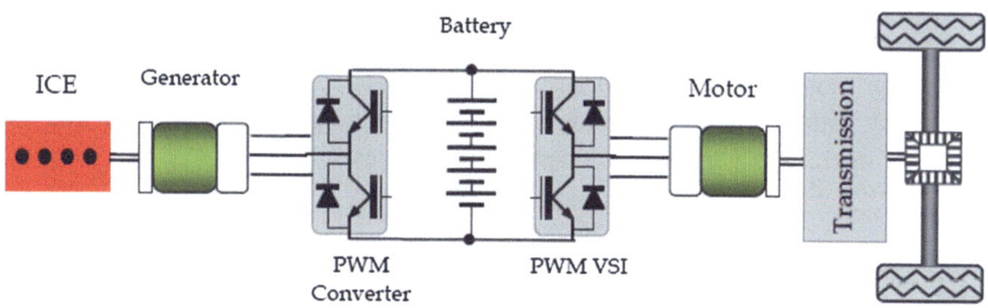

Fig. 22 Series hybrid configuration [54]

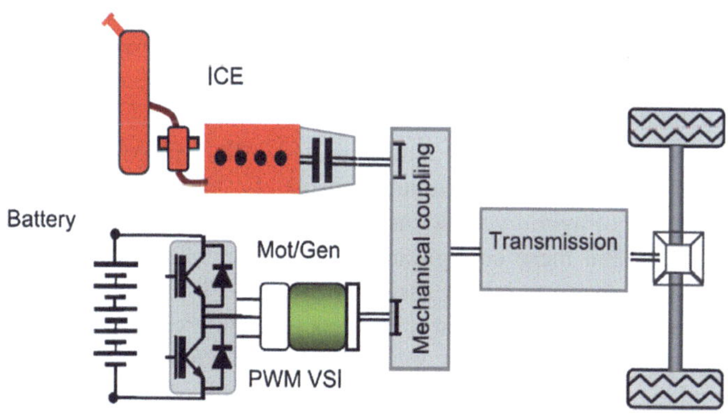

Fig. 23 Parallel hybrid configuration [55]

4.1.1 Series Hybrid

In series hybrid, electric generator is connected inline with an IC engine. Electric power generated by generators is stored in battery, which supplies power to electric motor. Mechanical power is transmitted to wheels through the transmission driven by electric motors. Series hybrid vehicles can operate in following modes:

1. Pure electric: IC engine is stopped, and the vehicle is propelled only by batteries. In this case, batteries have enough energy stored to propel the vehicle.
2. Pure engine mode: The vehicle is powered with energy provided by electric generator driven by engine. In this case, batteries' state of charge (SOC) is minimal and vehicle is propelled by power output of the generator. No energy is stored in battery.
3. Hybrid mode: The traction power is drawn from both the engine-generator and the batteries.
4. Engine traction and battery charging mode: The IC engine-generator provides the energy needed for the battery charging and propulsion of the vehicle.
5. Regenerative braking mode: The engine is turned off, the traction motor is operated as a generator, and the energy provided is used to charge the batteries.
6. Batteries charging mode: The engine-generator charges the batteries, and the traction motor is not supplied any current.
7. Hybrid batteries charging mode: Both the engine-generator and the traction motor operate as generator to charge the batteries.

4.1.2 Parallel Hybrid

In parallel hybrid, IC engine and motor are connected parallel via a mechanical coupling, as shown in Fig. 23. Power for propulsion is supplied by both engine and the motor. Depending on vehicle requirement, torque and speed are distributed by the engine and the motor. However, IC engine is operated at maximum efficiency zone most of the time. Engine and motor can be coupled as torque coupling, speed coupling, or both.

Different types of rechargeable batteries used in hybrid vehicles are lead-acid battery, nickel-cadmium battery, lithium-ion battery, or nickel-metal hydride battery. Other layouts of hybrid vehicle include plug-in hybrid (PHEV), microhybrids, and mild hybrids, but these configurations cannot operate in pure electric mode. Mild hybrid has a battery, which assists the vehicle for start–stop function and assists in propulsion sometimes, but it requires IC engine always in operation. Mild hybrid operates in parallel hybrid configuration. Start–stop condition arises during coasting, braking, or vehicle stoppage such as in traffic. Mild hybrid reduces the unnecessary operation of engine, thus reducing emissions and improving fuel economy. Mild hybrid can employ regenerative braking to charge battery. Microhybrid is barely considered as hybrid as it only allows start–stop of IC engine and sometimes can include battery to supply charge to onboard electric

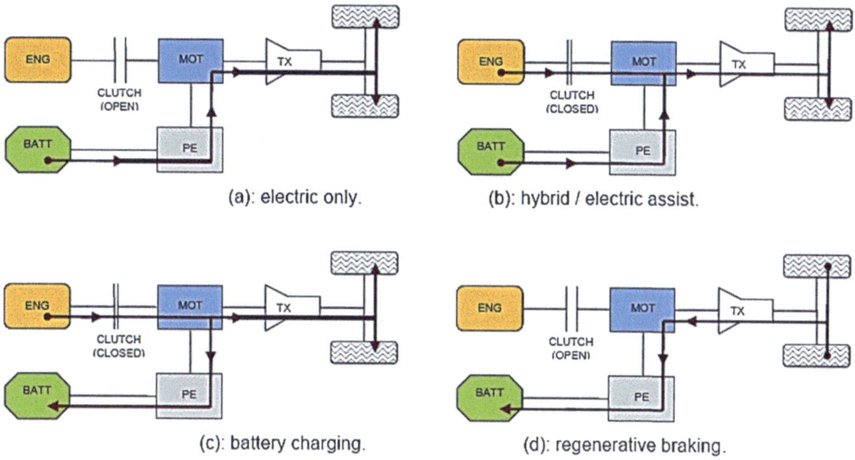

Fig. 24 Modes of operation of a typical hybrid vehicle [55]

components. Plug-in hybrid are full hybrid vehicles containing large batteries. These batteries can be charged through external electric grid so as to provide long duration EV mode operation of the vehicle. Figure 24 depicts different modes of operation of full hybrid electric vehicle (FHEV) during on-road movement.

Actuation of these modes is decided by vehicle control architecture. Control of hybrid vehicles is complex, and its architecture contains number of electronic control units (ECUs) as reflected in Fig. 25. Various ECUs support different functions such as hybrid ECU (HECU), engine ECU, transmission ECU, power electronic ECU, battery ECU, and electric motor ECU (EMECU).

HECU is the master ECU, which gathers driver inputs and commands the coordinating ECU through controlled area network (CAN) communication to respond and deliver output as per vehicle requirement. Hybrid vehicles have automated transmission; thus transmission ECU controls clutch operation and gear ratio determination according to torque and speed requirements. Battery ECU ensures proper functioning of battery such as controlling, overheating, and monitoring state of charge (SOC). All the ECUs provide feedback to the HECU and work in feed-loop control structure.

4.2 *Electric Vehicle (EVs)*

Electric vehicles are available in two formats: battery electric vehicle (BEV) and fuel-cell electric vehicle (FCEV). BEV utilizes stored chemical energy in rechargeable battery to propel the vehicle. Unlike hybrid vehicle, EVs do not have IC engine and purely operate by electricity either in stored form (BEV) or onboard generation (FCEV). EV propulsion system is simpler then IC engines and does not

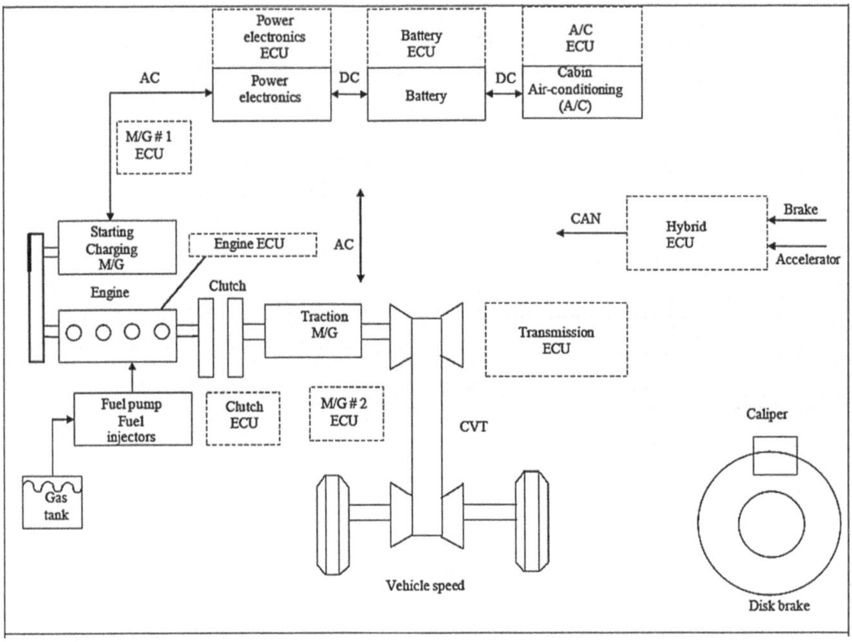

Fig. 25 Control architecture of a hybrid vehicle [56]

consist of any movable or rotary parts. EVs are zero-emission vehicles; however, well-to-wheel analysis reveals emissions at the time of electricity generation. BEV needs sufficient infrastructure for frequent battery charging. The requirement of charging stations can be avoided through implementation of fuel-cell electric vehicles (FCEVs), but FCEV requires hydrogen infrastructure as it utilizes hydrogen for onboard electricity production for charging of batteries.

FCEVs use a propulsion system similar to electric vehicles, where energy is stored as hydrogen, which gets converted to electricity by the fuel cell. Fuel cell in EV is more efficient and produces almost negligible pollutant compared to conventional internal combustion engines. From fuel cell, only water vapor and warm air are emitted as the product of electrochemical reaction. Figure 26 shows the layout and configuration of fuel-cell electric vehicle. Various components involved are hydrogen tank, battery stacks, cooling system, fuel-cell stacks, DC convertors, etc. Hydrogen stored in leak-proof tank is passed to fuel-cell stack. Electricity is generated in fuel-cell stack through electrochemical reactions. Fuel-cell stack converts the hydrogen gas and oxygen into electricity for powering the motor. Energy generated by fuel-cell stack is passed to run the motor for propelling the vehicle smoothly and quietly. A battery is used to store energy during regenerative braking. Power control unit governs the flow of electricity.

As mentioned, fuel cell involves reaction involving hydrogen to generate electricity. Figure 27 shows the reactions involved in a fuel cell. The most common

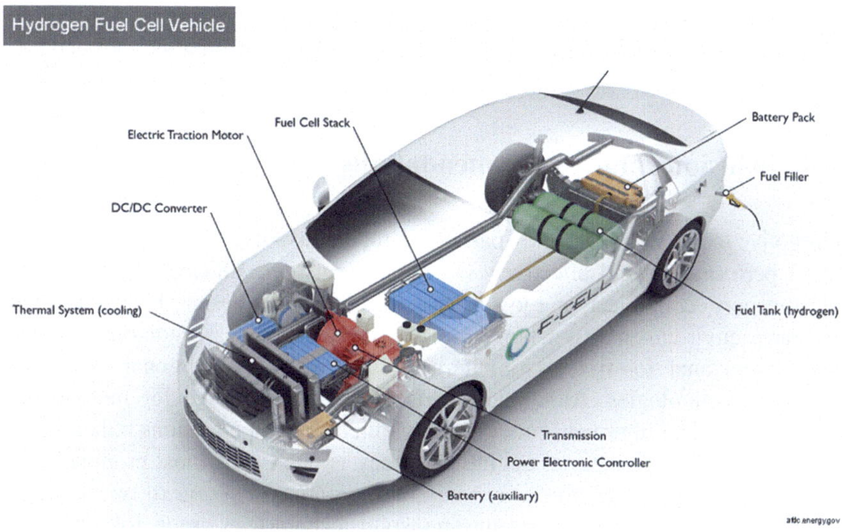

Fig. 26 Fuel-cell electric vehicle configuration [57]

Fig. 27 Working principle of a PEM fuel cell [58]

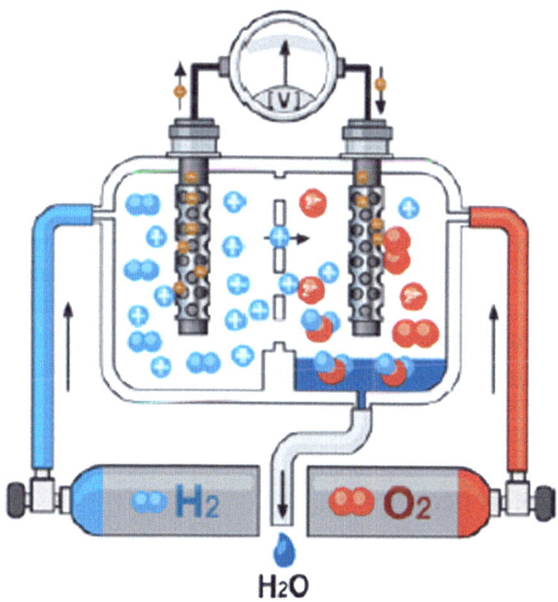

type of fuel cell for vehicle applications is the polymer electrolyte membrane (PEM) fuel cell, which is also called as proton-exchange membrane fuel cell.

5 Conclusions and Recommendations

Automotive technology is the key part of the road transport sector because it affects overall performance and effectiveness of road transport. Efficient vehicles make road transport economical, and lower emissions from vehicles are highly desirable for cleaner environment. To achieve these two key objectives, automotive engines either use cleaner fossil fuels or alternate fuels using conventional and newer propulsion technologies. Although IC engines have been used for more than a century, significant improvements in energy efficiency and emissions reduction are still required. Enhancement of engine efficiency is one of the most promising and cost-effective approaches for increasing fuel economy over next two decades. Important recommendations for future transport technologies are as follows:

- Fuel economy of existing engines should be increased by introducing advanced technologies such as engine downsizing, variable compression ratio, and lean-burn engines.
- Develop efficient combustion approaches for diesel combustion having near-zero emissions without complicated and expensive after-treatment systems.
- Develop technologies for energy recovery from engine's exhaust using thermoelectric devices, which convert heat into electricity, and can power vehicle and/or auxiliary devices.
- Research on existing advanced combustion technology such as HCCI, PCCI, RCCI, GDI is the way forward.
- Exploration of alternative/synthetic energy resource such as gaseous fuels, first-generation biofuels (alcohols, straight vegetable oils, etc.), second-generation biofuels (biodiesel from non-edible oils), and third-generation biofuels (algae-based), biomass-to-liquid (BTL) fuels, methanol, DME, biorefineries, etc., is a step in right direction.

References

1. Onishi S, Jo SH, Shoda K, Jo PD, Kato S (1979) Active thermo-atmosphere combustion (ATAC)—a new combustion process for internal combustion engines. SAE Technical Paper; 790501
2. Najt PM, Foster DE (1983) Compression-ignited homogeneous charge combustion. SAE Technical Paper; 830264
3. Thring RH (1989) Homogeneous-charge compression ignition (HCCI) engines. SAE Technical Paper; 892068

4. Stanglmaier RH, Roberts CE (1999) Homogeneous charge compression ignition (HCCI): benefits, compromise and future engine applications. SAE Technical Paper; 1999-01-3682
5. Maurya RK, Agarwal AK (2011) Experimental investigation on the effect of intake air temperature and air-fuel ratio on cycle-to-cycle variations of HCCI combustion and performance parameters. Appl Energy 88(4):1153–1163
6. Gan S, Ng HK, Pang KM (2011) Homogeneous charge compression ignition (HCCI) combustion: implementation and effects on pollutants in direct injection diesel engines. Appl Energy 88:559–567
7. Dec JE, Kelly-Zion PL (2000) The effects of injection timing and diluents addition on late-combustion soot burnout in a DI diesel engine based on simultaneous 2-D imaging of OH and soot. SAE Technical Paper; 2000-01-0238
8. Ryan III TW, Callahan TJ (1996) Homogeneous charge compression ignition of diesel fuel. SAE Technical Paper; 961160
9. Gray III AW, Ryan III TW (1997) Homogeneous charge compression ignition of diesel fuel. SAE Technical Paper; 971676
10. Singh AP, Agarwal AK (2012) Combustion characteristics of diesel HCCI engine: an experimental investigation using external mixture formation technique. Appl Energy 99: 116–125
11. Singh G, Singh AP, Agarwal AK (2014) Experimental investigations of combustion, performance and emission characterization of biodiesel fuelled HCCI engine using external mixture formation technique. Sustain Energy Technol Assess 6:116–128
12. Simescu S, Ryan TW, Neely GD, Matheaus AC, Surampudi B (2002) Partial pre-mixed combustion with cooled and uncooled EGR in a heavy-duty diesel engine. SAE Technical Paper; 2002-01-0963
13. Weiskirch C, Mueller E (2007) Advanced in diesel engine combustion: split combustion. SAE Technical Paper; 2007-01-0178
14. Benajes J, Molina S, García A, Monsalve-Serrano J (2015) Effects of low reactivity fuel characteristics and blending ratio on low load RCCI (reactivity controlled compression ignition) performance and emissions in a heavy-duty diesel engine. Energy 90:1261–1271
15. Li J, Yang WM, Zhou DZ (2017) Review on the management of RCCI engines. Renew Sustain Energy Rev 69:65–79
16. Li J, Yang WM, An H, Zhou DZ, Yu WB, Wang JX et al (2015) Numerical investigation on the effect of reactivity gradient in an RCCI engine fueled with gasoline and diesel. Energy Convers Manag 92:342–352
17. Nieman DE, Dempsey AB, Reitz RD (2012) Heavy-duty RCCI operation using natural gas and diesel. SAE Int J Eng 5:270–285
18. Salahi MM, Esfahanian V, Gharehghani A, Mirsalim M (2017) Investigating the reactivity controlled compression ignition (RCCI) combustion strategy in a natural gas/diesel fueled engine with a pre-chamber. Energy Convers Manag 132:40–53
19. Dahodwala M, Joshi S, Koehler F, Franke M, Tomazic D (2015) Experimental and computational analysis of diesel-natural gas RCCI combustion in heavy-duty engines. SAE Technical Paper; 2015-01-0849
20. Jia Z, Denbratt I (2015) Experimental investigation of natural gas-diesel dual-fuel RCCI in a heavy-duty engine. SAE Int J Eng 8:797–807
21. Pipitone E, Genchi G (2016) NO_x reduction and efficiency improvements by means of the Double Fuel HCCI combustion of natural gas–gasoline mixtures. Appl Therm Eng 102:1001–1010
22. Dempsey AB, Walker NR, Reitz R (2013) Effect of cetane improvers on gasoline, ethanol, and methanol reactivity and the implications for RCCI combustion. SAE Int J Fuels Lubr 6:170–187
23. Li Y, Jia M, Liu Y, Xie M (2013) Numerical study on the combustion and emission characteristics of a methanol/diesel reactivity controlled compression ignition (RCCI) engine. Appl Energy 106:184–197

24. Li Y, Jia M, Chang Y, Liu Y, Xie M, Wang T et al (2014) Parametric study and optimization of a RCCI (reactivity controlled compression ignition) engine fueled with methanol and diesel. Energy 65:319–332
25. Splitter DA, Reitz RD (2014) Fuel reactivity effects on the efficiency and operational window of dual-fuel compression ignition engines. Fuel 118:163–175
26. Curran S, Hanson R, Wagner R (2012) Effect of E85 on RCCI performance and emissions on a multi-cylinder light-duty diesel engine. SAE Technical Paper; 2012-01-0376
27. Zhang Y, Sagalovich I, Ojeda WD, Ickes A, Wallner T, Wickman DD (2013) Development of dual-fuel low temperature combustion strategy in a multi-cylinder heavy-duty compression ignition engine using conventional and alternative fuels. In: SAE technical paper, SAE international, pp 1481–1489
28. Qian Y, Ouyang L, Wang X, Zhu L, Lu X (2015) Experimental studies on combustion and emissions of RCCI fueled with n-heptane/alcohols fuels. Fuel 162:239–250
29. Qian Y, Wang X, Zhu L, Lu X (2015) Experimental studies on combustion and emissions of RCCI (reactivity controlled compression ignition) with gasoline/ n-heptane and ethanol/ n-heptane as fuels. Energy 88:584–594
30. DelVescovo D, Wang H, Wissink M, Reitz R (2015) Iso-butanol as both low reactivity and high reactivity fuels with addition of Di-Tert Butyl Peroxide (DTBP) in RCCI combustion. SAE Int J Fuels Lubr 8:329–343
31. Wang H, DelVescovo D, Yao M, Reitz RD (2015) Numerical study of RCCI and HCCI combustion processes using gasoline, diesel, iso-butanol and DTBP cetane improver. SAE Int J Eng 8(2):831–845
32. Liu H, Wang X, Zheng Z, Gu J, Wang H, Yao M (2014) Experimental and simulation investigation of the combustion characteristics and emissions using n-butanol/biodiesel dual-fuel injection on a diesel engine. Energy 74:741–752
33. Kokjohn SL, Hanson RM, Splitter DA, Reitz RD (2011) Fuel reactivity controlled compression ignition (RCCI): a pathway to controlled high-efficiency clean combustion. Int J Eng Res 12:209–226
34. Dempsey AB, Reitz RD (2011) Computational optimization of a heavy-duty compression ignition engine fueled with conventional gasoline. SAE Int J Eng 4:338–359
35. Jia M, Dempsey AB, Wang H, Li Y, Reitz RD (2015) Numerical simulation of cyclic variability in reactivity-controlled compression ignition combustion with a focus on the initial temperature at intake valve closing. Int J Eng Res 16:441–460
36. Molina S, García A, Pastor JM, Belarte E, Balloul I (2015) Operating range extension of RCCI combustion concept from low to full load in a heavy-duty engine. Appl Energy 143:211–227
37. Benajes J, Pastor JV, García A, Monsalve-Serrano J (2015) The potential of RCCI concept to meet EURO VI NOx limitation and ultra-low soot emissions in a heavy-duty engine over the whole engine map. Fuel 159:952–961
38. Agarwal AK, Singh AP, Thipse SS (Dec 2016) Technology vision 2015: technology roadmap transportation. TIFAC, New Delhi. https://www.tifac.org.in/vision2035/transportation/trans_roadmap1.pdf. Accessed on 17 July 2017
39. https://www.borgwarner.com/en/technologies. Accessed on 17 July 2017
40. http://articles.sae.org/13530. Accessed on 17 July 2017
41. http://www.popularmechanics.com/cars/how-to/a12765/4306310/. Accessed on 17 July 2017
42. http://www.automobile-sportive.com/technique/downsizing.php. Accessed on 17 July 2017
43. http://products.bosch-mobility-solutions.com/en/de/homepage/homepage_1.html. Accessed on 17 July 2017
44. http://pro.net.mk/speed/auto/fiats-invention-rules-world-diesel/. Accessed on 17 July 2017
45. http://www.fenderen.dk/forum/showthread.php?t=362. Accessed on 17 July 2017
46. http://www.felpro-only.com/blog/valve-stem-seals_/. Accessed on 17 July 2017
47. https://www.dieselnet.com/tech/engine_heavy-duty_aftertreatment.php. Accessed on 17 July 2017
48. https://www.bosch.com/ (Accessed on 17/07/2017)

49. https://www.autoevolution.com/news/how-the-diesel-particulate-filter-works-90866.html. Accessed on 17 July 2017
50. https://www.nettinc.com/information/emissions-faq/what-is-a-diesel-oxidation-catalyst. Accessed on 17 July 2017
51. Suresh M, Hari Prakash, Durga Prasad B (2015) Operation and developments of DISI engines—a review. Int R J Latest Trends Eng Technol (IJLTET) 5(4):31–37
52. https://www.continental-corporation.com/en-my. Accessed on 17 July 2017
53. Internal data, Daimler Chrysler, General Motors, Ricardo
54. Soylu S (ed) Electric vehicles-modelling and simulations. ISBN 978-953-307-477
55. http://www.wikiwand.com/en/Hybrid_vehicle_drivetrain. Accessed on 17 July 2017
56. Fuhs AE (2009) Hybrid vehicles and the future of personal transportation. CRC Press
57. www.afdc.energy.gov. Accessed on 17 July 2017
58. Basics of electric vehicle, Self-study program 820233, Volkswagen. http://www.natef.org/NATEF/media/NATEFMedia/VW%20Files/820233-Electric-Drives-7_9_2013_sm-(2).pdf. Accessed on 17 July 2017
59. Agarwal AK, Shukla PC, Patel C, Gupta JG, Sharma N, Prasad RK, et al (2016) Unregulated emissions and health risk potential from biodiesel (KB5, KB20) and methanol blend (M5) fuelled transportation diesel engines. Renew Energy
60. India: Heavy Duty: Emission. Source: http://transportpolicy.net/index.php?title=India:_Heavy-duty:_Emissions. Accessed on 17 July 2017

The manufacturer's authorised representative in the EU is Springer Nature Customer Service Centre GmbH, Europaplatz 3, 69115 Heidelberg, Germany. If you have any concerns regarding our products, please contact ProductSafety@springernature.com

Printed and bound by CPI Group (UK) Ltd, Croydon, CR0 4YY
25/09/2026
02222892-0013